APPLIED MATHEMATICS

APPLIED MATHEMATICS

Gerald Dennis Mahan

Pennsylvania State University
University Park, Pennsylvania

Kluwer Academic / Plenum Publishers
New York, Boston, Dordrecht, London, Moscow

Library of Congress Cataloging-in-Publication Data

Mahan, Gerald D.
    Applied mathematics/by Gerald D. Mahan.
        p.   cm.
    Includes bibliographical references and index.
    ISBN 0-306-46683-X
        1. Mathematics.   I. Title.

QA37.3 .M34 2001
510—dc21

2001038769

ISBN 0-306-46683-X

©2002 Kluwer Academic / Plenum Publishers, New York
233 Spring Street, New York, New York 10013

http://www.wkap.nl

10 9 8 7 6 5 4 3 2 1

A C.I.P. record for this book is available from the Library of Congress

Printed in the United States of America

Preface

This volume is a textbook for a year-long graduate level course in applied mathematics for scientists and engineers. All research universities have such a course, which could be taught in different departments, such as mathematics, physics, or engineering. I volunteered to teach this course when I realized that my own research students did not learn much in this course at my university. Then I learned that the available textbooks were too introductory. While teaching this course without an assigned text, I wrote up my lecture notes and gave them to the students. This textbook is a result of that endeavor. When I took this course many, many, years ago, the primary references were the two volumes of P. M. Morse and H. Feshbach, *Methods of Theoretical Physics* (McGraw-Hill, 1953). The present text returns the contents to a similar level, although the syllabus is quite different than given in this venerable pair of books.

My own research field is theoretical condensed matter physics. My first professional job was at the Research and Development Center of the General Electric Company. When I left there to become a college professor, they hired me back as a consultant. For many years I wrote analytic and computer models of all possible semiconductor devices for GE-CRD. In this latter activity I really learned applied mathematics. Most often I was solving the diffusion equation. Heat diffuses, as do minority carriers in devices. I once described consulting as "solving the diffusion equations in all possible dimensions with all possible initial and boundary conditions." Solving the transient diffusion equation is heavily emphasised in the chapters on partial differential equations. Similarly, the Laplace transform is most useful when solving transient problems, and is readily applied to diffusion. Conformal mapping is extraordinarily useful for solving problems in two dimensions, so this topic also gets emphasis.

Most of the syllabus is traditional. There are chapters on matrices, group theory, special functions, complex variables, and linear and nonlinear differential equations. Problems, which can be assigned as homework, can be found at the end of each chapter. There are several sections not found in most textbooks. Wavelets are included in the chapter on transforms, along with those of Fourier and Laplace. Wavelets are a new topic that have emerged in the past twenty years. It is applied to data streams that never end in time.

Another unusual topic is the method of Markov averaging of random systems. The solution proposed by S. Chandrasekhar is useful for many different types of problems.

I wish to thank my students in this course, who used the notes which became the preliminary draft. They let me know quickly if the first version was unclear. Also, I thank my editor at Kluwer-Academic-Plenum, Amelia McNamara, who agreed to publish another of my manuscripts. Her staff are always pleasant and efficient.

Contents

Applied Mathematics

Determinants

A determinant is a single number obtained by a particular evaluation of an $n \times n$ matrix of values. The number is real if the elements of the matrix are real, but could be complex if the elements are complex. Below are given some techniques for evaluating the determinant of a $n \times n$ matrix.

However, it is best to give some simple examples of matrix of small dimension:

1. $n = 1$: Here the matrix has a single element, and the determinant is the value of that number.
2. $n = 2$: Here the matrix has four elements, and the determinant is:

$$\det |a| = \det \begin{vmatrix} a_{11} & a_{12} \\ a_{21} & a_{22} \end{vmatrix}$$

$$= a_{11}a_{22} - a_{12}a_{21} \tag{1.1}$$

The two off-diagonal elements are multiplied together, and subtracted from the product of the diagonal elements.

3. $n = 3$: Here there are nine elements in the matrix. The determinant is

$$\det |a| = \det \begin{vmatrix} a_{11} & a_{12} & a_{13} \\ a_{21} & a_{22} & a_{23} \\ a_{31} & a_{32} & a_{33} \end{vmatrix}$$

$$= a_{11}a_{22}a_{33} + a_{12}a_{23}a_{31} + a_{13}a_{32}a_{21}$$
$$- a_{11}a_{23}a_{32} - a_{22}a_{13}a_{31} - a_{33}a_{12}a_{21} \tag{1.2}$$

There are six terms: three have plus signs and three have minus signs.

These three cases are the only ones that can be found in a simple fashion. For larger values of n it is necessary to employ minors.

Minors

The minor M_{ij} of an element a_{ij} of a determinant is the determinant one gets by crossing out row i and column j. The minor has dimension $(n - 1)$

if the determinant has dimension n. For example, in the 3×3 case above, we write down some different minors:

$$M_{11} = \begin{vmatrix} a_{22} & a_{23} \\ a_{32} & a_{33} \end{vmatrix}$$

$$= a_{22}a_{33} - a_{32}a_{23} \qquad (1.3)$$

$$M_{23} = \begin{vmatrix} a_{11} & a_{12} \\ a_{31} & a_{32} \end{vmatrix}$$

$$= a_{11}a_{32} - a_{12}a_{31} \qquad (1.4)$$

A determinant can be evaluated by taking any row, or any column, and summing the elements in that row (or column) times their minor. The terms in the summation have alternate sign, where the element a_{11} has positive sign and the rest alternate from there. The same result is obtained regardless of which row or column is used for the expansion in minor. The following results are equivalent

$$\det |a| = \sum_{i=1}^{n} (-1)^{i+j} a_{ij} M_{ij}$$

$$= \sum_{j=1}^{n} (-1)^{i+j} a_{ij} M_{ij} \qquad (1.5)$$

The first summation (over i) is for all of the elements of column (j). The second formula is a summation (over j) of the elements in row (i). The two formulas give the same result.

For example, if $n = 4$ then the minors are 3×3 and can be evaluated using the above prescription for this order. However, if $n = 5$ then the minors are 4×4. In this case, the minors also have to be evaluated by expanding them in minors. The procedure can be complicated and laborious once n exceeds a small number.

The expression $(-1)^{i+j} M_{ij}$ is called the *cofactor* of the element a_{ij}.

THEOREM. *The determinant of an $n \times n$ matrix has $n!$ terms.*

Proof. Let N_n denote the number of terms in the determinant of order n. The simple cases had $N_2 = 2$, $N_3 = 6$. If the determinant is evaluated using the expansion in minors, then from Eq. (1.5) the number of terms obeys the recursion $N_n = nN_{n-1}$, since there are n terms in the row or column, and the minor is a determinant with N_{n-1} terms. The relationship $N_n = nN_{n-1}$ defines the factorial function $N_n = n!$

Another way to think about the determinant is that each of the $n!$ terms contains n-elements multiplied together. Among these n-elements are one from each different row and one from each different column. How many combinations are there of one element from each row and from each column? The first row has n element, so that gives n-choices. In selecting an element from the second row, the column is different from the one in the first row, so there are only $(n - 1)$ choices. In the third row, the column is again different from the first two rows, so there are $(n - 2)$ choices. Continuing down the rows, the number of combinations is $n!$. Half of these terms are multiplied by a plus sign, and half by a minus sign.

1.1. Cramer's Rules

Some simple rules are very helpful in the evaluation of a general determinant.

1. A determinant is zero if all elements in a single row, or all elements in a single column, are zero.

2. The value of a determinant is unchanged if all rows are interchanged with all columns.

3. If two rows (or two columns) of a square matrix are interchanged, the sign of the determinant is changed: plus to minus or minus to plus.

4. If all elements of a single row, or all elements of a single column, are multiplied by the same constant k, the determinant is multiplied by k.

5. If the elements of two rows are identical, or differ only by a constant, the determinant is zero. Same applies to any two columns.

6. If each element in a single row (or in a single column) is expressed as the sum of two terms, then the determinant is equal to the sum of two determinants, in each of which one of the two terms is deleted in each element of that row (or column).

7. If one adds to the elements of any row (column), k times the corresponding elements of any other row (column), the determinant is unchanged.

These rules are easily proved. The expansion in minors is used to prove rules 1, 4, 6, and 7. Rule 7 also uses rule 6. Rule 5 follows from 3. If two identical rows are interchanged the determinant must change sign but is unchanged. The only number equal to negative is zero. Rule 3 is the hardest to prove for general n.

As an example of using these rules to evaluate determinants, consider the following one of order 4:

$$X = \det \begin{vmatrix} 1 & 2 & 3 & 4 \\ 5 & 6 & 7 & 8 \\ 9 & 10 & 11 & 12 \\ 13 & 14 & 15 & 16 \end{vmatrix} \tag{1.6}$$

Rather than expand it in minors, simplify using Cramer's rules. This process is mostly trial and error. Try something. If it works, fine. If not, try something else.

For the above determinant, subtract the first row from the other three. Use Rule 7 with $k = -1$ and the value is unchanged. This process gives

$$X = \det \begin{vmatrix} 1 & 2 & 3 & 4 \\ 4 & 4 & 4 & 4 \\ 8 & 8 & 8 & 8 \\ 12 & 12 & 12 & 12 \end{vmatrix} \tag{1.7}$$

Use Rule 4 to remove the factors of 4, 8, 12 from the last three rows. This step multiplies the determinant by these factors:

$$X = 4 \cdot 8 \cdot 12 \det \begin{vmatrix} 1 & 2 & 3 & 4 \\ 1 & 1 & 1 & 1 \\ 1 & 1 & 1 & 1 \\ 1 & 1 & 1 & 1 \end{vmatrix} \tag{1.8}$$

Now three rows are identical. According to Rule 5, then $X = 0$. The determinant is shown to be zero without having to expand in minors. In general, this type of manipulation is useful for evaluating many determinants.

1.2. Gaussian Elimination

Another example using Cramer's rules is the following:

$$X = \det \begin{vmatrix} 10 & 4 & 17 & 13 \\ 4 & 2 & 8 & 6 \\ 3 & -1 & 8 & 1 \\ 7 & 5 & 20 & 17 \end{vmatrix} \tag{1.9}$$

Use the following steps to find this value:

1. Divide each row by the first number in that row. This step results in the first column being all ones

$$X = (10 \cdot 4 \cdot 3 \cdot 7) \det \begin{vmatrix} 1 & 4/10 & 17/10 & 13/10 \\ 1 & 2/4 & 8/4 & 6/4 \\ 1 & -1/3 & 8/3 & 1/3 \\ 1 & 5/7 & 20/7 & 17/7 \end{vmatrix} \tag{1.10}$$

2. Next, subtract the first row from the next three. The first column is now mostly zeros:

$$X = (10 \cdot 4 \cdot 3 \cdot 7) \det \begin{vmatrix} 1 & 2/5 & 17/10 & 13/10 \\ 0 & 1/2 - 2/5 & 2 - 17/10 & 3/2 - 13/10 \\ 0 & -1/3 - 2/5 & 8/3 - 17/10 & 1/3 - 13/10 \\ 0 & 5/7 - 2/5 & 20/7 - 17/10 & 17/7 - 13/10 \end{vmatrix} \tag{1.11}$$

3. Now expand the first column in minors. Only the first element is nonzero. Combine fractions for the remaining elements:

$$X = (10 \cdot 4 \cdot 3 \cdot 7) \det \begin{vmatrix} 1/10 & 3/10 & 2/10 \\ -22/30 & 29/30 & -29/30 \\ 11/35 & 81/70 & 79/70 \end{vmatrix} \qquad (1.12)$$

4. The last step is to extract the common factors in the denominators in each row. Then evaluate the remaining (3×3)

$$X = \frac{(10 \cdot 4 \cdot 3 \cdot 7)}{(10 \cdot 30 \cdot 70)} \det \begin{vmatrix} 1 & 3 & 2 \\ -22 & 29 & -29 \\ 22 & 81 & 79 \end{vmatrix}$$

$$= \frac{1}{25} [1(29)79 + 3(-29)22 + 2(-22)81 - 2(29)22 - 3(-22)79 - 1(-29)81]$$

$$= \frac{3100}{25} = 124 \qquad (1.13)$$

Gaussian elimination is the process of eliminating the first column of a determinant by making all elements in the first row equal to one, and then subtracting the first row from the others. It effectively reduces the dimensionality of the determinant by one. Of course, if n is large, one has reduced it to a determinant of dimension $(n-1)$. Repeating the process reduces the dimension to $(n-2)$. Continue repeating until you get a dimension of one and the process is completed. Gaussian elimination is widely used in computer codes to evaluate determinants, since these steps are repetitive and easily programmed.

The computer program would take the Gaussian elimination down to a determinant of 1×1 before stopping. For example, continue the steps above, starting with (1.12)

5. Factor out the first element in each column of (1.12)

$$X = (10 \cdot 4 \cdot 3 \cdot 7) \cdot \left(\frac{1}{10} \cdot \frac{-22}{30} \cdot \frac{22}{70} \right) \det \begin{vmatrix} 1 & 3 & 2 \\ 1 & -29/22 & 29/22 \\ 1 & 81/22 & 79/22 \end{vmatrix} \qquad (1.14)$$

6. Simplify the prefactor. Subtract the first row from the others.

$$X = -\frac{(22 \cdot 22)}{(5 \cdot 5)} \det \begin{vmatrix} 1 & 3 & 2 \\ 0 & -29/22 - 3 & 29/22 - 2 \\ 0 & 81/22 - 3 & 79/22 - 2 \end{vmatrix}$$

$$= -\frac{(22 \cdot 22)}{(5 \cdot 5)} \det \begin{vmatrix} 1 & 3 & 2 \\ 0 & -95/22 & -15/22 \\ 0 & 15/22 & 35/22 \end{vmatrix} \qquad (1.15)$$

7. Expand the first column in minors. This step reduces the dimension by one

$$X = -\frac{(22 \cdot 22)}{(5 \cdot 5)} \det \begin{vmatrix} -95/22 & -15/22 \\ 15/22 & 35/22 \end{vmatrix} \tag{1.16}$$

8. Divide each row by the first element in that row.

$$X = \frac{(22 \cdot 22)}{(5 \cdot 5)} \left(\frac{95}{22} \cdot \frac{15}{22}\right) \det \begin{vmatrix} 1 & 3/19 \\ 1 & 7/3 \end{vmatrix} \tag{1.17}$$

9. Simplify the prefactor. Subtract the first row from the second one

$$X = (19 \cdot 3) \det \begin{vmatrix} 1 & 3/19 \\ 0 & 7/3 - 3/19 \end{vmatrix} \tag{1.18}$$

10. Expand in minors, which reduces the dimensionality to one

$$X = (19 \cdot 3) \left(\frac{7}{3} - \frac{3}{19}\right) = 124 \tag{1.19}$$

As successive operations reduce the dimensionality of the determinant, various constants are collected in front of the remaining determinant. These constants become the final answer. Gaussian elimination is a practical method for evaluating determinants.

Another technique for evaluating determinants is useful for analytical work. One just adds or subtracts multiples of one row from the others in order to produce a row or column that is mostly zeros. The choice of rows or columns is done by inspection. For example, consider again the evaluation of the determinant in Eq. (1.9). Perform the following steps:

- Multiply the third row by 4 and add it to the first row.
- Multiply the third row by 2 and add it to the second row.
- Multiply the third row by 5 and add it to the fourth row.

$$X = \det \begin{vmatrix} 22 & 0 & 49 & 17 \\ 10 & 0 & 24 & 8 \\ 3 & -1 & 8 & 1 \\ 22 & 0 & 60 & 22 \end{vmatrix} \tag{1.20}$$

- The second column has only one nonzero element. The answer is the minor of this term

$$X = \det \begin{vmatrix} 22 & 49 & 17 \\ 10 & 24 & 8 \\ 22 & 60 & 22 \end{vmatrix} \tag{1.21}$$

- Subtract the third column from the first column.

- Subtract thrice the third column from the second column.

$$X = \det \begin{vmatrix} 5 & -2 & 17 \\ 2 & 0 & 8 \\ 0 & -6 & 22 \end{vmatrix} = 2(-6)17 - [2(-2)22 + 5(-6)8]$$

$$= 124 \tag{1.22}$$

The last two steps were done to avoid multiplying many large numbers, that are prone to errors. This method is probably the easiest to do analytically. Gaussian elimination is easiest on the computer.

1.3. Special Determinants

Below are some special determinants that can be evaluated in a simple fashion.

1. A determinant with all zeros beneath the diagonal, or above the diagonal, is equal to the product of the diagonal elements.

$$D = \det \begin{vmatrix} a_{11} & a_{12} & a_{13} & \cdots & a_{1n} \\ 0 & a_{22} & a_{23} & \cdots & a_{2n} \\ 0 & 0 & a_{33} & \cdots & a_{3n} \\ \vdots & \vdots & \vdots & \ddots & \vdots \\ 0 & 0 & 0 & \cdots & a_{nn} \end{vmatrix}$$

$$= a_{11}a_{22}a_{33} \cdots a_{nn} \tag{1.23}$$

2. Vandermonde Determinant. The matrix is of order n. The elements are $a_{ij} = x_i^{j-1}$

$$D = \det \begin{vmatrix} 1 & x_1 & x_1^2 & \cdots & x_1^{n-1} \\ 1 & x_2 & x_2^2 & \cdots & x_2^{n-1} \\ 1 & x_3 & x_3^2 & \cdots & x_3^{n-1} \\ \vdots & \vdots & \vdots & \ddots & \vdots \\ 1 & x_n & x_n^2 & \cdots & x_n^{n-1} \end{vmatrix}$$

$$= \prod_{j=2}^{n} \prod_{i=1}^{j-1} (x_j - x_i) \tag{1.24}$$

- For $n = 1$ the answer is 1.
- For $n = 2$ the answer is $(x_2 - x_1)$.
- For $n = 3$ the answer is $(x_3 - x_2)(x_3 - x_1)(x_2 - x_1)$

The answer is the product of the difference between all different pairs of x_i. Note that this form is required. The general evaluation of the determinant is a polynomial of $n!$ terms. If any two values of x_i, x_j are alike, then the polynomial must vanish since two rows are identical. The answer must have a functional form of $(x_i - x_j)$ in order that it vanishes when $x_i = x_j$. Since x_i, x_j are any pair of rows, the polynomial must have the form of a product of $(x_i - x_j)$ for all pairs of rows. The final answer is a polynomial of degree $n(n-1)/2$. The degree of the polynomial is found by summing the exponents in a row

$$0 + 1 + 2 + \cdots + (n-1) = \sum_{j=0}^{n-1} j = \frac{n(n-1)}{2} \tag{1.25}$$

Since the product of all pairs is a polynomial of degree $n(n-1)/2$, and since the answer is a polynomial of the same order, then the final answer is the product of all pairs of differences. Of course, this line of reasoning does not eliminate the possibility that the answer is multiplied by a dimensionless number such as two, but that does not occur in the present example.

3. $a_{ij} = x_i - y_j$

$$D = \det \begin{vmatrix} x_1 - y_1 & x_1 - y_2 & x_1 - y_3 & \cdots & x_1 - y_n \\ x_2 - y_1 & x_2 - y_2 & x_2 - y_3 & \cdots & x_2 - y_n \\ x_3 - y_1 & x_3 - y_2 & x_3 - y_3 & \cdots & x_3 - y_n \\ \vdots & \vdots & \vdots & \ddots & \vdots \\ x_n - y_1 & x_n - y_2 & x_n - y_3 & \cdots & x_n - y_n \end{vmatrix} \tag{1.26}$$

- For $n = 1$ the answer is $(x_1 - y_1)$.
- For $n = 2$ the answer is $(x_1 - x_2)(y_1 - y_2)$.
- For $n \geqslant 3$ the answer is zero. This can be proved in a simple way. Subtract the first row from the other rows, which produces

$$D = \det \begin{vmatrix} x_1 - y_1 & x_1 - y_2 & x_1 - y_3 & \cdots & x_1 - y_n \\ x_2 - x_1 & x_2 - x_1 & x_2 - x_1 & \cdots & x_2 - x_1 \\ x_3 - x_1 & x_3 - x_1 & x_3 - x_1 & \cdots & x_3 - x_1 \\ \vdots & \vdots & \vdots & \ddots & \vdots \\ x_n - x_1 & x_n - x_1 & x_n - x_1 & \cdots & x_n - x_1 \end{vmatrix} \tag{1.27}$$

Each row except the first has the same element. The rows differ only by a constant, and the determinant vanishes according to Cramer's rules. This determinant, like the previous one, must vanish if any two values of x_i, or any two values of y_j, are equal. However, that observation only applies to the case for $n = 2$. For $n > 2$ it vanishes anyway. The vanishing of the result could have been anticipated on the basis of algebra. The determinant must vanish if any two values of x_i are alike, or if any two values of y_j are alike. The answer must be a polynomial with the factor of

$$\prod_{i>j} (x_i - x_j) \prod_{n>m} (y_n - y_m) \tag{1.28}$$

Each one of these products is a polynomial of degree $n(n-1)/2$, so the double product is a polynomial of degree $n(n-1)$. In the original determinant, each element is a polynomial of degree one, so each term in the expansion of the determinant is a polynomial of degree n. For these two statements to be consistent, then $n = n(n - 1)$. The only solution for positive n is that $n = 2$. That is the only order of the determinant that can have a consistent polynomial solution.

4. $a_{ij} = 1/(x_i - y_j)$

$$D = \det \begin{vmatrix} \dfrac{1}{x_1 - y_1} & \dfrac{1}{x_1 - y_2} & \dfrac{1}{x_1 - y_3} & \cdots & \dfrac{1}{x_1 - y_n} \\[2mm] \dfrac{1}{x_2 - y_1} & \dfrac{1}{x_2 - y_2} & \dfrac{1}{x_2 - y_3} & \cdots & \dfrac{1}{x_2 - y_n} \\[2mm] \dfrac{1}{x_3 - y_1} & \dfrac{1}{x_3 - y_2} & \dfrac{1}{x_3 - y_3} & \cdots & \dfrac{1}{x_3 - y_n} \\[2mm] \vdots & \vdots & \vdots & \ddots & \vdots \\[2mm] \dfrac{1}{x_n - y_1} & \dfrac{1}{x_n - y_2} & \dfrac{1}{x_n - y_3} & \cdots & \dfrac{1}{x_n - y_n} \end{vmatrix} \qquad (1.29)$$

Here the result is never zero as long as the x_i and y_j are all different. The numerator is the difference of all pairs $(x_j - x_i)(y_i - y_j)$:

- For $n = 1$ then $D = 1/(x_1 - y_1)$.
- For $n = 2$ the answer is:

$$D = \frac{(y_2 - y_1)(x_1 - x_2)}{(x_1 - y_1)(x_1 - y_2)(x_2 - y_1)(x_2 - y_2)} \qquad (1.30)$$

- For $n > 1$ the result is

$$D = \frac{\Pi_{j=2}^{n}\, \Pi_{i=1}^{j-1}\, (x_i - x_j)(y_j - y_i)}{\Pi_{i=1}^{n}\, \Pi_{j=1}^{n}\, (x_i - y_j)} \qquad (1.31)$$

The denominator is the product of all of the elements. The numerator is the product of all pair of values $(x_i - x_j)(y_j - y_i)$. The latter form is required for the determinant to vanish whenever $x_i = x_j$ or else $y_i = y_j$, since two rows, or two columns, are identical.

In this case the polynomials have the correct degree. The numerator of Eq. (1.30) is a polynomial of degree $n(n - 1)$ while the denominator has a degree of n^2. The net degree of the answer is $n(n-1)-n^2 = -n$. Since each element of the determinant has degree of -1, then each polynomial in the expansion has a degree of $-n$. The two methods of finding the degree are consistent.

5. Pfaffians are determinants of skew matrices: $a_{ji} = -a_{ij}$. This skewness also requires that the diagonal elements are zero ($a_{ii} = 0$). Pfaffians have the property that: (i) if n is an odd integer, the determinant vanishes; (ii) if n is an even integer, the determinant does not vanish, but is a perfect

square. Some examples are:

$$D_2 = \det \begin{vmatrix} 0 & a_{12} \\ -a_{12} & 0 \end{vmatrix} = a_{12}^2 \tag{1.32}$$

$$D_3 = \det \begin{vmatrix} 0 & a_{12} & a_{13} \\ -a_{12} & 0 & a_{23} \\ -a_{13} & -a_{23} & 0 \end{vmatrix} = 0 \tag{1.33}$$

$$D_4 = \det \begin{vmatrix} 0 & a_{12} & a_{13} & a_{14} \\ -a_{12} & 0 & a_{23} & a_{24} \\ -a_{13} & -a_{23} & 0 & a_{34} \\ -a_{14} & -a_{24} & -a_{34} & 0 \end{vmatrix} = (a_{12}a_{34} + a_{14}a_{23} - a_{13}a_{24})^2 \tag{1.34}$$

It is easy to prove that the Pfaffian vanish whenever n is odd. The proof proceeds in two steps. The first is to change the sign of every element. This step is equivalent to multiplying each row by (-1) so that the determinant is multiplied by $(-1)^n$. This action results in all elements above the diagonal having a minus sign, while those below have a plus sign. The original determinant is recovered by exchanging all rows and columns. This step does not change the answer. The derivation proves $D_n = (-1)^n D_n$. For an odd value of n then $D_n = -D_n$ that has the solution $D_n = 0$.

6. Let $a_{ij} = 1$ if $i \neq j$ and $a_{ii} = 1 + x$. The determinant has all ones except the diagonal elements, which are $1 + x$. This determinant is often encountered in physics problems, as discussed in Chapter 2.

$$D_n(x) = \det \begin{vmatrix} 1+x & 1 & 1 & \cdots & 1 \\ 1 & 1+x & 1 & \cdots & 1 \\ 1 & 1 & 1+x & \cdots & 1 \\ \vdots & \vdots & \vdots & \ddots & \vdots \\ 1 & 1 & 1 & \cdots & 1+x \end{vmatrix}$$

$$= (x)^{n-1}(x + n) \tag{1.35}$$

Of course, if $x = 0$ then the determinant must vanish, so it is expected that it is proportional to at least one power of x. However, it is proportional to the $n - 1$ power of x. This result can be proved by first subtracting the first row from the others:

$$D_n(x) = \det \begin{vmatrix} 1+x & 1 & 1 & \cdots & 1 \\ -x & x & 0 & \cdots & 0 \\ -x & 0 & x & \cdots & 0 \\ \vdots & \vdots & \vdots & \ddots & \vdots \\ -x & 0 & 0 & \cdots & x \end{vmatrix} \tag{1.36}$$

Expanding the last row in minors gives two determinants. One has the same form as D_n but has one less term on the diagonal, and is D_{n-1}. The other

$$\det \begin{vmatrix} 1 & 1 & \cdots & 1 & 1 \\ x & 0 & \cdots & 0 & 0 \\ 0 & x & \cdots & 0 & 0 \\ \vdots & \vdots & \vdots & \ddots & \vdots \\ 0 & 0 & \cdots & x & 0 \end{vmatrix} \qquad (1.37)$$

Expand the last row by minors that gives a single factor of x. Repeat this process that shows this minor equals x^{n-2}. The above steps provide the recursion relation

$$D_n = x^{n-1} + xD_{n-1} = x(x^{n-2} + D_{n-1}) \qquad (1.38)$$

The lowest few terms in the recursion are

$$D_1 = 1 + x \qquad (1.39)$$

$$D_2 = x[1 + (1 + x)] = x(2 + x) \qquad (1.40)$$

$$D_3 = x[x + x(2 + x)] = x^2(3 + x) \qquad (1.41)$$

The solution is (1.35).

References

1. F. B. Hildebrand, *Methods of Applied Mathematics* (Prentice Hall, 1952).

Problems

Evaluate the following determinants:

$$D_1 = \det \begin{vmatrix} 4 & 7 & 3 \\ -2 & 11 & 5 \\ -3 & 1 & -6 \end{vmatrix} \qquad (1.42)$$

$$D_2 = \det \begin{vmatrix} 10 & 4 & 3 & 7 \\ 4 & 2 & -1 & 5 \\ 17 & 8 & 8 & 20 \\ 13 & 6 & 1 & 17 \end{vmatrix} \qquad (1.43)$$

$$D_3 = \det \begin{vmatrix} 1 & 1 & 1 & 1 \\ 1 & 1+x & 1 & 1 \\ 1 & 1 & 1+y & 1 \\ 1 & 1 & 1 & 1+z \end{vmatrix} \qquad (1.44)$$

$$D_4 = \det \begin{vmatrix} 1+a & 1 & 1 & 1 \\ 1 & 1+b & 1 & 1 \\ 1 & 1 & 1+c & 1 \\ 1 & 1 & 1 & 1+d \end{vmatrix} \qquad (1.45)$$

$$D_5 = \det \begin{vmatrix} 1 & a & b & c \\ a & 1 & 0 & 0 \\ b & 0 & 1 & 0 \\ c & 0 & 0 & 1 \end{vmatrix} \tag{1.46}$$

$$D_6 = \det \begin{vmatrix} 0 & a & b & c \\ a & 1 & 0 & 0 \\ b & 0 & 1 & 0 \\ c & 0 & 0 & 1 \end{vmatrix} \tag{1.47}$$

$$D_7 = \det \begin{vmatrix} 1 & a & b+c \\ 1 & b & a+c \\ 1 & c & a+b \end{vmatrix} \tag{1.48}$$

$$D_8 = \det \begin{vmatrix} 1 & a & 2a^2 \\ 1 & b & 2b^2 \\ 1 & c & 2c^2 \end{vmatrix} \tag{1.49}$$

2

Matrices

Matrices are arrays of numbers. The arrays have a dimension of $n \times m$. The present discussion is limited to two cases: square arrays ($n \times n$) that are called matrices, and vectors ($1 \times n$). The matrices have elements a_{ij} where the indices i ($1 \leqslant i \leqslant n$) denote the rows while j ($1 \leqslant j \leqslant n$) denote the columns. The matrix is represented by the symbol $\mathscr{A}$. Some special definitions are:

- Real matrices have real numbers for all elements, these are summarized as $a_{ij}^* = a_{ij}$. The asterisk symbol denotes the complex conjugate.

- The transpose of a matrix is obtained by interchanging all rows and columns, that can be expressed by the symbol T such as $a_{ij}^T = a_{ji}$.

- A symmetric matrix equals its transpose: $a_{ij} = a_{ji}$, or, $\mathscr{A}^T = \mathscr{A}$.

- Hermitian matrices have $a_{ji} = a_{ij}^*$, that is expressed by a dagger ($\mathscr{A}^\dagger = \mathscr{A}$). In quantum mechanics the Hamiltonian operator is often represented by a matrix. This operator is always Hermitian.

- The identity matrix has all elements zero except for the diagonal elements that are one. It has the symbol $\mathscr{I}$

$$\mathscr{I} = \begin{bmatrix} 1 & 0 & 0 & \cdots & 0 \\ 0 & 1 & 0 & \cdots & 0 \\ 0 & 0 & 1 & \cdots & 0 \\ \vdots & \vdots & \vdots & \ddots & 0 \\ 0 & 0 & 0 & \cdots & 1 \end{bmatrix} \qquad (2.1)$$

The identity matrix commutes with all other matrices.

- A matrix could possess an inverse that is denoted by the exponent of (-1). The inverse of $\mathscr{A}$ is called $\mathscr{A}^{-1}$ and has the property $\mathscr{A}\mathscr{A}^{-1} = \mathscr{I} = \mathscr{A}^{-1}\mathscr{A}$. Not every matrix has an inverse. An inverse exists if the determinant of the matrix is nonzero. This criterion is proved below.

- A unitary matrix has its Hermitian matrix as its inverse: $\mathscr{A}\mathscr{A}^\dagger = \mathscr{I}$ so that $\mathscr{A}^{-1} = \mathscr{A}^\dagger$.

- Matrix multiplication is defined as $\mathscr{A}\mathscr{B} = \mathscr{C}$ that in terms of components is

$$\sum_{k=1}^{n} a_{ik} b_{kj} = c_{ij} \tag{2.2}$$

The element c_{ij} is obtained by multiplying the column j of $\mathscr{B}$ by the row i of $\mathscr{A}$. Here the word "multiplying" means that each element of one matrix is multiplied by the corresponding element of the other, and the results added.

- The *Trace* of a matrix is the summation of its diagonal elements. It is denoted by the symbol "Tr":

$$\text{Tr}[\mathscr{A}] = \sum_{i=1}^{n} a_{ii} \tag{2.3}$$

2.1. Several Theorems

Several theorems are proved regarding matrices. They are simple, yet important.

1. The transpose of a product of two matrices is the reverse product of their transposes: If $\mathscr{C} = \mathscr{A}\mathscr{B}$ then $\mathscr{C}^T = \mathscr{B}^T \mathscr{A}^T$. The proof is done by examining components

$$c_{ij} = \sum_{k=1}^{n} a_{ik} b_{kj} \tag{2.4}$$

$$c_{ij}^T = c_{ji} = \sum_{k=1}^{n} a_{jk} b_{ki}$$

$$= \sum_{k=1}^{n} b_{ik}^T a_{kj}^T \tag{2.5}$$

$$\mathscr{C}^T = \mathscr{B}^T \mathscr{A}^T \tag{2.6}$$

Take the complex conjugate and also prove that $\mathscr{C}^\dagger = \mathscr{B}^\dagger \mathscr{A}^\dagger$.

2. If $\mathscr{A}$ and $\mathscr{B}$ are unitary, then $\mathscr{C} = \mathscr{A}\mathscr{B}$ is unitary. The proof uses the associative law:

$$\mathscr{A}^\dagger \mathscr{A} = \mathscr{I} \tag{2.7}$$

$$\mathscr{B}^\dagger \mathscr{B} = \mathscr{I} \tag{2.8}$$

$$\mathscr{C}^\dagger \mathscr{C} = (\mathscr{A}\mathscr{B})^\dagger (\mathscr{A}\mathscr{B}) = (\mathscr{B}^\dagger \mathscr{A}^\dagger)(\mathscr{A}\mathscr{B})$$

$$= \mathscr{B}^\dagger (\mathscr{A}^\dagger \mathscr{A})\mathscr{B} = \mathscr{B}^\dagger \mathscr{B}$$

$$= \mathscr{I} \tag{2.9}$$

3. The trace of a product of matrices is unchanged by a cyclic rearrangement of the matrices that is expressed as:

$$\text{Tr}[\mathscr{A}\mathscr{B}\mathscr{C}\ldots\mathscr{X}\mathscr{Y}\mathscr{Z}] = \text{Tr}[\mathscr{B}\mathscr{C}\ldots\mathscr{X}\mathscr{Y}\mathscr{Z}\mathscr{A}] \tag{2.10}$$

where the matrix $\mathscr{A}$ has been moved from the far left to the far right. First, prove a simple version of the theorem that is $\text{Tr}[\mathscr{A}\mathscr{B}] = \text{Tr}[\mathscr{B}\mathscr{A}]$. In terms of components,

$$\text{Tr}(\mathscr{A}\mathscr{B}) = \sum_{i=1}^{n} \sum_{k=1}^{n} a_{ik} b_{ki} = \sum_{k=1}^{n} \sum_{i=1}^{n} b_{ki} a_{ik}$$
$$= \text{Tr}(\mathscr{B}\mathscr{A}) \tag{2.11}$$

The double summation can be achieved in either order, and the components can be moved, which proves the theorem. For three matrices

$$\text{Tr}(\mathscr{A}\mathscr{B}\mathscr{C}) = \sum_{ijk} a_{ij} b_{jk} c_{ki} = \sum_{ijk} b_{jk} c_{ki} a_{ij}$$
$$= \text{Tr}(\mathscr{B}\mathscr{C}\mathscr{A}) \tag{2.12}$$

The extension to N matrices is obvious.

2.2. Linear Equations

Linear equations can be solved easily. Assume there is a matrix $\mathscr{A}$ where all the elements are known. Also, there is a vector $\mathscr{C}$ whose elements are known. The task is to find the unknown elements of the vector $\mathscr{X}$ from the equation,

$$\mathscr{A}\mathscr{X} = \mathscr{C} \tag{2.13}$$

$$\sum_{k=1}^{n} a_{ik} x_k = c_i \tag{2.14}$$

The solution is rather trivial if the inverse exists of the matrix $\mathscr{A}$. Multiplying the above equation by this inverse gives

$$\mathscr{X} = \mathscr{A}^{-1}\mathscr{C} \tag{2.15}$$

$$x_i = \sum_{j=1}^{n} a_{ij}^{-1} c_j \tag{2.16}$$

The method of finding the inverse is given below.

Another way of solving these equations is using Gaussian elimination. The steps are:

1. Divide each row by the coefficient of the first term a_{i1}. If this coefficient is zero do not bother for this row.

$$x_1 + \frac{a_{12}}{a_{11}} x_2 + \frac{a_{13}}{a_{11}} x_3 + \cdots + = \frac{c_1}{a_{11}} \tag{2.17}$$

$$x_1 + \frac{a_{22}}{a_{21}} x_2 + \frac{a_{23}}{a_{21}} x_3 + \cdots + = \frac{c_2}{a_{21}} \tag{2.18}$$

$$x_1 + \frac{a_{32}}{a_{31}} x_2 + \frac{a_{33}}{a_{31}} x_3 + \cdots + = \frac{c_3}{a_{31}} \tag{2.19}$$

2. Subtract the first row from each of the other rows. This step eliminates the factor of x_1 from all but the first row.

$$x_1 + \frac{a_{12}}{a_{11}} x_2 + \frac{a_{13}}{a_{11}} x_3 + \cdots + = \frac{c_1}{a_{11}} \tag{2.20}$$

$$a'_{22} x_2 + a'_{23} x_3 + \cdots + = c'_2 \tag{2.21}$$

$$a'_{32} x_2 + a'_{33} x_3 + \cdots + = c'_3 \tag{2.22}$$

$$a'_{ij} = \frac{a_{ij}}{a_{i1}} - \frac{a_{1j}}{a_{11}} \tag{2.23}$$

$$c'_i = \frac{c_i}{a_{i1}} - \frac{c_1}{a_{11}} \tag{2.24}$$

3. The first row is now frozen and no longer manipulated. The remaining equations define a set of linear equations of dimension $(n-1)$. Solve it by Gaussian elimination. Eliminating the unknown x_2 in all but the first equation gives

$$x_1 + \frac{a_{12}}{a_{11}} x_2 + \frac{a_{13}}{a_{11}} x_3 + \cdots + = \frac{c_1}{a_{11}} \tag{2.25}$$

$$x_2 + \frac{a'_{23}}{a'_{22}} x_3 + \cdots + = \frac{c'_2}{a'_{22}} \tag{2.26}$$

$$a''_{33} x_3 + \cdots + = c''_3 \tag{2.27}$$

$$a''_{ij} = \frac{a'_{ij}}{a'_{i2}} - \frac{a'_{2j}}{a'_{22}} \tag{2.28}$$

$$c''_i = \frac{c'_i}{a'_{i2}} - \frac{c'_2}{a'_{22}} \tag{2.29}$$

4. Now the second equation is frozen, and Gaussian elimination is used to reduce further. This process is repeated n-times to get a set of equations of the form

$$x_1 + \frac{a_{12}}{a_{11}} x_2 + \frac{a_{13}}{a_{11}} x_3 + \cdots + = \frac{c_1}{a_{11}} \tag{2.30}$$

$$x_2 + \frac{a'_{23}}{a'_{22}} x_3 + \cdots + = \frac{c'_2}{a'_{22}} \tag{2.31}$$

$$x_3 + \frac{a''_{43}}{a''_{33}} + \cdots + = \frac{c''_3}{a''_{33}} \tag{2.32}$$

$$x_{n-1} + \frac{a^{(n-2)}_{n,n-1}}{a^{(n-2)}_{n-1,n-1}} x_n = \frac{c^{(n-2)}_{n-1}}{a^{(n-2)}_{n-1,n-1}} \tag{2.33}$$

$$x_n = \frac{c^{(n-1)}}{a^{(n-1)}_{nn}} \tag{2.34}$$

5. The equations are now solved. Equation (2.34) gives x_n. Equation (2.33) can be used to find x_{n-1} once x_n is known. Each of the equations gives one more of the x_j until the last which gives x_1. This simple procedure is simple to do on the computer, since one keeps doing the same operation over and over again.

Note that Gaussian elimination does not work if $\det[\mathscr{A}] = 0$. If the determinant vanishes, it is because two rows are proportional, or else two columns. Suppose that the first row of elements is proportional to another row r. When subtracting the first row from the others, the row r completely vanishes, since all the coefficients $a_{rj}/a_{r1} = a_{1j}/a_{11}$. If a complete row vanishes, there are no longer enough equations to provide a solution. In general, equation $\mathscr{A} \cdot X = C$ has no solution if $\det[\mathscr{A}] = 0$.

2.3. Inverse of a Matrix

The inverse of a matrix is obtained from its minors. Recall that in Chapter 1 on determinants, a minor M_{ij} of an element a_{ij} is the determinant of dimension $(n-1) \times (n-1)$ obtained by eliminating row i and column j from $\mathscr{A}$. Also recall that the determinant can be found by summing a row, or a column, of elements and their minors. Because of the sign alternation $(-1)^{i+j}$ it is useful to include it in the definition of another matrix

$$N_{ij} = (-1)^{i+j} M_{ij} \tag{2.35}$$

$$\det|\mathscr{A}| = \sum_{k=1}^{n} a_{ik} N_{ik}$$

$$= \sum_{k=1}^{n} a_{kj} N_{kj} \tag{2.36}$$

The two expressions show the summation over rows, and over columns.

Consider the result if the minors N_{ij} are summed using the elements from another row (or column)

$$\sum_{k=1}^{n} a_{rk} N_{ik} = \delta_{ri} \det|\mathscr{A}| \tag{2.37}$$

$$\sum_{k=1}^{n} a_{kr} N_{ki} = \delta_{ri} \det|\mathscr{A}| \tag{2.38}$$

The delta function expresses the result that the summation gives zero unless $r = i$. The zero results from Cramer's rules on determinants. The matrix N_{ik} is constructed using the elements of all rows except (ik). It uses the elements in row r if $r \neq i$, and uses the elements in column l if $l \neq k$. A determinant is zero if any two rows, or any two columns, are identical. If $r \neq i$ in the above summation, then one is evaluating a determinant that has row i identical to row r. It must vanish. The second expression gives the same result for columns.

These relations are the basis for the inverse matrix. Define a new matrix $\mathscr{L}$ as the transpose of $\mathscr{N}$ divided by the determinant

$$\mathscr{L} = \frac{\mathscr{N}^T}{\det|\mathscr{A}|} \tag{2.39}$$

$$L_{ij} = \frac{N_{ji}}{\det|\mathscr{A}|} \tag{2.40}$$

The transpose is taken because of the definition of matrix multiplication. The matrix $\mathscr{L}$ is the inverse of $\mathscr{A}$

$$\mathscr{I} = \mathscr{A}\mathscr{L} = \mathscr{L}\mathscr{A} \tag{2.41}$$

$$\delta_{ij} = \sum_{k=1}^{n} a_{ik} L_{kj} = \sum_{k=1}^{n} a_{ik} N_{jk}/\det|\mathscr{A}| \tag{2.42}$$

where the last line makes the result resemble the solution using minors. It is simple to construct the inverse of a matrix using the definition of the minor.

A necessary condition for the existence of an inverse matrix is that the determinant of the matrix be nonzero. If the determinant is zero, there is no inverse.

As an example, consider the inverse of the matrix

$$\mathscr{A} = \begin{pmatrix} 1 & 2 & 3 \\ 3 & 2 & 1 \\ 2 & 1 & 2 \end{pmatrix} \tag{2.43}$$

The first step is to construct the minor M_{ij} of each term, and then to multiply by $(-1)^{i+j}$ to get N_{ij}, and then transpose $\mathscr{N}$ to get the inverse.

$$\mathscr{M} = \begin{pmatrix} 3 & 4 & -1 \\ 1 & -4 & -3 \\ -4 & -8 & -4 \end{pmatrix} \tag{2.44}$$

$$\mathscr{N} = \begin{pmatrix} 3 & -4 & -1 \\ -1 & -4 & 3 \\ -4 & 8 & -4 \end{pmatrix} \tag{2.45}$$

$$\mathscr{A}^{-1} = -\frac{1}{8}\begin{pmatrix} 3 & -1 & -4 \\ -4 & -4 & 8 \\ -1 & 3 & -4 \end{pmatrix} \tag{2.46}$$

The prefix comes from $\det|\mathscr{A}| = -8$. The answer can be checked by explicit matrix multiplication.

2.4. Eigenvalues and Eigenvectors

The word *eigen* is German and means *self* or *own*. The words eigenvalue or eigenvectors are hybrids, with part of them from two languages.

For each $n \times n$ matrix $\mathcal{A}$, there are n sets of eigenvalues and eigenvectors. The eigenvalues are scalars that are denoted λ_α where the values of α run from $1 \leqslant \alpha \leqslant n$. The method of finding these eigenvalues is described below. The eigenvectors are vectors $X^{(\alpha)}$ with components $x_i^{(\alpha)}$. They have the property

$$\mathcal{A}X^{(\alpha)} = \lambda_\alpha X^{(\alpha)} \tag{2.47}$$

$$\sum_j a_{ij} x_j^{(\alpha)} = \lambda_\alpha x_i^{(\alpha)} \tag{2.48}$$

Multiplying an eigenvector times the matrix gives a multiple of the same eigenvector, and the multiple is the eigenvalue. Another way to write the above equation is

$$\mathcal{C} = \mathcal{A} - \lambda \mathcal{I} \tag{2.49}$$

$$\mathcal{C}X = 0 \tag{2.50}$$

This expression is the key to finding the eigenvalues. You may contemplate solving the above equation by multiplying it by the inverse of the matrix $\mathcal{C}$ that gives

$$\mathcal{C}^{-1}\mathcal{C}X = 0 \tag{2.51}$$

$$X = 0 \tag{2.52}$$

where $\mathcal{C}^{-1}\mathcal{C} = \mathcal{I}$. The last line shows that the eigenvector is identically zero. However, the eigenvector is not zero, so an error is made in this derivation. The result $X = 0$ is obtained as long as the matrix $\mathcal{C}$ has an inverse. The only way to have $X \neq 0$ is to assume the matrix $\mathcal{C}$ does not possess an inverse. The criteria for having an inverse is that $\det |\mathcal{C}| \neq 0$. The lack of an inverse requires that

$$0 = \det |\mathcal{C}| = \det |\mathcal{A} - \lambda \mathcal{I}| \tag{2.53}$$

This equation produces the eigenvalues. Expanding the determinant always gives an equation for λ that is an nth degree polynomial

$$0 = \lambda^n + b_{n-1}\lambda^{n-1} + \cdots + b_1\lambda + b_0$$

$$= \prod_{\alpha=1}^{n} [\lambda - \lambda_\alpha] \tag{2.54}$$

A polynomial equation of the nth degree always has n roots, called λ_α. They are the eigenvalues. The procedure for finding the eigenvalues is to reduce the determinant in Eq. (2.53) to a polynomial equation and then solve for its roots. A theorem in algebra states that the polynomial equation can always be written as the product of $(\lambda_\alpha - \lambda)$ of the eigenvalues. This latter result can be expressed as

$$\det|\mathscr{A} - \lambda\mathscr{I}| = \prod_{\alpha=1}^{n} [\lambda_\alpha - \lambda] \qquad (2.55)$$

$$\det|\mathscr{A}| = \prod_{\alpha=1}^{n} \lambda_\alpha \qquad (2.56)$$

The second line has a new result obtained by setting $\lambda = 0$. It is presented as a theorem.

THEOREM: *The determinant of a matrix is the product of all of the eigenvalues of that matrix.*

Two examples are presented for finding the eigenvalues and eigenvectors. The simplest way to find eigenvectors is to determine the vector X that satisfies $\mathscr{C}X = 0$. The rows of $\mathscr{C}$ should be identical within a constant factor.

1. The first matrix is

$$\mathscr{A} = \begin{bmatrix} 1 & 2 \\ 3 & 2 \end{bmatrix} \qquad (2.57)$$

$$0 = \det|\mathscr{A} - \lambda\mathscr{I}| = \det \begin{bmatrix} 1 - \lambda & 2 \\ 3 & 2 - \lambda \end{bmatrix}$$

$$= (1 - \lambda)(2 - \lambda) - 6 = (\lambda - 4)(\lambda + 1) \qquad (2.58)$$

The two eigenvalues and eigenvectors are:

- $\lambda_1 = -1$

$$\mathscr{C}_1 = \mathscr{A} - \lambda_1\mathscr{I} = \begin{bmatrix} 2 & 2 \\ 3 & 3 \end{bmatrix} \qquad (2.59)$$

$$X^{(1)} = \frac{1}{\sqrt{2}} \begin{bmatrix} 1 \\ -1 \end{bmatrix} \qquad (2.60)$$

- $\lambda_2 = 4$

$$\mathscr{C}_2 = \mathscr{A} - \lambda_2\mathscr{I} = \begin{bmatrix} -3 & 2 \\ 3 & -2 \end{bmatrix} \qquad (2.61)$$

$$X^{(2)} = \frac{1}{\sqrt{13}} \begin{bmatrix} 2 \\ 3 \end{bmatrix} \qquad (2.62)$$

The eigenvectors are always normalized so that $X^\dagger X = 1$, that explains the prefactors involving square roots.

2. The second matrix is chosen to show that real nonsymmetric matrices may have complex eigenvalues and eigenvectors.

$$\mathscr{A} = \begin{bmatrix} 1 & 2 \\ -2 & 3 \end{bmatrix} \tag{2.63}$$

$$0 = \det|\mathscr{A} - \lambda\mathscr{I}| = \det \begin{bmatrix} 1 - \lambda & 2 \\ -2 & 3 - \lambda \end{bmatrix}$$

$$= (1 - \lambda)(3 - \lambda) + 4 = (\lambda - \lambda_1)(\lambda - \lambda_2) \tag{2.64}$$

$$\lambda_1 = 2 + i\sqrt{3} \tag{2.65}$$

$$\lambda_2 = 2 - i\sqrt{3} \tag{2.66}$$

The two eigenvalues are complex. In fact, they are complex conjugates of each other. Complex eigenvalues of real matrices always occur in pairs which are mutual complex conjugates. This feature is guaranteed by Theorem (2.56) stating the determinant is the product of the eigenvalues. A real matrix has a real determinant, requiring that complex numbers occur in complex conjugate pairs. A real number results when they are multiplied together. $[\lambda_1\lambda_2 = (a + ib)(a - ib) = a^2 + b^2]$. The eigenvectors are

- $\lambda_1 = 2 + i\sqrt{3}$, and $\tan(\theta) = \sqrt{3}$ so that $\theta = 60°$

$$\mathscr{C}_1 = \mathscr{A} - \lambda_1\mathscr{I} = \begin{bmatrix} -1 - i\sqrt{3} & 2 \\ -2 & 1 - i\sqrt{3} \end{bmatrix}$$

$$= 2\begin{bmatrix} -e^{i\theta} & 1 \\ -1 & e^{-i\theta} \end{bmatrix} \tag{2.67}$$

$$X^{(1)} = \frac{1}{\sqrt{2}}\begin{bmatrix} 1 \\ e^{i\theta} \end{bmatrix} \tag{2.68}$$

- $\lambda_2 = 2 - i\sqrt{3}$,

$$\mathscr{C}_2 = \mathscr{A} - \lambda_2\mathscr{I} = 2\begin{bmatrix} -e^{-i\theta} & 1 \\ -1 & e^{i\theta} \end{bmatrix} \tag{2.69}$$

$$X^{(2)} = \frac{1}{\sqrt{2}}\begin{bmatrix} 1 \\ e^{-i\theta} \end{bmatrix} \tag{2.70}$$

In physics the Hamiltonian operator can often be represented by a matrix that is Hermitian. Two important theorems are proved regarding Hermitian matrices. These theorems also apply to real symmetric matrices.

1. *A Hermitian matrix has all of its eigenvalues being real.* The proof starts by taking the Hermitian conjugate of the eigenvalue equation and

remembering that $\mathscr{A}^\dagger = \mathscr{A}$

$$\mathscr{A} X^{(i)} = \lambda_i X^{(i)} \tag{2.71}$$

$$X^{(i)\dagger} \mathscr{A} = \lambda_i^* X^{(i)\dagger} \tag{2.72}$$

Next, multiply the first equation on the left by $X^{(i)\dagger}$ and the second from the right by $X^{(i)}$

$$X^{(i)\dagger} \mathscr{A} X^{(i)} = \lambda_i X^{(i)\dagger} X^{(i)} \tag{2.73}$$

$$X^{(i)\dagger} \mathscr{A} X^{(i)} = \lambda_i^* X^{(i)\dagger} X^{(i)} \tag{2.74}$$

The normalization of the eigenvectors is that $X^{(i)\dagger} X^{(i)} = 1$. When subtracting these two equations, the left-hand sides cancel, while the right side gives

$$0 = \lambda_i - \lambda_i^* \tag{2.75}$$

The only solution to this equation is that λ_i must be real.

2. *Different eigenvectors of a Hermitian matrix are orthogonal.* The proof starts out like the one above, except the conjugated equation is for a different eigenstate

$$\mathscr{A} X^{(i)} = \lambda_i X^{(i)} \tag{2.76}$$

$$X^{(j)\dagger} \mathscr{A} = \lambda_j X^{(j)\dagger} \tag{2.77}$$

Note it is assumed that the eigenvalues are real. Again, multiply the first equation from the left by $X^{(j)\dagger}$ and the second equation from the right by $X^{(i)}$ and then subtract these two equations

$$X^{(j)\dagger} \mathscr{A} X^{(i)} = \lambda_i X^{(j)\dagger} X^{(i)} \tag{2.78}$$

$$X^{(j)\dagger} \mathscr{A} X^{(i)} = \lambda_j X^{(j)\dagger} X^{(i)} \tag{2.79}$$

$$0 = (\lambda_i - \lambda_j) X^{(j)\dagger} X^{(i)} \tag{2.80}$$

The last line shows that if the eigenvalues are not equal, then the product of the eigenstates is zero. A zero product is called *orthogonal*. This theorem is expressed as

$$X^{(j)\dagger} X^{(i)} = \delta_{ij} \tag{2.81}$$

When $i = j$ the normalization gives the product as one. Many matrices have two or more eigenvalues that are equal. The proof fails for these cases: if $\lambda_i = \lambda_j$ then the above relationship does not require their eigenvectors to be orthogonal. However, in these cases, one can always construct orthogonal eigenvectors. The relationship Eq. (2.81) is of fundamental importance.

An earlier example found the eigenvalues and eigenvectors of two matrices that were 2 × 2. Their eigenvectors are *not* orthogonal. The original matrix was not symmetric, so the theorems do not apply. Matrices that are not Hermitian do not usually have eigenvectors that are orthogonal.

2.5. Unitary Transformations

Assume there is a Hermitian matrix $\mathscr{A}$ of dimension $n \times n$. It has n real eigenvalues λ_α, and each has an eigenvector $X^{(\alpha)}$ that is a column vector. Create an $n \times n$ matrix, called $\mathscr{S}$, whose columns are the n eigenvectors of $\mathscr{A}$.

$$\mathscr{S} = \begin{bmatrix} x_1^{(1)} & x_1^{(2)} & x_1^{(3)} & \cdots & x_1^{(n)} \\ x_2^{(1)} & x_2^{(2)} & x_2^{(3)} & \cdots & x_2^{(n)} \\ \vdots & \vdots & \vdots & \ddots & \vdots \\ x_n^{(1)} & x_n^{(2)} & x_n^{(3)} & \cdots & x_n^{(n)} \end{bmatrix} \tag{2.82}$$

Also define another matrix $\mathscr{T} = \mathscr{S}^\dagger$. The matrix $\mathscr{T}$ has its rows as the complex conjugate of the eigenvectors.

$$\mathscr{T} = \begin{bmatrix} x_1^{(1)*} & x_2^{(1)*} & x_3^{(1)*} & \cdots & x_n^{(1)*} \\ x_1^{(2)*} & x_2^{(2)*} & x_3^{(2)*} & \cdots & x_n^{(2)*} \\ \vdots & \vdots & \vdots & \ddots & \vdots \\ x_1^{(n)*} & x_2^{(n)*} & x_3^{(n)*} & \cdots & x_n^{(n)*} \end{bmatrix} \tag{2.83}$$

The matrices $\mathscr{S}$, $\mathscr{T}$ are unitary since $\mathscr{S}\mathscr{T} = \mathscr{I} = \mathscr{T}\mathscr{S}$. These identities follow from the orthogonality of the eigenvectors for Hermitian matrices.

As an example, consider the example of a real symmetric matrix

$$\mathscr{A} = \begin{bmatrix} 2 & 1 \\ 1 & 2 \end{bmatrix} \tag{2.84}$$

$$\det|\mathscr{A} - \lambda\mathscr{I}| = \det \begin{vmatrix} 2 - \lambda & 1 \\ 1 & 2 - \lambda \end{vmatrix} = (2 - \lambda)^2 - 1 = (\lambda - 3)(\lambda - 1) \tag{2.85}$$

The eigenvectors for the two eigenvalues are

1. $\lambda_1 = 1$

$$\mathscr{C}_1 = \mathscr{A} - \lambda_1\mathscr{I} = \begin{bmatrix} 1 & 1 \\ 1 & 1 \end{bmatrix} \tag{2.86}$$

$$X^{(1)} = \frac{1}{\sqrt{2}} \begin{pmatrix} 1 \\ -1 \end{pmatrix} \tag{2.87}$$

2. $\lambda_2 = 3$

$$\mathscr{C}_2 = \mathscr{A} - \lambda_2\mathscr{I} = \begin{bmatrix} -1 & 1 \\ 1 & -1 \end{bmatrix} \tag{2.88}$$

$$X^{(2)} = \frac{1}{\sqrt{2}} \begin{pmatrix} 1 \\ 1 \end{pmatrix} \tag{2.89}$$

The unitary matrices are

$$\mathscr{S} = \frac{1}{\sqrt{2}}\begin{bmatrix} 1 & 1 \\ -1 & 1 \end{bmatrix} \tag{2.90}$$

$$\mathscr{T} = \frac{1}{\sqrt{2}}\begin{bmatrix} 1 & -1 \\ 1 & 1 \end{bmatrix} \tag{2.91}$$

It is easy to show that $\mathscr{S}\mathscr{T} = \mathscr{T}\mathscr{S} = \mathscr{I}$. Other examples are given in the problems.

The matrices also have the property that

$$\mathscr{T}\mathscr{A}\mathscr{S} = \tilde{\lambda} = \begin{bmatrix} \lambda_1 & 0 & 0 & \cdots & 0 \\ 0 & \lambda_2 & 0 & \cdots & 0 \\ 0 & 0 & \lambda_3 & \cdots & 0 \\ \vdots & \vdots & \vdots & \ddots & \lambda_n \end{bmatrix} \tag{2.92}$$

The matrix product $\mathscr{A}\mathscr{S}$ gives a matrix whose columns are the eigenvectors multiplied by the eigenvalues. Multiplying $\mathscr{A}\mathscr{S}$ by $\mathscr{T}$ produces a diagonal matrix $\tilde{\lambda}$, whose diagonal elements are the eigenvalues. As an example, for the above matrix $\mathscr{A}$ then

$$\mathscr{T}\mathscr{A}\mathscr{S} = \frac{1}{2}\begin{bmatrix} 1 & -1 \\ 1 & 1 \end{bmatrix}\begin{bmatrix} 2 & 1 \\ 1 & 2 \end{bmatrix}\begin{bmatrix} 1 & 1 \\ -1 & 1 \end{bmatrix} = \begin{bmatrix} 1 & 0 \\ 0 & 3 \end{bmatrix} \tag{2.93}$$

Multiply the identity Eq. (2.92) on the left by $\mathscr{S}$, and on the right by $\mathscr{T}$, and produce the formula

$$\mathscr{S}(\mathscr{T}\mathscr{A}\mathscr{S})\mathscr{T} = \mathscr{S}\tilde{\lambda}\mathscr{T} \tag{2.94}$$

$$\mathscr{A} = \mathscr{S}\tilde{\lambda}\mathscr{T} \tag{2.95}$$

$$\mathscr{A}^2 = (\mathscr{S}\tilde{\lambda}\mathscr{T})(\mathscr{S}\tilde{\lambda}\mathscr{T})$$

$$= \mathscr{S}\tilde{\lambda}^2\mathscr{T} \tag{2.96}$$

$$\mathscr{A}^m = \mathscr{S}\tilde{\lambda}^m\mathscr{T} \tag{2.97}$$

$$\tilde{\lambda}^m = \begin{bmatrix} \lambda_1^m & 0 & 0 & \cdots & 0 \\ 0 & \lambda_2^m & 0 & \cdots & 0 \\ 0 & 0 & \lambda_3^m & \cdots & 0 \\ \vdots & \vdots & \vdots & \ddots & \lambda_n^m \end{bmatrix} \tag{2.98}$$

This derivation repeatedly used the fact that $\mathscr{S}$ and $\mathscr{T}$ are unitary. The matrix $\tilde{\lambda}^m$ is diagonal, with each element given by λ_α^m.

Any function that can be represented by a power series can be evaluated this way

$$f(\mathscr{A}) = \sum_j a_j \mathscr{A}^j = \mathscr{S}\left(\sum_j a_j \tilde{\lambda}^j\right)\mathscr{T}$$

$$= \mathscr{S}f(\tilde{\lambda})\mathscr{T} \tag{2.99}$$

As an example, consider the expression that is the exponential function of a matrix

$$f(\mathscr{A}) = \exp(\mathscr{A}) \tag{2.100}$$

What is meant by the exponential function of a matrix? The expression is defined by its power series that is always a proper function

$$f(\mathscr{A}) = \sum_{n=0}^{\infty} \frac{\mathscr{A}^n}{n!} = \mathscr{S} \sum_{n=0}^{\infty} \frac{\tilde{\lambda}^n}{n!} \mathscr{T} = \mathscr{S} \exp(\tilde{\lambda}) \mathscr{T} \tag{2.101}$$

$$\exp(\tilde{\lambda}) = \begin{bmatrix} e^{\lambda_1} & 0 & 0 & \cdots & 0 \\ 0 & e^{\lambda_2} & 0 & \cdots & 0 \\ 0 & 0 & e^{\lambda_3} & \cdots & 0 \\ \vdots & \vdots & \vdots & \ddots & e^{\lambda_n} \end{bmatrix} \tag{2.102}$$

The final expression is quite sensible. The result for $\exp(\mathscr{A})$ is obtained by multiplying together only three matrices.

Consider the trace of the matrix $\mathscr{A}$ as evaluated using the above representation

$$\mathrm{Tr}[\mathscr{A}] = \mathrm{Tr}[\mathscr{S}\tilde{\lambda}\mathscr{T}]$$
$$= \mathrm{Tr}[\tilde{\lambda}\mathscr{T}\mathscr{S}] = \mathrm{Tr}[\tilde{\lambda}] = \sum_{\alpha} \lambda_{\alpha} \tag{2.103}$$

The cyclic property of the trace are used to move $\mathscr{S}$ over by $\mathscr{T}$. Their product is the unit matrix, so that the final result is simple. The trace of a matrix equals the summation of its eigenvalues. Another theorem is:

THEOREM: *The sum of the eigenvalues of a matrix equals the sum of the diagonal elements.*

Another example is provided by the partition function in statistical mechanics that is the trace of the exponential of the Hamiltonian matrix

$$Z = \mathrm{Tr}[e^{-\beta\mathscr{H}}] = \mathrm{Tr}[\mathscr{S}e^{-\beta\tilde{\lambda}}\mathscr{T}]$$
$$= \mathrm{Tr}[e^{-\beta\tilde{\lambda}}\mathscr{T}\mathscr{S}] = \mathrm{Tr}[e^{-\beta\tilde{\lambda}}] = \sum_{\alpha=1}^{n} e^{-\beta\lambda_{\alpha}} \tag{2.104}$$

The matrix $\exp(-\beta\tilde{\lambda})$ has only diagonal terms, and the diagonals are $\exp(-\beta\lambda_{\alpha})$ for the values $\alpha = 1, 2, \ldots, n$. The trace is the summation of these terms. Every partition function can be evaluated this way.

2.6. Non-Hermitian Matrices

The above procedure is rather convenient, but only applies to Hermitian matrices. Similar expressions are derived for non-Hermitian matrices.

The first step is to define two different kinds of eigenvectors: right and left eigenvectors. The ones discussed so far are the right eigenvectors. They give a vector proportional to themselves when they multiply from the right side of the matrix.

$$\mathscr{A} X^{(Rj)} = \lambda_j X^{(Rj)} \tag{2.105}$$

$$X^{(Lj)} \mathscr{A} = \lambda_j X^{(Lj)} \tag{2.106}$$

$$X^{(Rj)} = \begin{pmatrix} x_{r1}^{(j)} \\ x_{r2}^{(j)} \\ x_{r3}^{(j)} \\ \vdots \\ x_{rn}^{(j)} \end{pmatrix} \tag{2.107}$$

$$X^{(Lj)} = (x_{l1}^{(j)}, x_{l2}^{(j)}, x_{l3}^{(j)}, \ldots, x_{ln}^{(j)}) \tag{2.108}$$

The left eigenvector is a row vector. When it is multiplied by the matrix $\mathscr{A}$ from the left side, it produces another row vector that is proportional to itself. The same eigenvalue λ_j gives the right and left eigenvectors.

For Hermitian matrices, the left eigenvector is the Hermitian conjugate of the right eigenvector. The proof is found by taking the Hermitian conjugate of the equation for the right eigenvector

$$[\mathscr{A} X^{(Rj)} = \lambda_j X^{(Rj)}]^\dagger \tag{2.109}$$

$$X^{(Rj)\dagger} \mathbf{A} = \lambda_j X^{(Rj)\dagger} \tag{2.110}$$

where for Hermitian matrices $\mathscr{A}^\dagger = \mathscr{A}$ and the eigenvalue is real. The above steps prove for Hermitian matrices $X^{(Lj)} = X^{(Rj)\dagger}$. This result is no longer true if the matrix is not Hermitian, since then $\mathscr{A}^\dagger \neq \mathscr{A}$ and the eigenvalue may not be real.

Find the left eigenvectors for the matrix considered before:

$$\mathscr{A} = \begin{bmatrix} 1 & 2 \\ 3 & 2 \end{bmatrix} \tag{2.111}$$

The two eigenvalues and related eigenvectors are:

- $\lambda_1 = -1$

$$\mathscr{C}_1 = \mathscr{A} - \lambda_1 \mathscr{I} = \begin{bmatrix} 2 & 2 \\ 3 & 3 \end{bmatrix} \tag{2.112}$$

$$X^{(R1)} = \frac{1}{\sqrt{2}} \begin{bmatrix} 1 \\ -1 \end{bmatrix} \tag{2.113}$$

$$X^{(L1)} = \frac{1}{\sqrt{13}} (3, -2) \tag{2.114}$$

$$\mathscr{C}_2 = \mathscr{A} - \lambda_2 \mathscr{I} = \begin{bmatrix} -3 & 2 \\ 3 & -2 \end{bmatrix} \qquad (2.115)$$

$$X^{(R2)} = \frac{1}{\sqrt{13}} \begin{bmatrix} 2 \\ 3 \end{bmatrix} \qquad (2.116)$$

$$X^{(L2)} = \frac{1}{\sqrt{2}} (1, 1) \qquad (2.117)$$

Next, examine the various orthogonality relations when taking the product of a left and right eigenvector

$$X^{(L1)} X^{(R1)} = \frac{5}{\sqrt{26}} \equiv \eta_1 \qquad (2.118)$$

$$X^{(L2)} X^{(R2)} = \frac{5}{\sqrt{26}} \equiv \eta_2 \qquad (2.119)$$

$$X^{(L1)} X^{(R2)} = 0 \qquad (2.120)$$

$$X^{(L2)} X^{(R1)} = 0 \qquad (2.121)$$

The left and right eigenvectors are orthogonal for different eigenvalues. The proof is provided below. The product of the left and right eigenvector for the same eigenvalue give a nonzero constant called η_α.

THEOREM. *The right and left eigenvectors for different eigenvalues are orthogonal.*

The theorem is expressed as

$$X^{(Li)} X^{(Rj)} = \delta_{ij} \eta_i \qquad (2.122)$$

In order to prove this theorem, a new matrix $\mathscr{Z}^{(\alpha\beta)}$ is constructed that is formed by the product of a left and right eigenvector. There are two ways to multiply together the left and right eigenvectors. The scalar product is the one used so far, and is written with the left eigenvector on the left

$$X^{(L\alpha)} X^{(R\beta)} = \sum_{i=1}^{n} x_{li}^{(\alpha)} x_{ri}^{(\beta)} = \delta_{\alpha\beta} \eta_\alpha \qquad (2.123)$$

The other way to multiply together the two vectors is to produce a matrix by multiplying together every element of one by every element of the other! It is

$$\mathscr{L}^{(\alpha\beta)} = X^{(R\alpha)}X^{(L\beta)} = \begin{pmatrix} x_{r1}^{(\alpha)} \\ x_{r2}^{(\alpha)} \\ \vdots \\ x_{rn}^{(\alpha)} \end{pmatrix} [x_{l1}^{(\beta)}, x_{l2}^{(\beta)}, \ldots, x_{ln}^{(\beta)}]$$

$$= \begin{bmatrix} x_{r1}^{(\alpha)}x_{l1}^{(\beta)} & x_{r1}^{(\alpha)}x_{l2}^{(\beta)} & x_{r1}^{(\alpha)}x_{l3}^{(\beta)} & \cdots & x_{r1}^{(\alpha)}x_{ln}^{(\beta)} \\ x_{r2}^{(\alpha)}x_{l1}^{(\beta)} & x_{r2}^{(\alpha)}x_{l2}^{(\beta)} & x_{r2}^{(\alpha)}x_{l3}^{(\beta)} & \cdots & x_{r2}^{(\alpha)}x_{ln}^{(\beta)} \\ \vdots & \vdots & \vdots & \ddots & \vdots \\ x_{rn}^{(\alpha)}x_{l1}^{(\beta)} & x_{rn}^{(\alpha)}x_{l2}^{(\beta)} & x_{rn}^{(\alpha)}x_{l3}^{(\beta)} & \cdots & x_{rn}^{(\alpha)}x_{ln}^{(\beta)} \end{bmatrix} \tag{2.124}$$

The columns are just the right eigenvector multiplied by a single element of the left eigenvector. The rows are just the left eigenvector multiplied by a single element of the right eigenvector. The matrix $\mathscr{L}^{(\alpha\beta)}$ has the following interesting properties

$$\mathscr{A}\mathscr{L}^{(\alpha\beta)} = \lambda_\alpha\mathscr{L}^{(\alpha\beta)} \tag{2.125}$$

$$\mathscr{L}^{(\alpha\beta)}\mathscr{A} = \lambda_\beta\mathscr{L}^{(\alpha\beta)} \tag{2.126}$$

The first result comes because the columns of $\mathscr{L}^{(\alpha\beta)}$ are the right eigenvectors that have the eigenvalue λ_α. The second relation is because the rows are the left eigenvectors, whose eigenvalue is λ_β. Take the difference of these two equations, and then take the trace

$$\text{Tr}[\mathscr{A}\mathscr{L}^{(\alpha\beta)} - \mathscr{L}^{(\alpha\beta)}\mathscr{A}] = (\lambda_\alpha - \lambda_\beta)\,\text{Tr}[\mathscr{L}^{(\alpha\beta)}] \tag{2.127}$$

$$0 = (\lambda_\alpha - \lambda_\beta)X^{(L\beta)}X^{(R\alpha)} \tag{2.128}$$

$$X^{(L\beta)}X^{(R\alpha)} = \eta_\alpha\delta_{\alpha\beta} \tag{2.129}$$

The left-hand side of Eq. (2.127) is zero because the trace of the product of two matrices is the same regardless of their order. On the right, the trace of $\mathscr{L}^{(\alpha\beta)}$ is the summation of the diagonal elements, that is also the scalar product of the two eigenvectors. It has been shown that if $\lambda_\alpha \neq \lambda_\beta$ then the left and right eigenvectors are orthogonal. If two eigenvalues are accidently degenerate, one can still construct their eigenvectors to be orthogonal.

Consider the following matrix that is composed of $\mathscr{L}^{(\alpha\alpha)}$ summed over all eigenvalues, and multiplied by λ/η:

$$\mathscr{B} = \sum_\alpha X^{(R\alpha)}\frac{\lambda_\alpha}{\eta_\alpha}X^{(L\alpha)} \tag{2.130}$$

$$\mathscr{B}X^{(Rj)} = \sum_\alpha X^{(R\alpha)} \frac{\lambda_\alpha}{\eta_\alpha} X^{(L\alpha)} X^{(Rj)}$$

$$= \sum_\alpha X^{(R\alpha)} \frac{\lambda_\alpha}{\eta_\alpha} \delta_{\alpha j} \eta_j$$

$$= \lambda_j X^{(Rj)} \tag{2.131}$$

$$X^{(Lj)} \mathscr{B} = \lambda_j X^{(Lj)} \tag{2.132}$$

When the matrix $\mathscr{B}$ operates on a right eigenvector, it gives the eigenvector times the eigenvalue. When it operates on a left eigenvector, it gives the eigenvector times the eigenvalue. This property is the same as that of the matrix $\mathscr{A}$. In fact, $\mathscr{B} = \mathscr{A}$.

$$\mathscr{A} = \sum_\alpha X^{(R\alpha)} \frac{\lambda_\alpha}{\eta_\alpha} X^{(L\alpha)} \tag{2.133}$$

Any matrix can be represented by this type of expansion. For a Hermitian matrix, the coefficients $\eta_j = 1$, and the left eigenvector is the Hermitian conjugate of the right eigenvector.

This representation of a matrix is convenient for finding powers of the matrix. Multiplying $\mathscr{A}$ by itself yields

$$\mathscr{A}^2 = \sum_\alpha X^{(R\alpha)} \frac{\lambda_\alpha}{\eta_\alpha} X^{(L\alpha)} \sum_\beta X^{(R\beta)} \frac{\lambda_\beta}{\eta_\beta} X^{(L\beta)}$$

$$= \sum_\alpha X^{(R\alpha)} \frac{\lambda_\alpha^2}{\eta_\alpha} X^{(L\alpha)} \tag{2.134}$$

$$\mathscr{A}^n = \sum_\alpha X^{(R\alpha)} \frac{\lambda_\alpha^n}{\eta_\alpha} X^{(L\alpha)} \tag{2.135}$$

$$f(\mathscr{A}) = \sum_\alpha X^{(R\alpha)} \frac{f(\lambda_\alpha)}{\eta_\alpha} X^{(L\alpha)} \tag{2.136}$$

P. A. M. Dirac is quite famous in physics for his development of relativistic quantum mechanics, among other things. He also wrote a famous book on quantum mechanics[1] in which he introduced a very useful notation for left and right eigenvectors. He took the word *bracket* and split it into two words, *bra* and *ket*. Bra is the left eigenvector and ket is the right eigenvector

$$X^{(Rj)} \equiv |j\rangle \tag{2.137}$$

$$X^{(Li)} \equiv \langle i| \tag{2.138}$$

$$X^{(Li)} X^{(Rj)} = \langle i|j\rangle = \delta_{ij} \eta_i \tag{2.139}$$

$$\mathscr{L}^{(\alpha\beta)} = X^{(R\alpha)} X^{(L\beta)} = |\alpha\rangle\langle\beta| \tag{2.140}$$

$$\mathscr{A} = \sum_\alpha |\alpha\rangle \frac{\lambda_\alpha}{\eta_\alpha} \langle\alpha| \tag{2.141}$$

The above formulas have been rewritten in the Dirac notation. His notation is quite easy to use, and has been widely adopted in physics.

Another matrix using the same notation is

$$\mathcal{I} = \sum_\alpha |\alpha\rangle \frac{1}{\eta_\alpha} \langle\alpha| \tag{2.142}$$

That the right-hand side is the identity matrix is proved by multiplying by itself

$$\mathcal{I} \cdot \mathcal{I} = \sum_{\alpha\alpha'} |\alpha\rangle \frac{1}{\eta_\alpha} \langle\alpha|\alpha'\rangle \frac{1}{\eta_{\alpha'}} \langle\alpha'|$$

$$= \sum_\alpha |\alpha\rangle \frac{1}{\eta_\alpha} \langle\alpha| = \mathcal{I} \tag{2.143}$$

The identity matrix has the feature that $\mathcal{I} \cdot \mathcal{I} = \mathcal{I}$. It is also the matrix that has all of its eigenvalues equal to one.

The trace of a matrix was defined as the sum over its diagonal elements. Later it was proved to be the sum of the eigenvalues. In fact the trace can be evaluated exactly as the summation over any set of eigenvectors

$$\text{Tr}[\mathcal{A}] = \sum_j \frac{\langle j|\mathcal{A}|j\rangle}{\eta_j} \tag{2.144}$$

The obvious choice for the eigenvectors are those of the matrix $\mathcal{A}$ itself

$$\text{Tr}[\mathcal{A}] = \sum_j \frac{1}{\eta_j} \langle j| \left[\sum_i |i\rangle \frac{\lambda_i}{\eta_i} \langle i| \right] |j\rangle = \sum_j \lambda_j \tag{2.145}$$

where $\langle i|j\rangle = \eta_i \delta_{ij}$. However, any other complete set of orthogonal vectors can be used and the same trace is obtained. The trace is independent of the basis set, as long as the basis set is complete and orthogonal.

Earlier, in Eq. (2.104), there was a discussion of the partition function, that is $Z = \text{Tr}[\exp(-\beta\mathcal{H})]$. It was shown that the partition function equals the summation over all of the eigenvalues. The same result can be obtained by using as the basis set, the eigenvectors of the matrix $\mathcal{H}$. Since $\mathcal{H}$ is Hermitian, all of the parameters $\eta_i = 1$:

$$\mathcal{H}|j\rangle = \lambda_j|j\rangle \tag{2.146}$$

$$\text{Tr}[e^{-\beta\mathcal{H}}] = \sum_j \langle j|e^{-\beta\mathcal{H}}|j\rangle$$

$$= \sum_j \langle j|e^{-\beta\lambda_j}|j\rangle = \sum_j e^{-\beta\lambda_j} \tag{2.147}$$

The last expression is the same as in Eq. (2.104).

The same techniques can be used to prove an interesting theorem.

THEOREM: *If $\mathscr{R}$ is any square matrix and $\mathscr{I}$ is the identity matrix, then:*

$$\ln(\det|\mathscr{I} + \mathscr{R}|) = \text{Tr}(\mathscr{R} - \tfrac{1}{2}\mathscr{R}\cdot\mathscr{R} + \tfrac{1}{3}\mathscr{R}\cdot\mathscr{R}\cdot\mathscr{R} - \cdots) \qquad (2.148)$$

The theorem is proved by showing that both sides equal the same quantity. Begin with the left-hand side. Write the matrices in terms of left and right eigenvectors of the matrix $\mathscr{R}$

$$\mathscr{I} + \mathscr{R} = \sum_\alpha |\alpha\rangle \frac{1 + \lambda_\alpha}{\eta_\alpha} \langle\alpha| \qquad (2.149)$$

The matrix $\mathscr{R}$ has eigenvalues λ_α, so $\mathscr{I} + \mathscr{R}$ has eigenvalues $1 + \lambda_\alpha$. The determinant of a matrix is the product of its eigenvalues. Therefore

$$\det|\mathscr{I} + \mathscr{R}| = \prod_\alpha (1 + \lambda_\alpha) \qquad (2.150)$$

$$\ln(\det|\mathscr{I} + \mathscr{R}|) = \sum_\alpha \ln(1 + \lambda_\alpha) \qquad (2.151)$$

Next consider the right-hand side of Eq. (2.148). Using the above formalism

$$\mathscr{R} - \frac{1}{2}\mathscr{R}\cdot\mathscr{R} + \frac{1}{3}\mathscr{R}\cdot\mathscr{R}\cdot\mathscr{R} - \cdots - = \sum_\alpha |\alpha\rangle \frac{1}{\eta_\alpha}\left\{ \lambda_\alpha - \frac{\lambda_\alpha^2}{2} + \frac{\lambda_\alpha^3}{3} - \cdots - \right\}\langle\alpha|$$

$$= \sum_\alpha |\alpha\rangle \frac{\ln(1 + \lambda_\alpha)}{\eta_\alpha}\langle\alpha| \qquad (2.152)$$

The series gives $\ln(1 + \lambda_\alpha)$. In order to evaluate the trace, use the basis states $|\alpha\rangle$ of the matrix $\mathscr{R}$.

$$\text{Tr}(O) = \sum_\alpha \frac{\langle\alpha|O|\alpha\rangle}{\eta_\alpha} \qquad (2.153)$$

$$\text{Tr}\left(\sum_\alpha |\alpha\rangle \frac{\ln(1 + \lambda_\alpha)}{\eta_\alpha}\langle\alpha|\right) = \sum_\alpha \ln(1 + \lambda_\alpha) \qquad (2.154)$$

Both sides of Eq. (2.148) equal the same expression, so the two expressions are equal.

The determinant of a matrix is the product of its eigenvalues. Using the above notation, this result can be expressed as

$$\det|\mathscr{A}| = \prod_j \lambda_j \qquad (2.155)$$

$$\ln[\det|\mathscr{A}|] = \sum_j \ln(\lambda_j) = \text{Tr}[\ln(\mathscr{A})] \qquad (2.156)$$

The above formula can be used to prove an important theorem:

If the square matrices A and B have the same dimension, then

$$\det|\mathscr{A}\mathscr{B}| = \det|\mathscr{A}|\det|\mathscr{B}| \tag{2.157}$$

The proof of the theorem uses Eq. (2.156) that are added for the two matrices

$$\ln[\det|\mathscr{A}|] + \ln[\det|\mathscr{B}|] = \text{Tr}[\ln(\mathscr{A}) + \ln(\mathscr{B})] \tag{2.158}$$

$$\ln[\det|\mathscr{A}|\det|\mathscr{B}|] = \text{Tr}[\ln(\mathscr{A}\mathscr{B})] = \ln[\det|\mathscr{A}\mathscr{B}|] \tag{2.159}$$

$$\det|\mathscr{A}|\det|\mathscr{B}| = \det|\mathscr{A}\mathscr{B}| \tag{2.160}$$

that completes the proof. The same method can be extended to the products of many matrices

$$\det|\mathscr{A}\mathscr{B}\cdots\mathscr{L}| = \det|\mathscr{A}|\det|\mathscr{B}|\cdots\det|\mathscr{L}| \tag{2.161}$$

2.7. A Special Matrix

This section considers a simple matrix that is encountered fairly often. It is a matrix of order n where all of the elements are identical. For example, say they are all 1.

$$\mathscr{A} = \begin{bmatrix} 1 & 1 & \cdots & 1 \\ 1 & 1 & \cdots & 1 \\ \vdots & \vdots & \ddots & \vdots \\ 1 & 1 & \cdots & 1 \end{bmatrix} \tag{2.162}$$

The eigenvalue equation is

$$0 = \det \begin{bmatrix} 1-\lambda & 1 & 1 & \cdots & 1 \\ 1 & 1-\lambda & 1 & \cdots & 1 \\ \vdots & \vdots & \vdots & \ddots & \vdots \\ 1 & 1 & 1 & \cdots & 1-\lambda \end{bmatrix}$$

$$= (-1)^n \lambda^{n-1}(\lambda - n) \tag{2.163}$$

The determinant was evaluated in Chapter 1, where $x = -\lambda$. There are $n-1$ eigenvalues of $\lambda = 0$ and one eigenvalue of $\lambda = n$. For $\lambda = n$ the eigenvector has all equal elements

$$X^{(Rn)} = \frac{1}{\sqrt{n}} \begin{pmatrix} 1 \\ 1 \\ 1 \\ \vdots \\ 1 \end{pmatrix} \tag{2.164}$$

The remaining eigenvectors are orthogonal to this one, and have a zero eigenvalue. Given $n - 1$ identical eigenvalues ($\lambda_j = 0$), the task is to construct $n - 1$ orthogonal eigenfunctions that have zero eigenvalue. There is more than one way to construct them. One method is presented here, and another method is in the next section.

The easiest way to write a general result is using complex numbers.

$$X^{(R\alpha)} = \frac{1}{\sqrt{n}} \begin{pmatrix} 1 \\ e^{i\alpha\theta} \\ e^{i2\alpha\theta} \\ e^{i3\alpha\theta} \\ \vdots \\ e^{i(n-1)\alpha\theta} \end{pmatrix}, \qquad \theta = \frac{2\pi}{n} \tag{2.165}$$

and α has integer values $\alpha = 1, 2, 3, \ldots, n - 1$. These eigenvectors are mutually orthogonal, and have a zero eigenvalue. All of these features depend on evaluating the summation

$$\sum_{j=0}^{n-1} e^{ij\alpha\theta} = \frac{1 - e^{in\alpha\theta}}{1 - e^{i\alpha\theta}} = 0 \tag{2.166}$$

Since $\theta = 2\pi/n$ then the exponent $n\alpha\theta = 2\pi\alpha$ and the exponent of this number times i is 1. The numerator vanishes. This summation, or something similar, always appears when proving the above eigenvectors are orthogonal. For example, to show two different eigenvectors of this type are orthogonal

$$X^{(R\alpha)\dagger} X^{(R\beta)} = \frac{1}{n} \sum_{j=0}^{n-1} e^{ij\theta(\beta - \alpha)} = \frac{1}{n} \frac{1 - e^{i\theta(\beta - \alpha)n}}{1 - e^{i\theta(\beta - \alpha)}} \tag{2.167}$$

Since $n\theta = 2\pi$ the numerator is zero if $\alpha \neq \beta$. If $\alpha = \beta$ the summation is n.

A problem where this determinant arises is a Hamiltonian for a particle on n sites. Each site has an energy E_0, and a matrix element V that transfers the particle to another site. A particle on each site can hop to all others with equal probability. Examples are the vertices of an equilateral triangle ($n = 3$) or a tetrahedron ($n = 4$). The Hamiltonian is the matrix

$$\mathcal{H} = \begin{bmatrix} E_0 & V & \cdots & V \\ V & E_0 & \cdots & V \\ \vdots & \vdots & \ddots & \vdots \\ V & V & \cdots & E_0 \end{bmatrix} \tag{2.168}$$

In the eigenvalue equation, let $\lambda = E_0 - V(1 + x)$ giving

$$0 = \det \begin{vmatrix} E_0 - \lambda & V & \cdots & V \\ V & E_0 - \lambda & \cdots & V \\ \vdots & \vdots & \ddots & \vdots \\ V & V & \cdots & E_0 - \lambda \end{vmatrix}$$

$$= V^n \det \begin{vmatrix} 1 + x & 1 & \cdots & 1 \\ 1 & 1 + x & \cdots & 1 \\ \vdots & \vdots & \ddots & \vdots \\ 1 & 1 & \cdots & 1 + x \end{vmatrix}$$

$$= V^n x^{n-1}(x + n) = (\lambda - E_0 + V)^{n-1}[\lambda - E_0 + (n - 1)V] \quad (2.169)$$

There are $(n - 1)$ eigenvalues of $E_0 - V$ and one of $\lambda = E_0 + (n - 1)V$. The eigenvectors are those constructed above.

2.8. Gram–Schmidt

The Gram–Schmidt process allows an orthogonal set of eigenvectors to be constructed from any set of vectors. By *orthogonal* is meant the property that

$$\langle \alpha | \beta \rangle = \sum_{i=1}^{n} x_{\alpha i}^* x_{\beta i} = \delta_{\alpha \beta} \quad (2.170)$$

where

$$|\alpha\rangle = \begin{bmatrix} x_{\alpha 1} \\ x_{\alpha 2} \\ \vdots \\ x_{\alpha n} \end{bmatrix} \quad (2.171)$$

$$|\beta\rangle = \begin{bmatrix} x_{\beta 1} \\ x_{\beta 2} \\ \vdots \\ x_{\beta n} \end{bmatrix} \quad (2.172)$$

$$\langle \alpha | = [x_{\alpha 1}^*, x_{\alpha 2}^*, \ldots, x_{\alpha n}^*] \quad (2.173)$$

The Dirac notation is used for the vectors.

Assume there are a set of vectors $|\alpha\rangle, |\beta\rangle, |\gamma\rangle$ that are not orthogonal and not normalized. The process for constructing a set that are normalized and orthogonal are:

1. Take one vector and normalize it by dividing by the square root of the length of the vector: normalized vectors are denoted $|\alpha)$:

$$|\alpha) = \frac{|\alpha\rangle}{\sqrt{\langle\alpha|\alpha\rangle}} \tag{2.174}$$

$$(\alpha|\alpha) = 1 \tag{2.175}$$

Any of the vectors can be chosen as the first one.

2. Make the second vector orthogonal to $|\alpha)$ and orthogonalize it:

$$|2\rangle = |\beta\rangle - |\alpha)(\alpha|\beta\rangle \tag{2.176}$$

$$|\beta) = \frac{|2\rangle}{\sqrt{\langle2|2\rangle}} \tag{2.177}$$

This vector is now normalized and orthogonal to $|\alpha)$

$$(\alpha|\beta) = 0 \tag{2.178}$$

$$(\beta|\beta) = 1 \tag{2.179}$$

3. The third vector is now made orthogonal to the first two, and then normalized

$$|3\rangle = |\gamma\rangle - |\alpha)(\alpha|\gamma\rangle - |\beta)(\beta|\gamma\rangle \tag{2.180}$$

$$|\gamma) = \frac{|3\rangle}{\sqrt{\langle3|3\rangle}} \tag{2.181}$$

$$(\alpha|\gamma) = 0 \tag{2.182}$$

$$(\beta|\gamma) = 0 \tag{2.183}$$

$$(\gamma|\gamma) = 1 \tag{2.184}$$

4. The next vector is made orthogonal to prior ones using the algorithm

$$|4\rangle = |\delta\rangle - \sum_{k=\alpha\beta\gamma} |k)(k|\delta\rangle \tag{2.185}$$

$$|\delta) = \frac{|4\rangle}{\sqrt{\langle4|4\rangle}} \tag{2.186}$$

The general procedure is now obvious. Each vector is made orthogonal to the prior ones using Eq. (2.185). It is normalized by dividing by its own length. This process generates a set orthonormal vectors from most beginning sets. The only problem that may arise is when two vectors are parallel. In this case, the number of orthogonal vectors is one less than the starting set.

As an example, consider the matrix

$$\mathcal{M} = \begin{pmatrix} 1 & 1 & 1 \\ 1 & 1 & 1 \\ 1 & 1 & 1 \end{pmatrix} \tag{2.187}$$

This matrix has three eigenvalues: $\lambda = 3, 0, 0$. The eigenvector for $\lambda = 3$ is

$$|1\rangle = \frac{1}{\sqrt{3}} \begin{bmatrix} 1 \\ 1 \\ 1 \end{bmatrix} \tag{2.188}$$

Note that the eigenvector is normalized properly. Earlier it was shown how to construct the eigenvectors for the degenerate eigenvalues $\lambda = 0$ in terms of complex numbers. Often it is useful to have the two eigenvectors expressed with real numbers. There is a simple way using Gram–Schmidt. The two eigenvectors must be orthogonal to $|1\rangle$. One of them can be chosen with just two elements that are ± 1

$$|2\rangle = \frac{1}{\sqrt{2}} \begin{bmatrix} 1 \\ -1 \\ 0 \end{bmatrix} \tag{2.189}$$

Again it is normalized. Now, how to choose the third eigenvector that also is an eigenvalue for $\lambda = 0$? Using Gram–Schmidt, start with any vector that is not obviously proportional to the first two, such as

$$|3\rangle = \begin{bmatrix} 1 \\ 2 \\ 3 \end{bmatrix} \tag{2.190}$$

Applying the Gram–Schmidt process gives

$$|3'\rangle = \begin{bmatrix} 1 \\ 2 \\ 3 \end{bmatrix} - \frac{6}{3} \begin{bmatrix} 1 \\ 1 \\ 1 \end{bmatrix} - \frac{(-1)}{2} \begin{bmatrix} 1 \\ -1 \\ 0 \end{bmatrix}$$

$$= -\frac{1}{2} \begin{bmatrix} 1 \\ 1 \\ -2 \end{bmatrix} \tag{2.191}$$

$$\langle 3'|3'\rangle = \frac{1}{4}[1 + 1 + 4] = \frac{3}{2} \tag{2.192}$$

$$|3\rangle = \sqrt{\frac{1}{6}} \begin{bmatrix} -1 \\ -1 \\ 2 \end{bmatrix} \tag{2.193}$$

It is easy to show that this eigenvector has a zero eigenvalue, and is orthogonal to the other two eigenvectors.

This process can be generalized to a vector of n dimension. If the starting vector has all elements equal to one, a set of mutually orthogonal vectors are:

$$|1) = \frac{1}{\sqrt{n}}\begin{bmatrix} 1 \\ 1 \\ 1 \\ \vdots \\ 1 \end{bmatrix}, \qquad |2) = \frac{1}{\sqrt{2}}\begin{bmatrix} 1 \\ -1 \\ 0 \\ \vdots \\ 0 \end{bmatrix} \qquad (2.194)$$

$$|3) = \frac{1}{\sqrt{6}}\begin{bmatrix} 1 \\ 1 \\ -2 \\ 0 \\ \vdots \\ 0 \end{bmatrix}, \qquad |4) = \frac{1}{\sqrt{12}}\begin{bmatrix} 1 \\ 1 \\ 1 \\ -3 \\ \vdots \\ 0 \end{bmatrix} \qquad (2.195)$$

$$|n) = \frac{1}{\sqrt{n(n-1)}}\begin{bmatrix} 1 \\ 1 \\ 1 \\ \vdots \\ 1 \\ -n-1 \end{bmatrix} \qquad (2.196)$$

If a matrix has dimension $n \times n$ the vector space has dimension of n. There are n different vectors in this space. That is, n different orthogonal vectors can be constructed that obey

$$X^{(iL)*} \cdot X^{(jR)} = \delta_{ij} \qquad (2.197)$$

Another feature of vector spaces is *completeness*. Any vector can be expressed exactly as a sum over these orthogonal vectors

$$X^{(R)} = \sum_{j=1}^{n} a_j X^{(jR)} \qquad (2.198)$$

$$a_j = X^{(jL)*} \cdot X^{(R)} \qquad (2.199)$$

This property of vector spaces is independent of the choice of the matrix $\mathscr{A}$. Any space of dimension n has a complete set of eigenvectors that are orthogonal.

The most trivial set of vectors are those with all elements zero except for one that is 1

$$X^{(jR)} = \begin{bmatrix} 0 \\ 0 \\ \vdots \\ 0 \\ 1 \\ 0 \\ \vdots \\ 0 \end{bmatrix} \tag{2.200}$$

$$X^{(jL)} = (0, 0, \ldots, 0, 1, 0, \ldots, 0) \tag{2.201}$$

The element "1" is placed in the jth row or column. These vectors form a complete set and are all orthogonal. For example, for $n = 3$ the completeness of a vector whose elements are $(1, 2, 3)$ is

$$\begin{bmatrix} 1 \\ 2 \\ 3 \end{bmatrix} = 1 \begin{bmatrix} 1 \\ 0 \\ 0 \end{bmatrix} + 2 \begin{bmatrix} 0 \\ 1 \\ 0 \end{bmatrix} + 3 \begin{bmatrix} 0 \\ 0 \\ 1 \end{bmatrix} \tag{2.202}$$

However, any three orthogonal vectors can be used for this expansion. Another possible set is

$$X^{(1R)} = \frac{1}{\sqrt{3}} \begin{bmatrix} 1 \\ 1 \\ 1 \end{bmatrix}, \qquad X^{(2R)} = \frac{1}{\sqrt{2}} \begin{bmatrix} 1 \\ 0 \\ -1 \end{bmatrix}, \qquad X^{(3R)} = \frac{1}{\sqrt{6}} \begin{bmatrix} 1 \\ -2 \\ 1 \end{bmatrix} \tag{2.203}$$

$$\begin{bmatrix} 1 \\ 2 \\ 3 \end{bmatrix} = 2 \begin{bmatrix} 1 \\ 1 \\ 1 \end{bmatrix} - 1 \begin{bmatrix} 1 \\ 0 \\ -1 \end{bmatrix} + 0 \begin{bmatrix} 1 \\ -2 \\ 1 \end{bmatrix} \tag{2.204}$$

Any set of n-orthogonal vectors can be used to span the vector space.

2.9. Chains

Another matrix has elements E_0 along the diagonal, and elements V that transfer the particle to adjacent sites. This Hamiltonian arises if the sites are a linear chain. The transfer takes place only to the first neighbor sites:

$$\mathcal{H} = \begin{bmatrix} E_0 & V & 0 & \cdots & 0 \\ V & E_0 & V & \cdots & 0 \\ 0 & V & E_0 & \cdots & 0 \\ \vdots & \vdots & \vdots & \ddots & \vdots \\ 0 & 0 & \cdots & V & E_0 \end{bmatrix} \tag{2.205}$$

This form of a matrix is called *tridiagonal*. In discussing problems in one dimension, physicists often take *periodic boundary conditions*. In that case, the chain wraps around on itself, so that the $n + 1$ site is back to being the first site. Note that the present boundary conditions are *not* periodic. The chain has n sites, it is linear, the first site is 1 and the last site is n.

The eigenfunction has the form $x_j^{(\alpha)} = \sin(jk_\alpha)$ where k_α denotes an eigenvalue to be determined.

$$X^{(R\alpha)} = \sqrt{\frac{2}{n}} \begin{pmatrix} \sin(k_\alpha) \\ \sin(2k_\alpha) \\ \sin(3k_\alpha) \\ \vdots \\ \sin(nk_\alpha) \end{pmatrix} \qquad (2.206)$$

Choose this form for the eigenfunction for a simple reason. The particle cannot get to sites 0 and $(n + 1)$ since they are not part of the chain. Therefore, $\psi_0 = \psi_{n+1} = 0$. The first is automatically satisfied by the choice of the sine function. It vanishes when $j = 0$. In order to satisfy the condition $\psi_{n+1} = \sin[(n + 1)k_\alpha] = 0$ choose

$$k_\alpha = \frac{\alpha\pi}{n + 1} \qquad (2.207)$$

where $\alpha = 1, 2, \ldots, n$ are positive integers. $\alpha = 0$ is not allowed since the eigenfunction vanishes. This also happens for $\alpha = n + 1$ and $(k_{n+1} = \pi)$. The case where $\alpha = n + 2$ gives the same eigenfunction and eigenvalue as $\alpha = n$. There are only n independent solutions that are $1 \leqslant \alpha \leqslant n$.

The eigenvalues for the matrix are

$$\lambda_\alpha = E_0 + 2V \cos(k_\alpha) \qquad (2.208)$$

Multiply this eigenfunction by each row of the original Hamiltonian, and show that it is a proper eigenvector. Except for the first and last rows, this step gives

$$\begin{aligned} 0 &= V \sin[(j - 1)k_\alpha] + (E_0 - \lambda) \sin(jk_\alpha) + V \sin[(j + 1)k_\alpha] \\ &= V\{\sin[(j - 1)k_\alpha] - 2\cos(k_\alpha)\sin(jk_\alpha) + \sin[(j + 1)k_\alpha]\} \\ &= 2V \cos(k_\alpha)\sin(jk_\alpha)[1 - 1] = 0 \end{aligned} \qquad (2.209)$$

where the trigometric identity is

$$\sin(a + b) + \sin(a - b) = 2\sin(a)\cos(b) \qquad (2.210)$$

Examine carefully the first and last rows. The first row is

$$0 = (E_0 - \lambda)\sin(k_\alpha) + V\sin(2k_\alpha) \qquad (2.211)$$

$$\begin{aligned} 0 &= V\{-2\cos(k_\alpha)\sin(k_\alpha) + \sin(2k_\alpha)\} \\ &= V\sin(2k_\alpha)[-1 + 1] = 0 \end{aligned} \qquad (2.212)$$

A similar result is found for the equation of the last row.

A different problem is the periodic chain. The particle on site n can hop to the site 1. The Hamiltonian has additional elements in the far corners

$$\mathscr{H} = \begin{bmatrix} E_0 & V & 0 & \cdots & V \\ V & E_0 & V & \cdots & 0 \\ 0 & V & E_0 & \cdots & 0 \\ \vdots & \vdots & \vdots & \ddots & \vdots \\ V & 0 & \cdots & V & E_0 \end{bmatrix} \tag{2.213}$$

This case is easy to solve. Take as the eigenfunction $\psi_j = \exp(ijk_\alpha)$. The periodicity is that $\psi_{j+n} = \psi_j$ which sets

$$1 = e^{ink_\alpha} \tag{2.214}$$

$$k_\alpha = \frac{2\pi\alpha}{n} \tag{2.215}$$

$$\alpha = 0,\ \pm 1,\ \pm 2, \ldots \tag{2.216}$$

$$\lambda_\alpha = E_0 + 2V\cos(k_\alpha) \tag{2.217}$$

$$X^{(R\alpha)} = \frac{1}{\sqrt{n}} \begin{pmatrix} 1 \\ \exp(ik_\alpha) \\ \exp(i2k_\alpha) \\ \exp(i3k_\alpha) \\ \vdots \\ \exp[i(n-1)k_\alpha] \end{pmatrix} \tag{2.218}$$

These eigenstates describe waves that run around the chain. The cases of $\alpha = \pm j$ can be combined to get sines or cosines. The periodic chain has the same equation for the eigenvalue as does the linear chain, but a different selection of wave vector values. There are only n independent eigenvectors. If the integer n is odd, then the independent values span the range of

$$-\frac{n-1}{2} \leqslant \alpha \leqslant \frac{n-1}{2} \tag{2.219}$$

For example, if $n=5$ the independent values are: $-2, -1, 0, 1, 2$. Values outside this set, such as $\alpha = 3$, have the same eigenfunction as one in this set of five: $\alpha = 3$ has the same eigenfunction and eigenvalue as does $\alpha = -2$. Similarly, if n is an even integer then $\alpha = \pm n/2$ give $k_\alpha = \pm\pi$ that have the same eigenvector and eigenvalue. For even values of n the independent eigenfunctions have values

$$-\frac{n}{2} \leqslant \alpha \leqslant \frac{n}{2} - 1 \tag{2.220}$$

A matrix of dimension n has only n independent eigenfunctions.

The last example solves for the vibrational modes of a linear chain of n masses. There is a harmonic spring of constant K between each mass. The classical equations of motion for the displacement x_j of the jth spring are

$$m\omega^2 x_1 = K(x_1 - x_2) \tag{2.221}$$

$$m\omega^2 x_2 = K(2x_2 - x_1 - x_3) \tag{2.222}$$

$$m\omega^2 x_j = K(2x_j - x_{j-1} - x_{j+1}) \tag{2.223}$$

$$m\omega^2 x_n = K(x_n - x_{n-1}) \tag{2.224}$$

where ω is the unknown frequency, and is the eigenvalue to be determined. The end masses, labeled 1 and n, obey a different equation since they are pulled by one spring. The other masses have two springs connected to them and obey a different equation. Define $\lambda = m\omega^2/K$ and the eigenvalue equation is now

$$\mathscr{A}X = \lambda X \tag{2.225}$$

$$\mathscr{A} = \begin{bmatrix} 1 & -1 & 0 & \cdots & 0 \\ -1 & 2 & -1 & \cdots & 0 \\ 0 & -1 & 2 & \cdots & 0 \\ \vdots & \vdots & \vdots & \ddots & \vdots \\ 0 & 0 & \cdots & -1 & 1 \end{bmatrix}, \qquad X = \begin{bmatrix} x_1 \\ x_2 \\ x_3 \\ \vdots \\ x_n \end{bmatrix} \tag{2.226}$$

Again the matrix has tridiagonal form. It is different from the previous case for electrons, since the first and last diagonal elements are different from the others. All diagonal elements are 2 except the first and last that are 1. The previous solution does not work in this case.

First note that there is a solution with $\lambda = 0$ and an eigenvector where $x_j = 1$. This represents a uniform translation of the entire chain. This solution is not interesting physically, although it is a valid eigenstate of the Hamiltonian. The other solutions must be orthogonal to this one, which means they have no center-of-mass motion. The other solutions have positive values of λ and represent collective vibrational modes.

The general solution has the form of $x_j^{(\alpha)} = \sin(jk_\alpha + \theta_\alpha)$. The phase angle θ_α has been introduced. The general solution to a tridiagonal matrix is always a sine or cosine function. Having an unspecified phase angle permits a solution that is any mixture of sines and cosines. For a mass not at the ends, the eigenvalue equation is

$$\lambda_\alpha \sin(jk_\alpha + \theta_\alpha) = 2\sin(jk_\alpha + \theta_\alpha) - \sin[(j+1)k_\alpha + \theta_\alpha] - \sin[(j-1)k_\alpha + \theta_2)]$$

$$= 2[1 - \cos(k_\alpha)]\sin(jk_\alpha + \theta_\alpha) \tag{2.227}$$

$$\lambda_\alpha = 2[1 - \cos(k_\alpha)] \tag{2.228}$$

which again used the trigometric identity Eq. (2.210). The parameters k_α and θ_α are determined by the equations of the two masses at the ends of the chain.

At the end $j = 1$

$$\lambda_\alpha \sin(k_\alpha + \theta_\alpha) = \sin(k_\alpha + \theta_\alpha) - \sin(2k_\alpha + \theta_\alpha) \tag{2.229}$$

Use the expression for λ_α and the trigometric identity to write the left-hand side as

$$\lambda_\alpha \sin(k_\alpha + \theta_\alpha) = 2[1 - \cos(k_\alpha)]\sin(k_\alpha + \theta_\alpha)$$
$$= 2\sin(k_\alpha + \theta_\alpha) - \sin(2k_\alpha + \theta_\alpha) - \sin(\theta_\alpha) \tag{2.230}$$

Put this expression into the left side of Eq. (2.229) and cancel some like factors. The remaining terms are

$$\sin(k_\alpha + \theta_\alpha) = \sin(\theta_\alpha) \tag{2.231}$$

At first it appears the solution is $k_\alpha = 0$. This solution does not work. There is another solution. There are different angles that have the same value for the sine. For example, $\sin(\theta) = \sin(\pi - \theta)$. Using this solution gives

$$k_\alpha + \theta_\alpha = \pi - \theta_\alpha \tag{2.232}$$
$$\theta_\alpha = \tfrac{1}{2}[\pi - k_\alpha] \tag{2.233}$$

This expression determines the phase angle θ_α. The last step is to determine k_α using the equation of motion of the last mass

$$\lambda_\alpha \sin(\Phi_n) = \sin(\Phi_n) - \sin(\Phi_{n-1}) \tag{2.234}$$
$$\Phi_n = nk_\alpha + \theta_\alpha. \tag{2.235}$$

Then using

$$\lambda_\alpha \sin(\Phi_n) = 2[1 - \cos(k_\alpha)]\sin(\Phi_n) = 2\sin(\Phi_n) - \sin(\Phi_{n+1}) - \sin(\Phi_{n-1}) \tag{2.236}$$

Combining the above two equations cancels many terms, and yields

$$\sin(\Phi_n) = \sin(\Phi_{n+1}) \tag{2.237}$$
$$\sin[nk_\alpha + \theta_\alpha] = \sin[(n+1)k_\alpha + \theta_\alpha] \tag{2.238}$$

Using Eq. (2.233), this last equation has the solution

$$\sin(\Phi_n) - \sin(\Phi_{n+1}) = \sin\left[\left(n - \frac{1}{2}\right)k_\alpha + \frac{\pi}{2}\right] - \sin\left[\left(n + \frac{1}{2}\right)k_\alpha + \frac{\pi}{2}\right]$$
$$= -2\sin(k_\alpha/2)\sin(nk_\alpha) = 0 \tag{2.239}$$
$$k_\alpha = \frac{\alpha\pi}{n} \tag{2.240}$$

This choice for k_α makes the equation at the upper end n identical to the one at the lower end 1. The integer α has positive values from zero to $(n-1)$. The choice $\alpha = 0$ gives $\lambda = 0$ and is the uniform translation.

Again the tridiagonal matrix has the eigenvalue $\lambda_\alpha = 2[1 - \cos(k_\alpha)]$. The choice of permissible values of k_α is different in this case, due to the change in the boundary conditions at the end of the chain.

A simple example is given for $n = 3$. The vibrational modes of this chain have the effective Hamiltonian of

$$\mathscr{A} = \begin{bmatrix} 1 & -1 & 0 \\ -1 & 2 & -1 \\ 0 & -1 & 1 \end{bmatrix} \tag{2.241}$$

The eigenvalues are found by using the old-fashioned method

$$0 = \det \begin{vmatrix} 1-\lambda & -1 & 0 \\ -1 & 2-\lambda & -1 \\ 0 & -1 & 1-\lambda \end{vmatrix}$$

$$= (1-\lambda)^2(2-\lambda) - 2(1-\lambda)$$

$$= -\lambda(\lambda - 1)(\lambda - 3) \tag{2.242}$$

The eigenfunctions for these three cases are

- $\lambda_1 = 0$

$$X^{(1)} = \frac{1}{\sqrt{3}} \begin{pmatrix} 1 \\ 1 \\ 1 \end{pmatrix} \tag{2.243}$$

This is the same solution as that with $\alpha = 0$, $k_\alpha = 0$, $\theta_\alpha = \pi/2$.

- $\lambda_2 = 1$

$$X^{(2)} = \frac{1}{\sqrt{2}} \begin{pmatrix} 1 \\ 0 \\ -1 \end{pmatrix} \tag{2.244}$$

This solution corresponds to the general case with $\alpha = 1$, $k_\alpha = \pi/3 = \theta_\alpha$.

- $\lambda_3 = 3$

$$X^{(3)} = \frac{1}{\sqrt{6}} \begin{pmatrix} 1 \\ -2 \\ 1 \end{pmatrix} \tag{2.245}$$

This solution corresponds to the general case with $\alpha = 2$, $k_\alpha = 2\pi/3$, $\theta_\alpha = \pi/6$.

References

1. P. A. M. Dirac, *Principles of Quantum Mechanics*, Oxford: Clarendon Press, 1958.

44

Problems

1. Find the inverse of

$$\mathscr{A} = \begin{bmatrix} 7 & -1 & 2 \\ -3 & 5 & 4 \\ 1 & 6 & 0 \end{bmatrix} \tag{2.246}$$

2. Use Gaussian elimination to solve the linear equations

$$x_1 + x_2 + x_3 + x_4 = 10 \tag{2.247}$$
$$x_1 - 2x_2 + 3x_3 - x_4 = 2 \tag{2.248}$$
$$4x_1 + 3x_2 + 2x_3 + x_4 = 20 \tag{2.249}$$
$$3x_2 - 7x_3 + 2x_4 = -7 \tag{2.250}$$

3. For the following matrix, find the eigenvalues, right eigenvectors, left eigenvectors, and the parameters η_i

$$\mathscr{A} = \begin{bmatrix} 1 & 2 & 3 \\ 3 & 2 & 1 \\ 1 & 1 & 1 \end{bmatrix} \tag{2.251}$$

4. Verify by explicit calculation, the two formulas below, using the eigenvectors and eigenvalues of the previous problem

$$\mathscr{A} = \sum_j |j\rangle \frac{\lambda_j}{\eta_j} \langle j| \tag{2.252}$$

$$\mathscr{I} = \sum_j |j\rangle \frac{1}{\eta_j} \langle j| \tag{2.253}$$

5. Construct the unitary matrices $\mathscr{S}$ and $\mathscr{T}$ for the Hermitian matrix

$$\mathscr{A} = \begin{pmatrix} 2 & -1 & 0 \\ -1 & 2 & -1 \\ 0 & -1 & 2 \end{pmatrix} \tag{2.254}$$

Then show that $\mathscr{T}\mathscr{A}\mathscr{S} = \tilde{\lambda}$.

6. Use Gram–Schmidt to find the real eigenvectors of the matrix

$$\mathscr{M} = \begin{pmatrix} 1 & 1 & 1 & 1 \\ 1 & 1 & 1 & 1 \\ 1 & 1 & 1 & 1 \\ 1 & 1 & 1 & 1 \end{pmatrix} \tag{2.255}$$

7. Derive the vibrational eigenvalues and eigenvectors of a chain of $n = 4$ atoms.

8. Derive the vibrational eigenvalues and eigenvectors of a chain of $(n + 2)$ atoms labeled by: $j = 0, 1, 2, \ldots, n, n+1$. Furthermore, the first ($j = 0$) and last ($j = n+1$)

ones are clamped and cannot move. Derive the matrix that describes the modes, and find its eigenvalue and eigenvectors.

9. Consider an electron hopping around an equilateral triangle. The Hamiltonian is

$$\mathscr{H} = \begin{bmatrix} E_0 & V & V \\ V & E_0 & V \\ V & V & E_0 \end{bmatrix} \tag{2.256}$$

Find the eigenvalues and eigenvectors.

10. An electron on a tetrahedron has the same energy E_0 on all four sites, and the same matrix element V for hopping to the other three sites. The Hamiltonian matrix is

$$\mathscr{H} = \begin{bmatrix} E_0 & V & V & V \\ V & E_0 & V & V \\ V & V & E_0 & V \\ V & V & V & E_0 \end{bmatrix} \tag{2.257}$$

Find the eigenvalues and eigenvectors.

11. Evaluate $Z = \text{Tr} \exp(-\beta \mathscr{H})$ where the Hamiltonian $\mathscr{H}$ is the matrix in the previous problem.

12. Consider an electron hopping around on the six sites of a hexagon, i.e., a benzene molecule.
    (a) Write down the Hamiltonian in the first neighbor approximation.
    (b) Find the eigenvalues and eigenvectors.

3

Group Theory

3.1. Basic Properties of Groups

Definition of a Group

A group is a collection of elements that obey certain multiplication properties. Here the word "multiplication" has a general meaning: it could entail multiplication of matrices, the adding of numbers, or the rotation of a figure. The elements of the group are denoted by the symbols $A, B, C \ldots$.

1. Closure. The product of any two elements in the group is also in the group. So if the product of A and B is denoted AB then AB is another element in the group.

2. The associative law holds. If A, B, C are all elements in the group, then $(AB)C = A(BC)$. That is, if we multiply B times C, and then this is multiplied by A, we obtain the same result as if B is multiplied by A, and then the product multiplies C.

3. The group contains the *unit element* E that has the property $EA = AE = A$.

4. Each element A has an inverse element A^{-1} that is also in the group. The inverse obeys the relations $AA^{-1} = E = A^{-1}A$. It matters not which order the inverse is taken.

As an example of a group, consider the elements to be integers. Multiplication is a simple addition. The identity is the integer zero. The inverse of n is $-n$. There is closure and the associative law holds. This group has an infinite number of elements.

Groups with a finite number of elements are called *finite groups*. The number of elements in finite groups is denoted by h, which is also called the *order* of the group. This chapter will only discuss finite groups.

Multiplication gives $AB = C$ where (A, B, C) are all elements of the group. To specify all possible multiplications of h elements take h^2 such results. They are conveniently collected in a *multiplication table*, which is an $h \times h$ array.

Table 3.1 Multiplication Table for D_3

	E	A	B	C	D	F
E	E	A	B	C	D	E
A	A	E	D	F	B	C
B	B	F	E	D	C	A
C	C	D	F	E	A	B
D	D	C	A	B	F	E
F	F	B	C	A	E	D

As an example, Table 3.1 contains the multiplication table for group D_3 that has six elements: Some comments follow:

- The element in the top row is multiplied by the element in the left column: e.g., $AB = D$, $BA = F$.

- A, B, and C are inverses of themselves.

- D and F are mutual inverses.

- D_3 is also the group of rotations that leave invariant an equilateral triangle as shown in Fig. 3.1:
  (a) Element E is no rotation;
  (b) Element D is a clockwise rotation by 120°;
  (c) Element F is a counterclockwise rotation by 120°;
  (d) Elements A, B, C are 180° rotations about axes in the plane of the triangle: Each axis goes through a vertice and bisects the opposite side.

Most of the groups in this chapter are for the rotations of simple polygons such as squares, hexagons, or cubes. Group elements are rotations about an axis. In

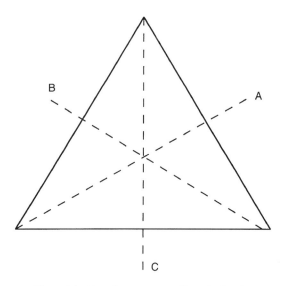

Figure 3.1 Rotations on an equilateral triangle.

Table 3.2 Angles in Degrees for C_n

ϕ (°)	C_n
0	E
60	C_6
90	C_4
120	C_3
180	C_2

this case, the rotations are represented by the symbol C_n where the value of n is determined by the angle of rotation $\phi = 360/n$ in degrees. A list of possibilities is shown in Table 3.2. For example, the group D_3 has elements: E, $2C_3$, $3C_{2'}$. The notation "$2C_3$" means there are two rotations by $\pm 120°$. Similarly, $3C_{2'}$ means there are three 180° rotations about axes in the plane. The prime on C_2 usually denotes the axis of rotation is in the plane. An unprimed rotation is about the axis perpendicular to the plane.

THEOREM. *Each element of the group appears only once in each row of the multiplication table, and only once in each column.*

Denote the elements of the group as A_i where $1 \leqslant i \leqslant h$ (and $A_1 \equiv E$). The proof of the theorem begins by multiplying each A_i by one of the group elements A_k. Denote the products as $B_i = A_i A_k$. The set of elements A_i are all different. The theorem is correct if the set of all B_i are different, regardless of the choice of A_k. The inverse of B_i is called $B_i^{-1} = A_k^{-1} A_i^{-1}$ since

$$B_i B_i^{-1} = A_i A_k A_k^{-1} A_i^{-1} = A_i A_i^{-1} = E \tag{3.1}$$

Then consider:

$$B_i B_j^{-1} = A_i A_k A_k^{-1} A_j^{-1} = A_i A_j^{-1} \neq E \tag{3.2}$$

If $i \neq j$ then $A_i \neq A_j$ and $A_i A_j^{-1} \neq E$. In this case $B_i B_j^{-1} \neq E$ and $B_i \neq B_j$. Since each pair of B_i and B_j are different, then none can appear twice in the product of $A_i A_k$. Each must appear once.

Other definitions are:

1. Abelian groups have all multiplications commutative. An example in Table 3.3 is Viergruppen, that in German means "four-group."

Table 3.3 Viergruppen

	E	A	B	C
E	E	A	B	C
A	A	E	C	B
B	B	C	E	A
C	C	B	A	E

Table 3.4 Multiplication Table of Group with Two Elements

	E	A
E	E	A
A	A	E

Note that this group is cyclic ($A^2 = E$) and Abelian.

Viergruppen is also the group of rotations that leave invariant a rectangle. E is no rotation. The elements A, B, C are 180° rotations about the three axes: x, y, z.

2. Cyclic groups have elements: E, A, A^2, A^3, ..., $A^n = E$ where the order is $h = n$. Cyclic groups are Abelian. The cyclic group with $h = 2$ is shown in Table 3.4.

3. Subgroups: Groups with a large number of elements usually have subgroups. They are a subset of elements that form a group by themselves. The element E is always a subgroup of order one. For example, the other subgroups of D_3 are:

   - E, A,

   - E, B,

   - E, C,

   - E, D, F.

In finding the subgroups, a useful theorem is due to LaGrange:

THEOREM. *The order h_j of every subgroup is an integer divisor of the order h of the original group.*

For example, if $h = 6$ then subgroups can have orders: 1, 2, or 3, but not 4 or 5.

4. The groups with small values of h are:

   - $h = 1$: Only element is E.
   - $h = 2$: Is the group with (E, A) given in Table 3.4.
   - $h = 3$: Is the cyclic group given in Table 3.5 as (E, D, F).

Table 3.5 Multiplication Table of Group with Three Elements

	E	D	F
E	E	D	F
D	D	F	E
F	F	E	D

Note that this group is actually cyclic since $D^2 = F$, $D^3 = E$.

Table 3.6 Multiplication Table of Cyclic Group with $A^4 = E$

	E	A	A^2	A^3
E	E	A	A^2	A^3
A	A	A^2	A^3	E
A^2	A^2	A^3	E	A
A^3	A^3	E	A	A^2

- $h = 4$: Here there are two groups: the Viergruppen in Table 3.3 and the cyclic group in Table 3.6.

5. Conjugate elements of a group: B is conjugate to A if $B = XAX^{-1}$ where X is any group element.

- *If B is conjugate to A, then A is conjugate to B.*

    Proof. If B is conjugate to A then there is a group element X such that $B = XAX^{-1}$. Multiply by X^{-1} from the left and by X from the right:

$$X^{-1}BX = X^{-1}(XAX^{-1})X = A \tag{3.3}$$

Let $Y = X^{-1}$ and Y is also an element of the group since it is the inverse of X. Then $B = YAY^{-1}$ and B is conjugate to A.

- *If B and C are conjugate to A, then B is conjugate to C.*

    Proof. Say that $B = XAX^{-1}$ and $C = YAY^{-1}$. Then the latter definition is equivalent to $A = Y^{-1}CY$ and

$$\begin{aligned}
B &= X(Y^{-1}CY)X^{-1} = (XY^{-1})C(YX^{-1}) \\
  &= (XY^{-1})C(XY^{-1})^{-1} = ZCZ^{-1}
\end{aligned} \tag{3.4}$$

if $Z = XY^{-1}$ which is an element of the group.

- Elements which are mutually conjugate are in a **class**.
- E is always in a class by itself.
- In Abelian and cyclic groups, each element is in a class by itself.
- In D_3, the classes are: (i) E, (ii) (A, B, C), and (iii) (D, F). See Table 3.7.

The classes of D_3 are written in the C_n notation as: $(A, B, C) = 3C_{2'}$ and $(D, F) = 2C_3$. Rotations by identical angles are often in the same class. This fact often allows one to guess the classes of a group of rotations.

3.2. Group Representations

There are many different ways to represent the elements of the group. Most representations are only matrices. A scalar is a 1×1 matrix. For example,

Table 3.7 Conjugate Elements in D_3

X	XAX^{-1}	X	XDX^{-1}
E	A	E	D
A	A	A	F
B	C	B	F
C	B	C	F
D	B	D	D
F	C	F	D

consider the group D_3. Some of its representations are

- Represent each element by one. Then $AB = D$ is just $1 \cdot 1 = 1$ which is trivially true. All groups have this representation.
- Let E, D, F be represented by $+1$ and A, B, C be represented by -1. This works, as you can check using the multiplication tables.
- The operations can be represented by 2×2 matrices:

$$E = \begin{pmatrix} 1 & 0 \\ 0 & 1 \end{pmatrix} \tag{3.5}$$

$$A = \begin{pmatrix} -1 & 0 \\ 0 & 1 \end{pmatrix} \tag{3.6}$$

$$B = \begin{pmatrix} 1/2 & -\sqrt{3}/2 \\ -\sqrt{3}/2 & -1/2 \end{pmatrix} \tag{3.7}$$

$$C = \begin{pmatrix} 1/2 & \sqrt{3}/2 \\ \sqrt{3}/2 & -1/2 \end{pmatrix} \tag{3.8}$$

$$D = \begin{pmatrix} -1/2 & \sqrt{3}/2 \\ -\sqrt{3}/2 & -1/2 \end{pmatrix} \tag{3.9}$$

$$F = \begin{pmatrix} -1/2 & -\sqrt{3}/2 \\ \sqrt{3}/2 & -1/2 \end{pmatrix} \tag{3.10}$$

- The next representation is based on the determinant of a matrix. Use the theorem that if $AB = D$ then $\det |A| \det |B| = \det |D|$. The determinant obeys all of the multiplication table that is actually the second representation listed above.
- These three are the only *irreducible representations* of D_3 and are called: Γ_j for $j = (1, 2, 3)$. $\Gamma_3(A)$ is the 2×2 matrix for A listed above.
- The representations obey the same multiplication tables as the elements of the group that is represented by: $\Gamma_i(A)\Gamma_i(B) = \Gamma_i(D)$ if $AB = D$,

Reducible representations are those that can be "reduced" to irreducible representations. An example of a reducible representation is given for the group D_3. It is then reduced to the three irreducible ones.

The group D_3 describes the rotations that leave invariant the equilateral triangle. Label the three vertices as $(1, 2, 3)$. The rotation by A changes these to $(2, 1, 3)$ that can be represented by a 3×3 matrix that is the representation called Γ_4

$$\begin{pmatrix} 2 \\ 1 \\ 3 \end{pmatrix} = \begin{pmatrix} 0 & 1 & 0 \\ 1 & 0 & 0 \\ 0 & 0 & 1 \end{pmatrix} \begin{pmatrix} 1 \\ 2 \\ 3 \end{pmatrix} \tag{3.11}$$

$$\Gamma_4(A) = \begin{pmatrix} 0 & 1 & 0 \\ 1 & 0 & 0 \\ 0 & 0 & 1 \end{pmatrix} \tag{3.12}$$

In the same way, construct the other representations of Γ_4:

$$\Gamma_4(E) = \begin{pmatrix} 1 & 0 & 0 \\ 0 & 1 & 0 \\ 0 & 0 & 1 \end{pmatrix} \tag{3.13}$$

$$\Gamma_4(B) = \begin{pmatrix} 1 & 0 & 0 \\ 0 & 0 & 1 \\ 0 & 1 & 0 \end{pmatrix} \tag{3.14}$$

$$\Gamma_4(C) = \begin{pmatrix} 0 & 0 & 1 \\ 0 & 1 & 0 \\ 1 & 0 & 0 \end{pmatrix} \tag{3.15}$$

$$\Gamma_4(D) = \begin{pmatrix} 0 & 0 & 1 \\ 1 & 0 & 0 \\ 0 & 1 & 0 \end{pmatrix} \tag{3.16}$$

$$\Gamma_4(F) = \begin{pmatrix} 0 & 1 & 0 \\ 0 & 0 & 1 \\ 1 & 0 & 0 \end{pmatrix} \tag{3.17}$$

The next step is to perform a unitary transformation on this representation. First define a matrix S and its inverse S^{-1} that are

$$S = \begin{pmatrix} 1/\sqrt{3} & 1/\sqrt{2} & -1/\sqrt{6} \\ 1/\sqrt{3} & -1/\sqrt{2} & -1/\sqrt{6} \\ 1/\sqrt{3} & 0 & \sqrt{2/3} \end{pmatrix} \tag{3.18}$$

$$S^{-1} = \begin{pmatrix} 1/\sqrt{3} & 1/\sqrt{3} & 1/\sqrt{3} \\ 1/\sqrt{2} & -1/\sqrt{2} & 0 \\ -1/\sqrt{6} & -1/\sqrt{6} & \sqrt{2/3} \end{pmatrix} \tag{3.19}$$

$$SS^{-1} = S^{-1}S = E = \begin{pmatrix} 1 & 0 & 0 \\ 0 & 1 & 0 \\ 0 & 0 & 1 \end{pmatrix} \tag{3.20}$$

Examine the properties of a new representation called Γ_4' that is formed from a unitary transformation on Γ_4:

$$\Gamma_4'(X) = S^{-1}\Gamma_4(X)S \tag{3.21}$$

Note that Γ_4' is also a valid representation of the group since

$$\Gamma_4'(A)\Gamma_4'(B) = (S^{-1}\Gamma_4(A)S)(S^{-1}\Gamma_4(B)S) = S^{-1}\Gamma_4(A)\Gamma_4(B)S$$
$$= S^{-1}\Gamma_4(D)S = \Gamma_4'(D) \tag{3.22}$$

All group multiplications are maintained. Doing the indicated multiplications for three group elements gives the following results:

$$\Gamma_4'(E) = S^{-1}\Gamma_4(E)S = \begin{pmatrix} 1 & 0 & 0 \\ 0 & 1 & 0 \\ 0 & 0 & 1 \end{pmatrix} \tag{3.23}$$

$$\Gamma_4'(A) = S^{-1}\Gamma_4(A)S = \begin{pmatrix} 1 & 0 & 0 \\ 0 & -1 & 0 \\ 0 & 0 & 1 \end{pmatrix} \tag{3.24}$$

$$\Gamma_4'(B) = S^{-1}\Gamma_4(B)S = \begin{pmatrix} 1 & 0 & 0 \\ 0 & 1/2 & -\sqrt{3}/2 \\ 0 & -\sqrt{3}/2 & -1/2 \end{pmatrix} \tag{3.25}$$

Note that they can be given by the matrix

$$\Gamma_4'(X) = S^{-1}\Gamma_4(X)S = \begin{pmatrix} \Gamma_1(X) & 0 \\ 0 & \Gamma_3(X) \end{pmatrix} \tag{3.26}$$

The representation $\Gamma_4'(X)$ is a matrix in *block diagonal form*. Its first element is the representation $\Gamma_1(X)$. The other nonzero parts are the 2×2 matrix $\Gamma_3(X)$. The representation $\Gamma_4(X)$ has been "reduced" to a matrix that is composed Γ_1 and Γ_3. The matrix Γ_4 has no new information, it contains only the known information in Γ_1, Γ_3. Therefore, Γ_4 is "reducible" to Γ_1, Γ_3. A representation that can be reduced is not of fundamental importance. Only the "irreducible" representations are important, so only they are listed in character tables. In group theory the common notation is to write: $\Gamma_4 = \Gamma_1 + \Gamma_3$. The + sign does not mean addition, instead, it means that Γ_4 can be put into block diagonal form, composed of representations Γ_1 and Γ_3. This notation will be used often.

3.3. Characters

Characters are the trace of the matrix representation: for scalars, the trace is just the scalar. The trace is the summation of the diagonal elements of the matrix.

Table 3.8 Character Table for D_3

	E	$3C_{2'}$	$2C_3$
Γ_1	1	1	1
Γ_2	1	−1	1
Γ_3	2	0	−1

- *All elements in the same class have the same trace and the same character for each representation.*

  Proof. The character is the trace of the matrix representation.

$$\chi_i(A) = \sum_{\alpha=1}^{l_i} \Gamma_{\alpha,\alpha,i}(A) = \mathrm{Tr}(A) \tag{3.27}$$

  Let B be a matrix that is in the same class as A, so there is an element X such that $B = X^{-1}AX$. Take the trace of B:

$$\chi(B) = \mathrm{Tr}(B) = \mathrm{Tr}(X^{-1}AX) = \mathrm{Tr}(XX^{-1}A) = \mathrm{Tr}(A) = \chi(A)$$

  The cyclic properties of the trace were used to get to $XX^{-1} = I$. The characters of all elements in a class are identical, for the same representation.

- The first representation is denoted Γ_1 that is equal to 1 for all classes. Its characters are all 1 since the representation is all 1s.
- The characters for $\Gamma_j(E) = l_j$, where l_j is the dimension of the representation. The representations for E are just the identity matrix that has all 1s along the diagonal, and 0s elsewhere. The trace is just the dimension of the matrix.
- A *character table* is constructed that lists the characters in each class, for each representation. For D_3 it is shown in Table 3.8

The character table has an important role in group theory. Many properties of the group can be deduced from its character table. Therefore it is useful to develop a method of finding the character table quickly, without having to construct the multiplication tables for the group. This procedure is outlined below.

Rules for Constructing Character Tables

1. The number of irreducible representations equals the number of classes. It is denoted by p.
2. Some definitions

   - h is the dimension of group.
   - p is the number of classes and the number of irreducible representations.
   - l_i is the dimension of a representation, i.e., the matrix is $l_i \times l_i$. l_i are nonzero positive integers.

- N_k is the number of elements in a class C_k of rotations. (*Note:* h is divisible by N_k.) Classes can be found by inspection and grouping together like elements. All 90° (C_4) and 60° (C_6) rotations usually go together, the notable exceptions are 180° rotations. They are in the same class if they have similar axes. All C_2 about an edge go together, but are in a separate class from all C_2 rotations about a corner.

- The character $\chi_i(X)$ of an element X for the representation $\Gamma_i(X)$ is the trace of the matrix $\Gamma_i(X)$.

3. The set of values l_i can be found using

$$h = \sum_{i=1}^{p} l_i^2 \tag{3.28}$$

This identity is proved below. Note that there is only one way of selecting the values l_i to satisfy this equation. For example, if $h = 8$ and $p = 5$ then the values of l_i are (1, 1, 1, 1, 2). None of the l_i can be 0.

4. The identity E is always diagonal so $\chi_i(E) = l_i$.

5. The first representation has all elements 1 so that $\chi_1(C_k) = 1$.

6. A representation of dimension 1 has all of its characters with a magnitude of 1 and could be phase factors such as $\exp(i\alpha)$ including -1, but they cannot be 0 or a real number that is not ± 1.

7. The columns of the character table are orthogonal vectors. This orthogonality is given by the first of the two equations below. The summation i is over representations.

$$\sum_{i=1}^{p} \chi_i(C_k)^* \chi_i(C_{k'}) = \delta_{k,k'} \frac{h}{N_k} \tag{3.29}$$

$$\sum_{k=1}^{p} N_k \chi_i(C_k)^* \chi_j(C_k) = h\delta_{ij} \tag{3.30}$$

As an example, if $k = k' = 1$ and $C_k = C_{k'} = E$, $N_1 = 1$, $\chi_i(E) = l_i$, then Eq. (3.29) gives the identity Eq. (3.28).

8. The rows of the character table are also orthogonal vectors if the multiplication is weighted by the factor N_k. This orthogonality is the second of the above two equations, where the symbol k denotes classes.

9. The above rules are sufficient to uniquely construct the character table without ever constructing the multiplication table.

As an example, the character table for the group that has $h = 8$, $p = 5$ is shown in Table 3.9. The rotations $C_2, 2C_4$ in Fig. 3.2 are about the axis perpendicular to the plane. The other four C_2 rotations are about axes in the plane; two along the (x, y) axes and two about body diagonals. The four different kinds of rotations are in separate classes. This group of rotations (D_4) leaves a square invariant.

Table 3.9 Character Table for D_4

	E	C_2	$2C_4$	$2C_{2'}$	$2C_{2''}$
Γ_1	1	1	1	1	1
Γ_2	1	1	1	-1	-1
Γ_3	1	1	-1	1	-1
Γ_4	1	1	-1	-1	1
Γ_5	2	-2	0	0	0

Representations of Cyclic Groups

Cyclic groups are Abelian. The number of classes equals the order of the group $h = p$. All values of $l_i = 1$. Define a complex number (where $i = \sqrt{-1}$)

$$\omega = e^{2\pi i/h}, \qquad \omega^h = 1 \tag{3.31}$$

The representations, and the characters since $l_i = 1$, are multiples of this complex number.

- The first representation has $\Gamma_1 = 1$.
- The second representation has $A = \omega, A^2 = \omega^2, \ldots$.
- The third representation has $A = \omega^2, A^2 = \omega^4, \ldots$.
- The fourth representation has $A = \omega^3, A^2 = \omega^6, \ldots$.
- The cyclic group with $h = 3$ is shown in Table 3.10.
- Note that for $h = 3$ then $\omega^4 = \omega$ that simplifies the last entry in the table.
- One can show that these rows and columns obey the orthogonality rules for character tables. Take the complex conjugate of one vector in multiplying vectors. In proving orthogonality, it is useful to note the

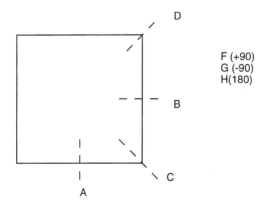

Figure 3.2 Rotations that leave a square invariant.

Table 3.10 Representation of the Cyclic Group $h = 3$

	E	A	A^2
Γ_1	1	1	1
Γ_2	1	ω	ω^2
Γ_3	1	ω^2	ω

simple identity

$$1 + \omega + \omega^2 + \cdots + \omega^{h-1} = \sum_{j=0}^{h} \omega^j = \frac{1 - \omega^h}{1 - \omega} = 0 \qquad (3.32)$$

where $\omega^h = 1$ is used to make the numerator vanish.

3.4. Direct Product Groups

A product group is formed by taking two different groups and combining them in a way that involves a "product."

- One group $\mathcal{G}_a$ has order h_a and elements $E, A_2, A_3, \ldots, A_{h_a}$.
- Another group $\mathcal{G}_b$ has order h_b and elements $E, B_2, B_3, \ldots, B_{h_b}$.
- The only common elements between the two groups is E.
- All operations of one group commute with those of the other.

If these conditions are satisfied, one can form a *direct product group* denoted as $\mathcal{G}_a \times \mathcal{G}_b$ that has the following properties

- The number of elements is $h_a h_b$.
- The elements are all products $A_i B_j$.
- One can show it is a group. For example, it has closure, since multiplying two elements together produce another group element:

$$(A_i B_j)(A_{i'} B_{j'}) = (A_i A_{i'})(B_j B_{j'}) = (A_k B_l) \qquad (3.33)$$

The first equation invokes the fact that the element of $\mathcal{G}_a$ commutes with those of $\mathcal{G}_b$, so the order can be rearranged. The second identity uses the fact that $A_i A_{i'} = A_k$ and $B_j B_{j'} = B_l$. Since $\mathcal{G}_a$ and $\mathcal{G}_b$ are both groups, then A_k and B_l are both in these two groups. Then their product must also be in the product group.   QED

In the same way, one can prove the other requirements to be a group, such that there is an inverse for each element.

3.4.1. Representations

The representations of the direct product group are just the "product" of the representations of the individual groups. Here the word "product" has a precise and unusual meaning. If a representation of $\mathcal{G}_a$ has dimensionality l_{ja}

Table 3.11 Character Table of
Any Group with $h = 2$

	E	σ_h
Γ_1	1	1
Γ_2	1	-1

and the representation of $\mathscr{G}_b$ has dimensionality l_{jb}, then the representation of the direct product group has dimension $l_{ia}l_{jb}$. For example, if $l_{ia} = 2$, $l_{jb} = 3$ then the product group representation has dimension of six. It is formed by multiplying every element of $\Gamma_{ia}(A_k)$ by every element of $\Gamma_{jb}(B_m)$. This is expressed as

$$\Gamma_{ij}^{(a \times b)}(A_k B_m) = \Gamma_{ia}(A_k) \times \Gamma_{jb}(B_m) \tag{3.34}$$

It is not matrix multiplication. The character table is constructed as the direct product of the character table of the two groups. This new character table still obeys all of the rules of character tables. The rows and columns are mutually orthogonal and the sum of the squares of the dimensions equals $h_a h_b$. For example,

$$h_a h_b = \sum_{ij} (l_{ai}l_{bj})^2 = \sum_i (l_{ai})^2 \sum_j (l_{bj})^2 \tag{3.35}$$

As an example, consider the operations that leave unchanged a three dimensional block that is in the shape of an equilateral triangle when viewed from the top, but the sides are straight. Introduce a reflection σ_h in the horizontal plane through the center of the block, that leaves it unchanged. Rotational operations are identical to those in D_3 for the equilateral triangle. The twelve operations that leave the block unchanged are E, $2C_3$, $3C_{2'}$, σ_h, $2\sigma_h C_3$, $3\sigma_h C_{2'}$. This group is the direct product group of $D_{3h} = (E, \sigma_h) \times D_3$. The character table of the group (E, σ_h) is shown in Table 3.11. The character table of the direct product group of the triangular block is shown in Table 3.12. Note that this character table has four quadrants, three are alike and are identical to the character table of D_3. The fourth quadrant is the negative of the characters in the other three. This placement of plus and minus signs

Table 3.12 Character Table of the Direct Product Group
$(E, \sigma_h) \times D_3$

D_{3h}	E	$3C_{2'}$	$2C_3$	σ_h	$(3\sigma_h C_{2'})$	$(2\sigma_h C_3)$
Γ_1	1	1	1	1	1	1
Γ_2	1	-1	1	1	-1	1
Γ_3	2	0	-1	2	0	-1
Γ_4	1	1	1	-1	-1	-1
Γ_5	1	-1	1	-1	1	-1
Γ_6	2	0	-1	-2	0	1

corresponds to the four elements in the character table of (E, σ_h). The character table can also be expressed as

$$
\begin{bmatrix}
1 & 1 & 1 \\
1 & -1 & 1 \\
2 & 0 & -1
\end{bmatrix}
\times
\begin{bmatrix}
1 & 1 \\
1 & -1
\end{bmatrix}
\tag{3.36}
$$

which is a simpler notation.

3.5. Basis Functions

Basis functions are functions of $\mathbf{r} = (x, y, z)$ that transform according to individual representations of a group. Usually they are given for low order polynomials of the three components of $\mathbf{r}$.

As an example, consider the group D_3. The group operations can be considered to be the rotations on an equilateral triangle. Assign an (x, y, z) coordinate system to the triangle, where (x, y) are in the plane of the triangle, while the z-axis points upward. First assign z to a group representation. When an operation leaves the z-axis still pointing up, that is called $+z$, and assign $+1$ to the operation. That is the case for E plus the two rotations $(D, F) = C_{\pm 3}$ around the z-axis. However, the rotations $(A, B, C) = C_2$, flip $+z$ to $-z$ and are assigned a representation of -1. The coordinate z is assigned to the representation shown in Table 3.13. This representation is identical to the one we have been calling Γ_2. Therefore z transform according to the representation Γ_2. Similarly z^2 has a representation in which every element is $+1$, that is the representation Γ_1.

The next step is to consider how the rotations change the unit vectors in the plane (x, y). Consider the operator D, that is a rotation by an angle $2\pi/3$ clockwise around the z-axis. This rotates the (x, y) axes to a new set we call (x', y'). The relationship is $(\theta = 2\pi/3)$

$$
x' = x \cos \theta + y \sin \theta = -\frac{1}{2} x + \frac{\sqrt{3}}{2} y
\tag{3.37}
$$

$$
y' = -x \sin \theta + y \cos \theta = -\frac{\sqrt{3}}{2} x - \frac{1}{2} y
\tag{3.38}
$$

$$
\begin{pmatrix} x' \\ y' \end{pmatrix} =
\begin{pmatrix}
-\dfrac{1}{2} & \dfrac{\sqrt{3}}{2} \\
-\dfrac{\sqrt{3}}{2} & -\dfrac{1}{2}
\end{pmatrix}
\begin{pmatrix} x \\ y \end{pmatrix}
\tag{3.39}
$$

Table 3.13 Representation of z and z^2

	E	$3C_{2'}$	$2C_3$	
z	1	-1	1	Γ_2
z^2	1	1	1	Γ_1

Table 3.14 Some Basis Functions for D_3

		E	$2C_3$	$3C_{2'}$
$z^2, x^2 + y^2$	Γ_1	1	1	1
z	Γ_2	1	1	-1
$(x, y), (yz, -xz), [xy, \frac{1}{2}(x^2 - y^2)]$	Γ_3	2	-1	0

The operation by D can be represented by the 2×2 matrix in Eq. (3.9). This matrix should be familiar, since it is the same matrix introduced earlier in Γ_3. The two functions (x, y) together form a basis for the Γ_3 representation. As another example, consider the operation A in D_3 that is a $C_{2'}$ rotation around the y axis. This rotation changes $x' = -x, y' = y$ so the matrix is

$$A = \begin{pmatrix} -1 & 0 \\ 0 & 1 \end{pmatrix} \tag{3.40}$$

It is the same matrix $\Gamma_3(A)$. All of the six 2×2 matrices just state how x and y are changed during the rotations. In fact, that is how these matrices were constructed.

A representation with l_i has that number of basis functions in a set. For example, the Γ_3 representation of D_3 has as one basis (x, y). The two functions must go together in one basis, since the variables x and y get scrambled together as a basis. However, the combination $x^2 + y^2$ is unaffected by any rotation, and so belongs to Γ_1. Table 3.14 shows some of the basis functions for the representations of D_3. Note that the representation Γ_3 has three different pairs of basis functions. These functions account for all polynomials up to second order.

Another feature of basis functions is that they must be orthogonal. The best way to think of this is in spherical coordinates: Set

$$x = r \sin(\theta) \cos(\phi) \tag{3.41}$$

$$y = r \sin(\theta) \sin(\phi) \tag{3.42}$$

$$z = r \cos(\theta) \tag{3.43}$$

$$\langle xz \rangle = r^2 \int_0^\pi \sin(\theta) \, d\theta \int_0^{2\pi} d\phi \cos(\theta) \sin(\theta) \cos(\phi) = 0 \tag{3.44}$$

The last integral gives an example of "orthogonality." One integrates over the angles in spherical coordinates, and the functions are orthogonal if the integral is zero.

3.6. Angular Momentum

In three dimensions the Cartesian coordinates are (x, y, z). Often it is convenient to work in spherical coordinates where the three variables become (r, θ, ϕ). Here, θ is the polar angle and is the angle that the vector makes with the z axis. The angle ϕ is the azimuthal angle and is the angle from the x axis when the vector is projected onto the (x, y) plane. The three variables each have an eigenvalue (n, l, m). In Chapter 12 the associated Legendre polynomials $P_l^m(\theta)$ are introduced and by using them, the eigenfunctions have the form

$$\phi_{nlm}(\mathbf{r}) = R_{nl}(r)P_l^{|m|}(\theta)e^{im\phi} \tag{3.45}$$

where the radial function is $R_{nl}(r)$. Values of l run from zero to infinity. Values of m span the range $-l \leqslant m \leqslant l$, so there are $2l + 1$ values. An important feature of this problem is that the eigenvalues often do not depend on the quantum number m. There are $2l + 1$ states of the same energy with basis states $\exp(im\phi)$. The group representation has a dimension $2l + 1$ of the form

$$\Gamma = \begin{vmatrix} e^{il\phi} & 0 & 0 & \cdots & 0 \\ 0 & e^{i(l-1)\phi} & 0 & \cdots & 0 \\ 0 & 0 & e^{i(l-2)\phi} & \vdots & 0 \\ 0 & 0 & 0 & \ddots & 0 \\ 0 & 0 & 0 & \cdots & e^{-il\phi} \end{vmatrix} \tag{3.46}$$

The character is the sum of the diagonal elements

$$\chi_l(\phi) = \sum_{m=-l}^{l} e^{im\phi} = \frac{\sin[(l + 1/2)\phi]}{\sin(\phi/2)} \tag{3.47}$$

$$\chi_l(0) = 2l + 1 \tag{3.48}$$

This formula is important and will be used often. The angle ϕ is the angle of the rotations and C_n is the angle $\phi = 2\pi/n$. It is useful to have a table of χ_l or various angles of rotation. The simple examples are $\chi_0(\phi) = 1$, $\chi_l(0) = 2l + 1$. Other examples are shown in Table 3.15. The results in the table are universal, i.e., they apply to all groups.

As an example, consider how the angular momentum characters behave in D_3. The numbers in the first three columns of Table 3.16 for χ_l are merely the evaluation of Eq. (3.47) for the various angles: 120° and 180° for C_3 and $C_{2'}$, respectively. The assignment of the angular momentum states to the irreducible representations Γ_j is easy in some cases. The $l = 0$ case is isotropic, and must belong to Γ_1. Similarly, the three $l = 1$ functions are (x, y, z), that belong to $\Gamma_2(z)$ and $\Gamma_3(x, y)$. Similarly, the $l = 2$ functions are the quadratic polynomials that all belong to either Γ_1 or Γ_3. The angular momentum decomposition is useful in cases where the basis functions have not been

Table 3.15 Angular Momentum Characters for Different Angles

$\phi =$	0	60	90	120	180
χ_1	3	2	1	0	−1
χ_2	5	1	−1	−1	1

Table 3.16 Angular Momentum Basis Functions for D_3

	E	$2C_3$	$3C_{2'}$	
Γ_1	1	1	1	
Γ_2	1	1	-1	
Γ_3	2	-1	0	
χ_0	1	1	1	$= \Gamma_1$
χ_1	3	0	-1	$= \Gamma_2 + \Gamma_3$
χ_2	5	-1	1	$= \Gamma_1 + 2\Gamma_3$

worked out previously. A listing such as $\Gamma_2 + \Gamma_3$ means block diagonal form.

How do we know how to decompose the angular functions to the irreducible representations? To be correct, the characters in each column must add correctly. It is necessary to find the integers n_{li} that form the decomposition

$$\chi_l(C_k) = \sum_i^p n_{li}\chi_i(C_k) \tag{3.49}$$

where C_k is any class. The summation is over irreducible representations. The integers n_{li} are determined by considering the following expression while using the orthogonality rules for character tables

$$\sum_k^p N_k \chi_j^*(C_k)\chi_l(C_k) = \sum_i^p n_{li} \sum_k^p N_k \chi_j^*(C_k)\chi_i(C_k) \tag{3.50}$$

$$= \sum_i n_{li}(h\delta_{ij}) = hn_{lj} \tag{3.51}$$

$$n_{lj} = \frac{1}{h}\sum_k^p N_k \chi_j^*(C_k)\chi_l(C_k) \tag{3.52}$$

This formula is useful for larger values of l where one cannot decompose by inspection.

Another example of the characters for angular momentum is given in Table 3.17 for the group D_6. Note that all rotations by the same angle $(C_2, C_{2'}, C_{2''})$ have the same characters.

Table 3.17 Angular Momentum Basis Functions for D_6.

	E	C_2	$2C_3$	$2C_6$	$3C_{2'}$	$3C_{2''}$	
χ_0	1	1	1	1	1	1	$= \Gamma_1$
χ_1	3	-1	0	2	-1	-1	$= \Gamma_2 + \Gamma_5$
χ_2	5	1	-1	1	1	1	$= \Gamma_1 + \Gamma_5 + \Gamma_6$

3.7. Products of Representations

Suppose there is a set of basis functions B_{ni} belonging to representation n. Perhaps we wish to multiply these basis functions by another set of basis functions B_{mj} that belong to representation m. To which representation does the product $B_{ni}B_{mj}$ belong? The answer is that this product may be a basis set of one of the same representations (n or m), of another representation, or possibly a linear combination of other representations. These different results are obtained from the character tables. One can merely multiply the characters of these representations, and then decompose the product into irreducible representations. The decomposition uses the same formula, Eq. (3.52), for decomposing angular momentum. An example from the group D_3 is shown in Table 3.18. It shows the result of multiplying the characters $\chi_i(C_k) \times \chi_j(C_k)$ for the different irreducible representations. For the six different possible arrangements, the first five produce a single representation. The last case ($\Gamma_3 \times \Gamma_3$) produces the sum of all three irreducible representations. This simple example illustrates two theorems:

THEOREM 1. *The direct product of a representation by itself always contains the representation Γ_1.*

THEOREM 2. *The direct product $\Gamma_i \times \Gamma_j$ contains Γ_1 if, and only if, $i = j$.*

In the example for D_3, the representation Γ_1 is only found for the cases of $\Gamma_1 \times \Gamma_1, \Gamma_2 \times \Gamma_2, \Gamma_3 \times \Gamma_3$. These results are the basis for the next section.

3.8. Quantum Mechanics

3.8.1. Eigenvalues

An important problem in quantum mechanics is to find the eigenstates $\phi_n(\mathbf{r})$ of a Hamiltonian H for a single particle moving in a potential $V(\mathbf{r})$

$$H\phi_n(\mathbf{r}) = E_n\phi_n(\mathbf{r}) \tag{3.53}$$

$$H = -\frac{\hbar^2\nabla^2}{2m} + V(\mathbf{r}) \tag{3.54}$$

Table 3.18 Products of Representations in D_3

	E	$2C_3$	$3C_{2'}$	
Γ_1	1	1	1	
Γ_2	1	1	-1	
Γ_3	2	-1	0	
$\Gamma_1 \times \Gamma_1$	1	1	1	$= \Gamma_1$
$\Gamma_1 \times \Gamma_2$	1	1	-1	$= \Gamma_2$
$\Gamma_1 \times \Gamma_3$	2	-1	0	$= \Gamma_3$
$\Gamma_2 \times \Gamma_2$	1	1	1	$= \Gamma_1$
$\Gamma_2 \times \Gamma_3$	2	-1	0	$= \Gamma_3$
$\Gamma_3 \times \Gamma_3$	4	1	0	$= \Gamma_1 + \Gamma_2 + \Gamma_3$

The group symmetry properties of the Hamiltonian are determined by the potential $V(\mathbf{r})$. For example, suppose the potential consisted of three classical charges at equal distance from each other and are at the corners of an equilateral triangle. Then all of the rotations of the group D_3 leave the potential unchanged, since all rotations return the charges to the same arrangement. The Hamiltonian is said to be *invariant* under the operations of group D_3. The kinetic energy term is always invariant under group operations, since ∇^2 has the same symmetry as r^2 that is usually Γ_1. If X is any of the six elements in D_3 then $XH = HX$. The group operations commute with the Hamiltonian. For example, multiply Eq. (3.53) by X and get

$$XH\phi_n = HX\phi_n = E_n X\phi_n \tag{3.55}$$

The first equality states that the Hamiltonian commutes with X. The second equality proves an important result:

THEOREM. *If ϕ_n is an eigenstate of energy E_n, then $X\phi_n$ is also an eigenstate of energy E_n.*

Take an eigenstate, operate by any group element, and get another function that is also an eigenstate of the same energy.

This theorem is actually a trivial result of group theory. The eigenstates ϕ_n must belong to a single group representation Γ_j. They must be basis functions of one representation or linear combinations of basis functions from one representation. The operations of the group merely scramble the basis functions from the same representations. So all basis functions in the same set have the same eigenvalue. In group D_3, the two functions in the basis set of Γ_3 such as (x, y) must have the same eigenvalue if they are part of the same eigenstate. Note that any function of r^2 such as $\xi(r^2)$ belongs to Γ_1 since both z^2 and $x^2 + y^2$ belong to this representation. Therefore, $\xi(r^2)$ is unchanged by any group representation. Any function of the form $[x\xi(r^2), y\xi(r^2)]$ also belongs to Γ_3 and these two have the same eigenvalue. Note that $z\xi(r^2)$ belongs to Γ_2 and will have a different eigenvalue. Different pairs in the same representation, such as $[yz\xi(r^2), -xz\xi(r^2)]$ will have a different energy than $[x\xi(r^2), y\xi(r^2)]$.

In quantum mechanics, states are *degenerate* if they have the same energy. All basis set groups are degenerate.

3.8.2. Representations

Chapter 2 discussed the method of finding the eigenvalues and eigenvectors of various matrices and some of these matrices were formulated as problems in quantum mechanics. For example, the matrix that describes the hopping of an electron between three equidistant sites is

$$\mathcal{H} = \begin{bmatrix} E_0 & V & V \\ V & E_0 & V \\ V & V & E_0 \end{bmatrix} \tag{3.56}$$

The eigenvalues and eigenvectors are known for this matrix. How are they assigned to the various group representations? Group D_3 describes the states of an equilateral triangle, so the eigenstates of this matrix belong to the representations of this group.

A simple model for the eigenstates are linear combinations of atomic orbitals (LCAO), where one orbital is centered at each atomic site

$$\psi_j(\mathbf{r}) = \sum_i a_i(j)\phi(\mathbf{r} - \mathbf{R}_i) \tag{3.57}$$

The eigenvectors of the matrix Eq. (3.56) are just the coefficients $a_i(j)$.

- For $\lambda_1 = E_0 + 2V$

$$\Psi_1(\mathbf{r}) = \frac{1}{\sqrt{3}}[\phi(\mathbf{r} - \mathbf{R}_1) + \phi(\mathbf{r} - \mathbf{R}_2) + \phi(\mathbf{r} - \mathbf{R}_3)] \tag{3.58}$$

- For $\lambda_2 = E_0 - V$

$$\Psi_2(\mathbf{r}) = \frac{1}{\sqrt{2}}[\phi(\mathbf{r} - \mathbf{R}_1) - \phi(\mathbf{r} - \mathbf{R}_3)] \tag{3.59}$$

- For $\lambda_3 = E_0 - V$

$$\Psi_3(\mathbf{r}) = \frac{1}{\sqrt{6}}[\phi(\mathbf{r} - \mathbf{R}_1) - 2\phi(\mathbf{r} - \mathbf{R}_2) + \phi(\mathbf{r} - \mathbf{R}_3)] \tag{3.60}$$

The next step is to assign these solutions to the representations of group D_3. The result depends on the symmetry of the atomic orbitals. Orbitals are assigned rotational quantum numbers (l, m): Those with $(l = 0, m = 0)$ are s states, those with $(l = 1, -1 \leqslant m \leqslant 1)$ are p states. S-state orbitals are unchanged by rotations of the atom. If the orbitals $\phi(\mathbf{r})$ have s symmetry, then the state Ψ_1 belongs to representation Γ_1 since it is unchanged by any of the group rotations. On the other hand, if the orbital has p_z symmetry $(l = 1, m = 0)$, then the orbitals are perpendicular to the plane of the molecule. The orbital changes sign under a rotation $C_{2'}$. In this case $\Psi_1(\mathbf{r})$ belongs to the representation Γ_2.

The two states Ψ_2, Ψ_3 belong to the representation Γ_3. They are paired basis functions for this representation, similar to (x, y). The fact that they are degenerate suggests this assignment. However, the true test is how the states change under a group rotation. A rotation by C_3 changes the atoms $(1 \rightarrow 2, 2 \rightarrow 3, 3 \rightarrow 1)$.

$$C_3\Psi_2 = \frac{1}{\sqrt{2}}[\phi(\mathbf{r} - \mathbf{R}_2) - \phi(\mathbf{r} - \mathbf{R}_1)] = -\frac{1}{2}\Psi_2 - \frac{\sqrt{3}}{2}\Psi_3$$

$$C_3\Psi_3 = \frac{1}{\sqrt{6}}[\phi(\mathbf{r} - \mathbf{R}_2) - 2\phi(\mathbf{r} - \mathbf{R}_3) + \phi(\mathbf{r} - \mathbf{R}_1)] = \frac{\sqrt{3}}{2}\Psi_2 - \frac{1}{2}\Psi_3$$

$$\tag{3.61}$$

$$C_3\begin{pmatrix}\Psi_2 \\ \Psi_3\end{pmatrix} = \begin{pmatrix}-1/2 & -\sqrt{3}/2 \\ \sqrt{3}/2 & -1/2\end{pmatrix}\begin{pmatrix}\Psi_2 \\ \Psi_3\end{pmatrix} \tag{3.62}$$

The rotations of the group are represented by 2×2 matrices that belong to Γ_3. The above matrix is called F in Section 3.2. The states $\Psi_{2,3}$ belong to Γ_3 when the orbitals $\phi(\mathbf{r})$ are either s- or p_z-states. Assigning eigenstates to group representations is an important step in any group theory analysis.

3.8.3. Matrix Elements

Often in quantum mechanics there is a need to evaluate integrals such as

$$\int d^3r \psi_i^*(\mathbf{r}) \psi_j(\mathbf{r}) = ? \tag{3.63}$$

The Hamiltonian H describes the properties of one or several particles. It has eigenstates ψ_i such that $H\psi_j = E_j \psi_j$ where E_j are the eigenvalues. In these instances, what can be stated about the integral in Eq. (3.63)? The Hamiltonian belongs to some group, and this group has representations Γ_v. The eigenfunctions ψ_i must belong to one of the group representations, for example, Γ_n, while ψ_j belongs to Γ_m. The product of the two wave functions $\psi_i^*(\mathbf{r}) \psi_j(\mathbf{r})$ is an example of multiplying together two group representations $\Gamma_n \times \Gamma_m$. According to the theorem, this product contains Γ_1 only if $n = m$ and only if the eigenfunctions belong to the same group representations. Another theorem, along with those in Section 3.7 is

THEOREM 3. *The integral is nonzero only if the argument belongs to the representation Γ_1.*

This theorem is true since the basis functions of the other representations are always functions such as x, y, z, $x^2 - y^2$, $zx, \ldots$, that always average to zero when integrated over all angular directions. The representation Γ_1 has the basis functions such as x^2, y^2, z^2 that average to a nonzero answer. Only the basis functions of Γ_1 have a nonzero result when averaged over all angular directions. The integral in Eq. (3.63) is zero unless $n = m$ and unless the two eigenfunctions belong to the same group representation.

$$\int d^3r \psi_i^*(\mathbf{r}) \psi_j(\mathbf{r}) = \delta_{ij} \tag{3.64}$$

Another example is to consider an integral such as

$$\int d^3r \psi_i^*(\mathbf{r}) H \psi_j(\mathbf{r}) = E_i \delta_{ij} \tag{3.65}$$

where the Hamiltonian H has been inserted between the eigenfunctions. However, since initially $H\psi_j = E_j \psi_j$, the constant E_j can be removed from the integral. What is left is the prior integral that is zero unless $i = j$.

The same result can be proven using group theory. The Hamiltonian always belongs to the representation Γ_1. The above matrix element is the direct product of three representations, of which one is $\Gamma_1 : \Gamma_n \times (\Gamma_1 \times \Gamma_m)$. However,

Table 3.19 Optical Matrix Elements for D_3

$\Gamma_1 \times (\Gamma_2 \times \Gamma_1)$	Γ_2
$\Gamma_1 \times (\Gamma_2 \times \Gamma_2)$	Γ_1
$\Gamma_1 \times (\Gamma_2 \times \Gamma_3)$	Γ_3
$\Gamma_2 \times (\Gamma_2 \times \Gamma_2)$	Γ_2
$\Gamma_2 \times (\Gamma_2 \times \Gamma_3)$	Γ_3
$\Gamma_3 \times (\Gamma_2 \times \Gamma_3)$	$\Gamma_1 + \Gamma_2 + \Gamma_3$
$\Gamma_1 \times (\Gamma_3 \times \Gamma_1)$	Γ_3
$\Gamma_1 \times (\Gamma_3 \times \Gamma_2)$	Γ_3
$\Gamma_1 \times (\Gamma_3 \times \Gamma_3)$	$\Gamma_1 + \Gamma_2 + \Gamma_3$
$\Gamma_2 \times (\Gamma_3 \times \Gamma_2)$	Γ_3
$\Gamma_2 \times (\Gamma_3 \times \Gamma_3)$	$\Gamma_1 + \Gamma_2 + \Gamma_3$
$\Gamma_3 \times (\Gamma_3 \times \Gamma_3)$	$\Gamma_1 + \Gamma_2 + \Gamma_3$

since Γ_1 has characters that are all unity, then $(\Gamma_1 \times \Gamma_m) = \Gamma_m$ and the representation Γ_1 has no effect on the outcome. Again, one finds the result is nonzero unless $n = m$.

Another example is to consider some other operator in the matrix element

$$\int d^3 r \psi_i^*(\mathbf{r}) \mathscr{L}(\mathbf{r}) \psi_j(\mathbf{r}) = ? \tag{3.66}$$

Let the function $\mathscr{L}$ belong to the group representation Γ_L. The product in the integrand is now $\Gamma_n \times (\Gamma_L \times \Gamma_m)$. The integral is nonzero only if this triple product contains Γ_1 as one of its representations. The answer depends on which representation is involved, so one has to work out each case.

An important example in physics is to have $\mathscr{L} = \mathbf{p} = -i\hbar \nabla$. This matrix element occurs in the rate of optical transitions (absorption or emission). The group symmetry of the derivative is the same as that of $\mathbf{r}$. In group D_3 the derivative p_z belongs to Γ_2 while (p_x, p_y) belong to the representation Γ_3. Table 3.19 gives the products of two other representations with these two: The multiplication table shows that for light polarized in the z direction, that has representation Γ_2, transitions are allowed only between the representations (Γ_1, Γ_2) and (Γ_3, Γ_3). For light polarized in the (p_x, p_y) direction, all transitions are allowed involving Γ_3. Group theory provides the optical selection rules. In atomic physics, where the optical transitions are between energy states of atoms, the selection rules involve angular momentum. The operator $\mathbf{p}$ has angular momentum $l = 1$.

Group theory shows which matrix elements are zero by symmetry. If group theory allows the matrix element to have a nonzero value, it does not require it to be nonzero. One must actually evaluate the integral to obtain the value that could vanish for other reasons.

3.9. Double Groups

Fermions have halfinteger angular momentum from their spin. Examples are electrons, protons, and neutrons. The wave functions of spin one-half have an interesting property under rotations: a rotation by 360°, or 2π radians, causes the wave function to change sign. One must rotate by 720° in order to return the wave function to its initial state. This behavior is evident in the character for rotations. Examine how it behaves under a rotation by 2π

$$\chi_l(\theta) = \frac{\sin[(l + 1/2)\theta]}{\sin(\theta/2)} \tag{3.67}$$

$$\chi_l(\theta + 2\pi) = (-1)^{2l} \frac{\sin[(l + 1/2)\theta]}{\sin(\theta/2)} \tag{3.68}$$

Replace the angular momentum l by j, let j be half-integer, and $(-1)^{2J} = -1$. The character changed sign. Introduce another group rotation R that is a rotation by 2π. For every rotation X in the single group, there is another group element RX giving a different result. The number of elements in the group has doubled. The new set of elements is called the *double group*. If there were h elements before there are $2h$ in the double group.

The important question is whether the number of classes also doubled, in which case it is a direct product group of the original group times the group (E, R). In deciding whether it is a double group, there are two important rules stated by Opechowski:

1. If C_n denotes a rotation by angle $2\pi/n$, the group elements C_n and RC_n belong in different classes unless $n = 2$.

2. For $n = 2$, RC_2 is in the same class as C_2 if, and only if, there is another twofold axis of rotation that is perpendicular to the one in question.

COROLLARY. *If all RC_2, C_2 combinations are in different classes, the double group is a direct product group of the original group times group (E, R). Recall that this direct product group has twice the number of classes and representations.*

Two Examples. The first is D_3. This group has one C_2 rotation around the z axis. Since there is no other C_2 rotation, C_2 and RC_2 are in different classes and it is a direct product group.

The second example is the group of rotations that leave a cube invariant. It has many twofold rotations and each has another twofold rotation whose axis is perpendicular. All of the rotations RC_2 are in the same classes as C_2. The double group has $h = 48$ and $p = 8$. The single group has five classes ($h = 24$, $p = 5$). The character table of the double group is given in Table 3.20. The double group representations are easy to find since they obey some simple rules:

• The character table still obeys the rules that rows and columns are orthogonal, and the summation of $\Sigma \, l_i^2 = 2h$, where h is the number of elements in the single group.

• The characters in the first p representations of the double group are the same as in the single group representations.

Table 3.20 Double Group of O

	E	R	$8C_3$	$8RC_3$	$3C_2, 3RC_2$	$6C_2, 6RC_2$	$6C_4$	$6RC_4$
Γ_1	1	1	1	1	1	1	1	1
Γ_2	1	1	1	1	1	-1	-1	-1
Γ_3	2	2	-1	-1	2	0	0	0
Γ_4	3	3	0	0	-1	-1	1	1
Γ_5	3	3	0	0	-1	1	-1	-1
Γ_6	2	-2	1	-1	0	0	$\sqrt{2}$	$-\sqrt{2}$
Γ_7	2	-2	1	-1	0	0	$-\sqrt{2}$	$\sqrt{2}$
Γ_8	4	-4	-1	1	0	0	0	0

- For these single group representations, the double group characters are the same: $\chi_i(RC_n) = \chi_i(C_n)$. For example, RC_4 has the same character as C_4 for the first five representations $\Gamma_1 \ldots \Gamma_5$.

- The new representations ($\Gamma_6, \Gamma_7, \Gamma_8$) are called *the double group representations*. For them, the characters $\chi(RX)$ and $\chi(X)$ have the opposite sign.

- When RC_2 and C_2 are in the same class, they can have opposite sign in the double group representations only when the character is zero.

- The rotation characters are still given by $\chi_J(\theta) = \sin[(J + 1/2)\theta]/\sin(\theta/2)$. In Table 3.20, one can use this formula to find that $\chi_{1/2} = \Gamma_6$ while $\chi_{3/2} = \Gamma_8$.

The Fermion wave functions will generally belong to only the double group representations, since they are the ones that have half-integer angular momentum.

Problems

1. The rotations that leave a square invariant are of order eight.

   - Define all eight elements.
   - Construct the multiplication table.
   - List all subgroups.
   - List all classes.
   - Give the character table.

2. The rotations that leave a pentagon invariant are of order ten.

   - Define all ten elements.
   - Construct the multiplication table.
   - List all subgroups.
   - List all classes.

3. The rotations that leave a cube invariant are of order $h = 24$.

   - Define all elements.

- List all classes.
- Give the character table.

4. Give the character table of the cyclic group with $h = 6$.

5. If p is any prime number, the group of order $(p - 1)$ contains the integers from 1 to $(p - 1)$ where multiplication is ordinary multiplication modulus p (i.e., if $p = 7$ then $4 \times 5 = 20 - 2 \times 7 = 6$). Construct the multiplication tables for $p = 3$ and $p = 5$. Is each element in every row and column? Show that this multiplication does not give a group for $p = 4$.

6. Show that the elements in the prior problem is a group for any prime value of p.

7. Find a group of rotations that leave a tetrahedron invariant. There are twelve and the group is called T. (*Hint:* Inscribe the tetrahedron in a cube, and which of the 24 rotations of the cube also leave the tetrahedron unchanged?)
   (a) Divide the group into classes.
   (b) Construct the character table.

8. What is the group of operations that leaves invariant a square block of thickness not equal to an edge? Show that it can be given as a product group.

9. Find the group representations of the basis functions up to second order of the group D_4 that leaves a square invariant.

10. For the group of rotations D_6 that leaves a hexagon invariant:
    (a) List the 12 elements.
    (b) List the classes.
    (c) Give the character table.
    (d) Decompose the angular momentum χ_l for $l = 0, 1, 2, 3$ into its irreducible representations.

11. In Chapter 2 there was a problem finding the eigenvalues and eigenvectors of a benzene molecule for the model of an electron that can hop from each carbon to the nearest neighbor carbon. Assign these eigenvectors to the representations of the group D_6 for the cases that, (a) the orbitals have s symmetry and (b) the orbitals have p_z symmetry.

12. For the group of rotations D_4 that leave a square invariant, multiply each irreducible representation by the others, and then decompose these products. Optical transitions are allowed between which representations of the group of the square?

13. Construct the double group of the rotations that leave a square invariant. Decompose the angular momentum χ_l into irreducible representations for values of $l = 0, 1/2, 1, 3/2, 2$.

14. Consider the character table for the double group of the rotations that leave a pentagon invariant. Show that it is a direct product group.

Complex Variables

4.1. Introduction

Complex variables use the imaginary constant $i = \sqrt{-1}$. Before discussing the variables, we will review some properties of imaginary numbers. First of all, note that

$$i^2 = -1 \qquad (4.1)$$

$$i^3 = -i \qquad (4.2)$$

$$i^4 = 1 \qquad (4.3)$$

A fundamental relationship is

$$e^{i\theta} = \cos(\theta) + i\sin(\theta) \qquad (4.4)$$

Using this formula, some formulas at key angles are

$$e^{i\pi/2} = i \qquad (4.5)$$

$$e^{i\pi} = i^2 = -1 \qquad (4.6)$$

$$e^{3i\pi/2} = i^3 = -i \qquad (4.7)$$

$$e^{2i\pi} = i^4 = 1 \qquad (4.8)$$

$$e^{i\pi/4} = \sqrt{i} = \frac{1+i}{\sqrt{2}} \qquad (4.9)$$

$$e^{i\pi/6} = i^{1/3} = \frac{\sqrt{3}+i}{2} \qquad (4.10)$$

In two dimensions a vector can be represented using polar coordinates. An example is shown in Fig. 4.1. The horizontal axis is labeled x and the vertical axis is labeled y. A vector in two dimensions goes from $(0,0)$ to (x, y). In polar coordinates, the vector is labeled by (r, θ) where

$$r = \sqrt{x^2 + y^2} \qquad (4.11)$$

$$\tan(\theta) = \frac{y}{x} \qquad (4.12)$$

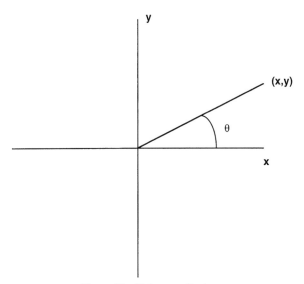

Figure 4.1 Polar coordinates.

The length of the vector is r, while θ is the angle as measured from the x axis. Using the polar coordinates, consider the function

$$re^{i\theta} = r[\cos(\theta) + i\sin(\theta)] \tag{4.13}$$

$$re^{i\theta} = x + iy \tag{4.14}$$

$$x = r\cos(\theta) \tag{4.15}$$

$$y = r\sin(\theta) \tag{4.16}$$

Complex variables are used in two dimensions. The standard symbol is to write

$$z = x + iy \tag{4.17}$$

Here x denotes the x axis, y denotes the y axis, and z does not denote the z axis. Instead, the symbol z represents this variable which is a mix of real and imaginary parts.

The *complex conjugate* operation on an imaginary number changes the sign of i. It is denoted by an asterisk: $i* = -i$. Some examples are

$$(3 + 2i)* = 3 - 2i \tag{4.18}$$

$$z* = (x + iy)* = (x - iy) \tag{4.19}$$

$$(e^{i\theta})* = e^{-i\theta} = \cos(\theta) - i\sin(\theta) \tag{4.20}$$

Combining the last equation with Eq. (4.4) gives

$$\cos(\theta) = \tfrac{1}{2}(e^{i\theta} + e^{-i\theta}) \tag{4.21}$$

$$\sin(\theta) = \frac{1}{2i}(e^{i\theta} - e^{-i\theta}) \tag{4.22}$$

This representation of the sine and cosine in terms of complex numbers is very useful.

The *absolute magnitude* of a complex number is denoted by parallel lines such as $|z|$. If $z = x + iy$ then the absolute magnitude is just

$$|z| = \sqrt{x^2 + y^2} = r \tag{4.23}$$

the length of the polar vector r. This length can be expressed in terms of complex conjugates as

$$z^*z = (x - iy)(x + iy) = x^2 + y^2 \tag{4.24}$$

$$|z| = \sqrt{z^*z} \tag{4.25}$$

This formula is often useful. If one has a complicated formula for a complex function, its absolute magnitude is found by multiplying it by its complex conjugate, and then taking the square root of the resulting mess. The absolute magnitude just removes the phase angle

$$|z| = |x + iy| = |re^{i\theta}| = r \tag{4.26}$$

Finally, as a last exercise in this section, consider how one would solve the equation of $z^n = 1$, in order to find all of the roots. First, write z in polar form and then take its absolute magnitude

$$(z^n)^*z^n = (r^n e^{-in\theta})(r^n e^{in\theta}) = r^{2n} = 1 \tag{4.27}$$

Since the length r is always real, the equation $r^{2n} = 1$ has its only solution as $r = 1$. The equation has been reduced to

$$z^n = e^{in\theta} = 1 \tag{4.28}$$

The net phase angle $(n\theta)$ must be a multiple of 2π radians according to Eq. (4.8). The solution to $z^n = 1$ is

$$z = e^{2i\pi l/n} \tag{4.29}$$

where the integers l range in value from 0 to $n - 1$. There are n independent choices of l giving n roots to the equation. Note that setting $l = n$ gives the same root ($z = 1$) as $l = 0$. Similarly, choosing l to be negative gives the same solution as one of the positive values, i.e., $l = -2$ has the same root as $l = n - 2$. There are exactly n independent solutions.

4.2 Analytic Functions

An analytic function is a mathematical function of x and y that depends on these variables only in the combination of $f(z) = f(x + iy)$. Some examples of analytic functions are: z^2, $\exp(z)$, $\ln(z)$, and $\cos(z)$. Functions involving complex conjugates are never analytic and some examples of nonanalytic functions are: z^*z, $\ln(z^*/z)$.

The real and imaginary parts of an analytic function are denoted as $u(x, y)$ and $v(x, y)$. Both of these functions are real. Some examples are

$$f(z) = u(x, y) + iv(x, y) \tag{4.30}$$

$$z^2 = (x + iy)^2 = x^2 - y^2 + i2xy, \ u = x^2 - y^2, \ v = 2xy \tag{4.31}$$

$$e^z = e^{x+iy} = e^x[\cos(y) + i\sin(y)], \ u = e^x\cos(y), \ v = e^x\sin(y) \tag{4.32}$$

$$\cos(z) = \cos(x + iy) = \cos(x)\cos(iy) - \sin(x)\sin(iy)$$

$$= \cos(x)\cosh(y) - i\sin(x)\sinh(y) \tag{4.33}$$

$$u = \cos(x)\cosh(y), \ v = -\sin(x)\sinh(y) \tag{4.34}$$

The definitions of sine and cosine in terms of phase angle are used to show that they become hyperbolic functions when the arguments are imaginary

$$\cos(iy) = \tfrac{1}{2}(e^{i(iy)} + e^{-i(iy)}) = \tfrac{1}{2}(e^{-y} + e^y) = \cosh(y) \tag{4.35}$$

$$\sin(iy) = \frac{1}{2i}(e^{i(iy)} - e^{-i(iy)}) = \frac{1}{2i}(e^{-y} - e^y) = i\sinh(y) \tag{4.36}$$

Often it is difficult to decide whether a function is analytic. For example, if you were given

$$\cos(x)\cosh(y) - i\sin(x)\sinh(y) \tag{4.37}$$

you might not immediately recognize that this formula represents $\cos(z)$. There is a test that one can perform on any complex function in order to ascertain whether it is analytic. This test is called the *Cauchy–Riemann conditions*, that are now derived. The test involves taking derivatives. The derivative with respect to z can be done as either a derivative with respect to x, or as a derivative with respect to iy. These two derivatives should give the same answer on an analytic function

$$\frac{\partial}{\partial z} f(z) = \frac{\partial}{\partial x} f(x + iy)$$

$$= \frac{\partial}{\partial(iy)} f(x + iy) = -i\frac{\partial}{\partial y} f(x + iy) \tag{4.38}$$

In terms of the components $f = u + iv$ in the above formula gives

$$\frac{\partial}{\partial x}[u(x, y) + iv(x, y)] = -i\frac{\partial}{\partial y}[u(x, y) + iv(x, y)] \tag{4.39}$$

In Eq. (4.39) equate the real components, and then the imaginary components, that leads to the following two equations. They are the Cauchy–Riemann

$$\frac{\partial}{\partial x} u(x, y) = \frac{\partial}{\partial y} v(x, y) \qquad (4.40)$$

$$\frac{\partial}{\partial x} v(x, y) = -\frac{\partial}{\partial y} u(x, y) \qquad (4.41)$$

An analytic function must satisfy the conditions. Some examples are given using some of the analytic functions listed above.

- $f = z^2$, $u = x^2 - y^2$, $v = 2xy$

$$\frac{\partial}{\partial x} u(x, y) = 2x = \frac{\partial}{\partial y} v(x, y) \qquad (4.42)$$

$$\frac{\partial}{\partial x} v(x, y) = 2y = -\frac{\partial}{\partial y} u(x, y) \qquad (4.43)$$

- $f = \exp(z)$, $u = e^x \cos(y)$, $v = e^x \sin(y)$

$$\frac{\partial}{\partial x} u(x, y) = e^x \cos(y) = \frac{\partial}{\partial y} v(x, y) \qquad (4.44)$$

$$\frac{\partial}{\partial x} v(x, y) = e^x \sin(y) = -\frac{\partial}{\partial y} u(x, y) \qquad (4.45)$$

- $f = \cos(z)$, $u = \cos(x) \cosh(y)$, $v = -\sin(x) \sinh(y)$

$$\frac{\partial}{\partial x} u(x, y) = -\sin(x) \cosh(y) = \frac{\partial}{\partial y} v(x, y) \qquad (4.46)$$

$$\frac{\partial}{\partial x} v(x, y) = -\cos(x) \sinh(y) = -\frac{\partial}{\partial y} u(x, y) \qquad (4.47)$$

It is also easy to show that the Cauchy–Riemann conditions do not work on nonanalytic functions.

The Cauchy–Riemann conditions presume that the derivative exists. This assumption, that they have derivatives that exist for the values of (x, y), is another condition on analytic functions under consideration. This condition can be stated more formally as

- *A function is analytic at a point $z_0 = x_0 + iy_0$ if the derivative $\partial f / \partial z$ exists at every point in the vicinity of z_0.*
- *A region is analytic if it is analytic at each point in the region.*

For example, consider the function

$$f(z) = \frac{1}{z^2} \qquad (4.48)$$

It is not analytic at the point $z = 0$ ($x = 0$, $y = 0$) since the derivative diverges at this point. The function is analytic at all points except the origin. Other examples are $f = \sqrt{z}, \ln(z), z \ln(z)$.

A *pole* is a point where a function $f(z)$ diverges as a power law

- *A pole at the point z_0 occurs if*

$$\lim_{z \to z_0} f(z) \to \frac{R_0}{(z - z_0)^n} \tag{4.49}$$

where n is a positive integer.

A *simple pole* has the exponent $n = 1$. The constant R_0 is called the *residue* of the pole. It is the part of the function that is not diverging.

As an example, consider the poles of the function

$$f(z) = \frac{1}{\sin(z)} \tag{4.50}$$

The function $\sin(z)$ vanishes at $z = 0$ and $f(z)$ has a pole at that point. The proof uses the standard expansion for the sine of an angle

$$\sin(z) = z - \frac{z^3}{3!} + \frac{z^5}{5!} - \cdots -$$

$$= z \left[1 - \frac{z^2}{6} + \frac{z^4}{120} - \cdots - \right] \tag{4.51}$$

At small value of z, $f(z) \to 1/z$ so that it has a simple pole with a residue equal to one. However, the function has other poles. The function $\sin(\theta) = 0$ at angles $\theta = l\pi$ where l is any integer. The above case had $l = 0$, so now do the case that $l \neq 0$. To find out the nature of these poles, set $z = l\pi + \delta z$, where $\delta z = z - l\pi$ is small and will be set equal to zero as the pole is approached. Expand the sine function in terms of the small variable δz

$$\sin(z) = \sin(l\pi + \delta z) = \sin(l\pi) \cos(\delta z) + \sin(\delta z) \cos(l\pi)$$
$$\approx \delta z \cos(l\pi) + O(\delta z)^3 = \delta z (-1)^l \tag{4.52}$$

$$\lim_{z \to l\pi} f(z) \to \frac{(-1)^l}{\delta z} \tag{4.53}$$

$$R_l = (-1)^l \tag{4.54}$$

The residue is plus or minus one, depending on whether l is an even or odd integer. The function $f(z) = 1/\sin(z)$ has an infinite number of simple poles, which are strung out along the real axis at the point $x_l = l\pi$, $y_l = 0$.

Another feature of analytic functions is demonstrated by graphing lines where, (i) $u = $ constant and (ii) $v = $ constant. An example is shown in Fig. 4.2 for one quadrant for the function $w = z^2$, $u = x^2 - y^2$, $v = 2xy$. When lines of constant u cross those of constant v they are locally perpendicular.

- *Lines of constant u are always perpendicular to lines of constant v at the points where they cross.*

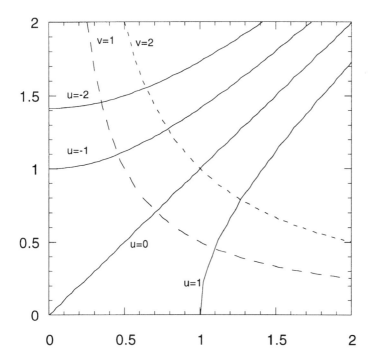

Figure 4.2 Lines of constant u (solid) and lines of constant v (dashed) for the function z^2, $u = x^2 - y^2$, $v = 2xy$.

The proof of this theorem uses the Cauchy–Riemann conditions. Along a line of constant u then

$$du = dx\frac{du}{dx} + dy\frac{du}{dy} = 0 \tag{4.55}$$

$$dy = -dx\frac{du}{dx}\bigg/\frac{du}{dy} \tag{4.56}$$

A unit vector $\hat{n}_u$ along the line of constant u is in the direction

$$\hat{n}_u = \frac{(dx, dy)}{\sqrt{(dx)^2 + (dy)^2}} = \frac{(du/dy, -du/dx)}{\sqrt{(du/dx)^2 + (du/dy)^2}} \tag{4.57}$$

Similarly, the line of constant v has

$$dv = dx\frac{dv}{dx} + dy\frac{dv}{dy} = 0 \tag{4.58}$$

$$dy = -dx\frac{dv}{dx}\bigg/\frac{dv}{dy} \tag{4.59}$$

$$\hat{n}_v = \frac{(dx, dy)}{\sqrt{(dx)^2 + (dy)^2}} = \frac{(dv/dy, -dv/dx)}{\sqrt{(dv/dx)^2 + (dv/dy)^2}} \tag{4.60}$$

Using the Cauchy–Riemann conditions gives for the unit vector $\hat{n}_v$

$$\hat{n}_v = \frac{(du/dx, du/dy)}{\sqrt{(du/dx)^2 + (du/dy)^2}} \tag{4.61}$$

$$\hat{n}_u \cdot \hat{n}_v = \frac{(du/dx)(du/dy)(1 - 1)}{(du/dx)^2 + (du/dy)^2} = 0 \tag{4.62}$$

The two unit vectors $\hat{n}_u$ and $\hat{n}_v$ are orthogonal. The lines of constant u are perpendicular to the lines of constant v.

4.3 Multivalued Functions

Consider the properties of the function $f(z) = \ln(z)$. If the value of $z = 3 (x = 3, y = 0)$ then the evaluation is rather trivial $\ln(3) = 1.0986$. However, the evaluation is less trivial than it appears. Employ the polar definition of the complex variable

$$z = \rho e^{i\theta} \tag{4.63}$$

$$\ln(z) = \ln(\rho e^{i\theta}) = i\theta + \ln(\rho) \tag{4.64}$$

At the point $(x = 3, y = 0)$ the angle $\theta = 0$, so that $\ln(\rho) = \ln(3)$ is unchanged. However, the angle θ may be increased by 2π radians that is equivalent to rotating the point z around the origin. The rotation returns to the same point on the real axis. However, after this rotation

$$f(z) = 2i\pi + \ln(3) \tag{4.65}$$

The value of the function has been given an imaginary part. Rotating the variable z around the origin and returning it to the same point changes the evaluation of the function $\ln(z)$. In fact, rotating completely around the origin n times gives

$$f(z) = 2i\pi n + \ln(\rho) \tag{4.66}$$

where n is any integer. Including negative integers means rotating around the origin in a clockwise fashion, since plus angles are counterclockwise rotations.

The functions $f(z) = \ln(z)$ is multivalued and has many different values. Each different value of n is called a *branch*. The function has many different values with each one valid.

All of the rotations for $\ln(z)$ were about the origin. This point is where $\ln(z)$ is nonanalytic and the derivative of $\ln(z)$ diverges at the point $z = 0$. The word *branch point* is applied to the pivoting point of a multivalued function with more than one branch. The branch point is always a point of nonanalyticity, however, nonanalyticity does not require a branch point. The function $f(z) = z^{-\alpha}$ is nonanalytic at $z = 0$, but a branch point is required only if α is not an integer. A pole is not a branch point. In thinking about a multivalued function, the first step is to always identify the branch point(s).

Another important concept is the *branch cut*. It is a line in the (x, y) plane that separates the various branches. As the point z is rotated around the branch points, some signal is needed to determine where to change branches. This switching occurs at the branch cuts and the branch cuts have one end at a branch point. The other end can be at another branch point or at infinity. An important feature of branch cuts is that they can be selected by the user to fit the problem. There is no absolute right choice for many problems.

As an example, consider the branch cut appropriate for the problem of $f(z) = \ln(z)$. For this function, the cut must start at the origin, and extend to infinity. One choice is to have the cut along the positive real axis $(x > 0, y = 0)$. This choice would fit the earlier discussion, since when rotating θ by $2\pi n$ we kept returning to the real axis with a different value. A branch cut along the real axis would be a possible choice for the branch cut. A function $f(z)$ has a different value on the two sides of a branch cut.

Perhaps a better choice is to put the branch cut along the negative real axis $(x < 0, y = 0)$. The reason for this choice is that the actual problem is with negative numbers. Any algebra student can evaluate $\ln(3)$. The problem comes when asked to evaluate $\ln(-3)$. The factor of (-1) can be either $-1 = \exp(i\pi)$ or $-1 = \exp(-i\pi)$. The first choice gives $f(-3) = \ln(-3) = i\pi + \ln(3)$ while the second choice gives $\ln(-3) = -i\pi + \ln(3)$. The ambiguity is along the negative real axis. When putting the branch cut there, it appears as the wiggly line in Fig. 4.3(a). The function $\ln(z)$ is chosen to be real along the real axis at point A. One gets to point B by rotating the angle θ by π radians, so that at point B $\ln z = i\pi + \ln(3)$. One gets to point C below the branch cut by starting at point A and rotating the angle θ by $-\pi$ radians. At point C the result is $\ln(-3) = -i\pi + \ln(3)$. The branch cut makes sense along the negative real axis, since that is where the actual ambiguity occurs. In practice, one branch cut separating points B and C is all that is needed.

This example is concluded with a remark that the branch cut for $\ln(z)$ can be drawn to infinity in any direction from the origin. The choices discussed here are not the only ones.

Another example is $f(z) = \sqrt{z + 3}$. Here the branch point is at $z = -3 (x = -3, y = 0)$. The multivalued nature of this function is explored by going to polar coordinates with the branch point as the origin

$$z + 3 = \rho_3 e^{i\theta_3} \tag{4.67}$$

The subscript "3" on the ρ_3 and θ_3 reminds us that these coordinates are not from the origin in space, but are from the branch point. Again, draw the branch cut to infinity, along the negative real axis, starting at the branch point. It is shown as the wiggly line in Fig. 4.3(b). Along the real axis, for $x > -3$, define $\sqrt{z + 3}$ as a real function. The problem is when the value of z is in the domain $x < -3$. Then the value of the function depends on the path to get there. To get to point B, one sets $\theta_3 = \pi$ so that

$$f(z) = \sqrt{\rho_3} \, e^{i\pi/2} = i\sqrt{\rho_3} \tag{4.68}$$

At point B we find that $f(z) = i\sqrt{\rho_3}$ since $i = \exp(i\pi/2)$. Point C is reached by

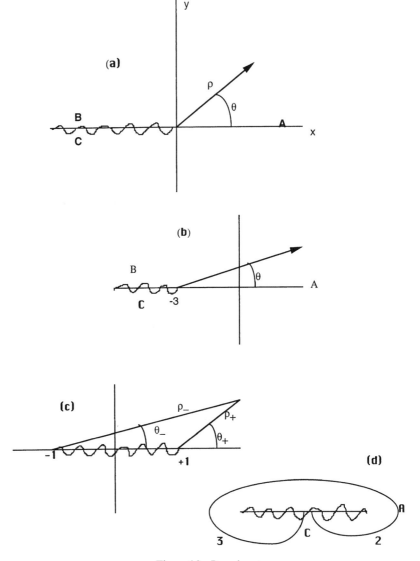

Figure 4.3 Branch cuts.

taking $\theta_3 = -\pi$ and then $f(z) = -i\sqrt{\rho_3}$. It is the complex conjugate of the value at **B**. The function $\sqrt{z+3}$ is multivalued along the negative real axis, and requires a branch cut as shown. In this case there are only two branches. Going around the branch point in angles $\theta_3 = 2\pi n$ gives the result that $\sqrt{z+3}$ has only two possible values. So there are only two different branches to this function.

The next example is the function $f(z) = \sqrt{z^2 - 1}$. There are two branch points at $z = \pm 1$. In this case the branch cut can extend between them. It is not necessary to extend them to infinity. In discussing this case, each point z is represented by two radial vectors, one from each branch point. They are shown in Fig. 4.3(c).

$$z - 1 = \rho_+ e^{i\theta_+} \tag{4.69}$$

$$z + 1 = \rho_- e^{i\theta_-} \tag{4.70}$$

$$f(z) = \sqrt{z^2 - 1} = \sqrt{\rho_+ \rho_-} \; e^{i(\theta_+ + \theta_-)/2} \tag{4.71}$$

Now consider the value of this function at various points. In each case, the point is reached by moving z along a path. Keep track of the angles for each path:

- Real axis, $x > 1$. Here both angles are set equal to zero, and $f = \sqrt{\rho_+ \rho_-}$. The same real constant is obtained by rotating both angles simultaneously by $2\pi n$. So the real axis for $x > 1$ is singlevalued, and does not need a branch cut.
- The negative real axis for $x < -1$. This point is reached two possible ways: (i) set $\theta_+ = \theta_- = \pi$. Then $f = -\sqrt{\rho_+ \rho_-}$. The value of the function is negative but real. (ii) The choice $\theta_+ = \theta_- = -\pi$ gives the same functions as in part (i). So the value of the function for the negative real axis $x < -1$ is single valued, and a branch cut is not needed. That is why the branch cut needs only to extend along a finite segment.
- Real axis for $-1 < x < 1$. It can be reached by several possible paths as shown in Fig. 4.3(d):

1. $\theta_+ = \pi,\ \theta_- = 0$ that gives $f = i\sqrt{\rho_+ \rho_-}$
2. $\theta_+ = -\pi,\ \theta_- = 0$ that gives $f = -i\sqrt{\rho_+ \rho_-}$
3. $\theta_+ = \pi,\ \theta_- = 2\pi$ that gives $f = -i\sqrt{\rho_+ \rho_-}$

The first case ends above the branch cut by rotating the point z clockwise from the real axis. The last two cases end below the branch cut — the second case by rotating the point z counterclockwise from the real axis. Both give the same answer. The multivalued function has one value beneath the branch cut and another above it. Both are independent of the path used to get there.

The next example is the function $f(z) = [z(z - 1)]^{1/3}$. Again there are two branch points, that are now at $z = 0$ and $z = 1$. One cannot use a finite branch cut that goes between them, since the value along the negative real axis is multivalued. So the branch cut goes from $z = 1$ to the left, along the negative axis all the way to infinity. On the other hand, the following function can be described with a finite branch cut

$$f(z) = \left(\frac{z-1}{z}\right)^\alpha = \left(\frac{\rho_1}{\rho}\right)^\alpha e^{i\alpha(\theta_1 - \theta)} \tag{4.72}$$

$$z - 1 = \rho_1 e^{i\theta_1} \tag{4.73}$$

$$z = \rho e^{i\theta} \tag{4.74}$$

The constant α is neither an integer nor one-half integer. This function can be evaluated with a branch cut that only goes between the two branch points. The values are real along the real axis except in the interval $0 < x < 1$.

4.4 Contour Integrals

The word *contour* denotes a line. *Contour integrals* are integrals along a line. Most of the time the contours will be complete loops.

When an integral is created in two dimensional space, there are two ways to do it. One can integrate over an area $\int dx\,dy$. The other possibility is to integrate on a line along a path. Contour integrals are the latter case. They will be written as

$$\int_{\text{path}} dzf(z) \tag{4.75}$$

The "path" will signify a starting point, an end point, and the route. An example is shown in Fig. 4.4. The integral goes from z_1 to z_2 along two possible paths labeled A and B. These two integrals are

$$\text{path A} \quad \int_{y_1}^{y_2} idyf(x_1 + iy) + \int_{x_1}^{x_2} dxf(x + iy_2) \tag{4.76}$$

$$\text{path B} \quad \int_{x_1}^{x_2} dxf(x + iy_1) + \int_{y_1}^{y_2} idyf(x_2 + iy) \tag{4.77}$$

Note that $dz = idy$ when integrating along the y axis. Sometimes these two paths give the same result for the integral, while at other times they give different results. The differences are explained next.

Consider the integral

$$I_1 = \oint \frac{dz}{z} \tag{4.78}$$

The circle on the integral denotes a path that is a complete circle. The path of integration is a circle about the origin, as shown in Fig. 4.5(a).

This integral is easy to do. Use the polar form for the variable $z = \rho \exp(i\theta)$. A circular contour has a constant value of ρ, and the angle θ is

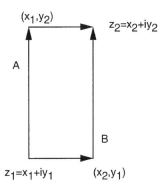

Figure 4.4 Two integration contours.

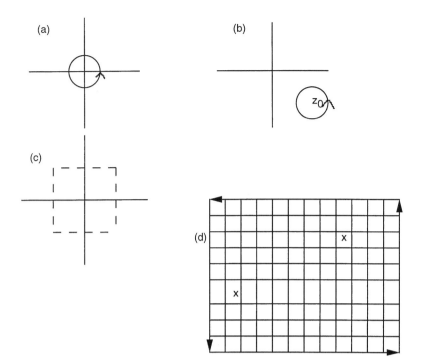

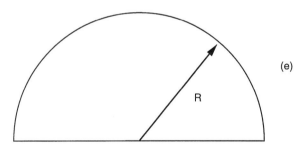

Figure 4.5 Paths of contour integrals.

varying over the interval $0 < \theta < 2\pi$. Therefore write

$$dz = \frac{dz}{d\theta} d\theta = i\rho e^{i\theta} d\theta = zi \, d\theta \tag{4.79}$$

$$I_1 = i \int_0^{2\pi} d\theta = 2\pi i \tag{4.80}$$

This result is of fundamental importance in the theory of contour integrals. The integral which is a complete circle around a simple pole gives $2\pi i$. In order to stress the importance of this result, several variations are done on the same integral:

1. The result is the same regardless of the location of the simple pole. Let the pole be at the point z_0 and consider the contour integral around this pole, as shown in Fig. 4.5(b)

$$I_2 = \oint \frac{dz}{z - z_0} = 2\pi i \qquad (4.81)$$

The circle is centered at the point z_0. Let $z = z_0 + \rho \exp(i\theta)$ and then $dz = (z - z_0)i\, d\theta$, where z_0, ρ are constants. The integral is now the same as Eq. (4.80) and one finds the same result $I_2 = 2\pi i$.

2. The above integrals have circled the pole in a counterclockwise direction, that is the conventional direction. What is the result if the path is the circle in the clockwise direction? The steps in the derivation are the same, except that the angle θ varies from 0 to -2π radians. The integral in this case is $-2\pi i$.

3. Consider the result when the polynomial is not a pole

$$I_3 = \oint dz z^m \qquad (4.82)$$

The contour is a circle around the origin. Again set $dz = iz\, d\theta$ and in polar coordinates

$$I_3 = i\rho^{m+1} \int_0^{2\pi} d\theta e^{i(m+1)\theta} = \frac{\rho^{m+1}}{m+1} [e^{i(m+1)2\pi} - 1] = 0 \qquad (4.83)$$

The result is zero since $\exp(2\pi i) = 1$ and the same result is obtained for $\exp(i2n\pi) = 1$ where $n = m+1$ is any integer not equal to zero ($m \neq -1$). This result is only valid for integer exponents. If the exponent is not an integer, then introduce a branch cut. Paths of integrals can never cross a branch cut. This topic is treated below. The circular contour gives zero for any polynomial z^m as long as m is an integer that does not equal minus one.

4. Consider the case of multiple poles, say at the point z_0

$$I_4 = \oint \frac{dz}{(z - z_0)^n} \qquad (4.84)$$

The path of integration is a circle around the point z_0. Again change variables of integration to the angle θ, where $z = z_0 + \rho \exp(i\theta)$, $dz = (z - z_0)i\, d\theta$, resulting in

$$I_4 = i\rho^{1-n} \int_0^{2\pi} d\theta e^{-i(n-1)\theta} = -\frac{1}{\rho^{n-1}(n-1)} [e^{-i2\pi(n-1)} - 1] = 2\pi i \delta_{n=1} \qquad (4.85)$$

The integral is zero unless $n = 1$ since the two terms in the bracket equal one and cancel. Only the simple pole ($n = 1$) has a nonzero integral.

Consider the integral around the simple pole at z_0

$$I_5 = \oint \frac{dz z^3}{z - z_0} \tag{4.86}$$

Again change variables of integration to the angle θ, where

$$z = z_0 + \rho \exp(i\theta), \qquad dz = (z - z_0) i \, d\theta$$

so that

$$I_5 = i \int_0^{2\pi} d\theta [z_0 + \rho e^{i\theta}]^3 \tag{4.87}$$

$$I_5 = i \int_0^{2\pi} d\theta [z_0^3 + 3z_0^2 \rho e^{i\theta} + 3z_0 \rho^2 e^{2i\theta} + \rho^3 e^{i3\theta}]$$

$$= 2\pi i z_0^3 \tag{4.88}$$

Only the first term in the expansion (z_0^3) gives a nonzero result. The other terms contain exponents of the form $\exp(in\theta)$ that integrate to zero. The final answer is

$$I_5 = \oint \frac{dz z^3}{z - z_0} = 2\pi i z_0^3 \tag{4.89}$$

$$I_6 = \oint \frac{dz z^n}{z - z_0} = 2\pi i z_0^n \tag{4.90}$$

The second equation has extended the result to any value of n.

6. Assume there is a function $f(z)$ that is a polynomial in z^n with only nonnegative values of the exponent n:

$$f(z) = a_0 + a_1 z + a_2 z^2 + \cdots + \tag{4.91}$$

$$I_7 = \oint \frac{dz f(z)}{z - z_0} = \oint \frac{dz}{z - z_0} [a_0 + a_1 z + a_2 z^2 + \cdots +]$$

$$= 2\pi i [a_0 + a_1 z_0 + a_2 z_0^2 + \cdots +] = 2\pi i f(z_0) \tag{4.92}$$

$$f(z_0) = \oint \frac{dz}{2\pi i} \frac{f(z)}{z - z_0} \tag{4.93}$$

This last integral is called *Cauchy's theorem*. It is valid for any function $f(z)$ that has neither poles nor branch cuts inside the contour of integration. The function $f(z)$ must be analytic within the contour of integration.

7. The same result $(2\pi i)$ is obtained if the contour of integration is not a circle but is another shape. It must still encircle the simple pole. Do the integral for a contour in the shape of a square box as shown in Fig. 4.5(c).

The four corners of the box are at the points

$$z_1 = 1 + i = \sqrt{2}\, e^{i\pi/4} \tag{4.94}$$

$$z_2 = -1 + i = \sqrt{2}\, e^{3i\pi/4} \tag{4.95}$$

$$z_3 = -1 - i = \sqrt{2}\, e^{5i\pi/4} \tag{4.96}$$

$$z_4 = 1 - i = \sqrt{2}\, e^{-i\pi/4} \tag{4.97}$$

The integral has four segments, one for each side of the square. Each segment gives the same result $i\pi/2$. The sum of the four segments is therefore $2\pi i$. To show that the segments give $i\pi/2$ consider one of them

$$\int_{z_1}^{z_2} \frac{dz}{z} = \int_1^{-1} \frac{dx}{x+i} = \ln\left[\frac{z_2}{z_1}\right] = \ln(e^{i\pi/2}) = i\pi/2 \tag{4.98}$$

The other segments give the same result.

8. The last example is a circular contour which does *not* encircle a pole, but misses it entirely

$$I_8 = \oint \frac{dz}{z - z_0} \tag{4.99}$$

The contour is a circle about the origin, so $dz = iz\, d\theta$. Since the pole at z_0 is not in the circle of integration, then $\rho < |z_0|$. Therefore expand the denominator in a power series using the small parameter $\rho/|z_0| < 1$

$$I_8 = -i\frac{\rho}{z_0} \int_0^{2\pi} e^{i\theta} \frac{d\theta}{1 - \rho e^{i\theta}/z_0}$$

$$= -i\frac{\rho}{z_0} \int_0^{2\pi} e^{i\theta}\, d\theta[1 + \rho e^{i\theta}/z_0 + \rho^2 e^{2i\theta}/z_0^2 + \cdots +] = 0 \tag{4.100}$$

The integral is zero since all of the integrals have factors of $\exp(in\theta)$ that integrate to zero when $n \neq 0$. A nonzero integral is obtained only when the pole is inside of the circular path of integration.

These results show that the contour integral which completely encircles a pole gives a nonzero result. The above examples have proved an important theorem:

THEOREM. *If the function $f(z)$ has only simple poles at the points z_j with residues R_j, the contour integral around any closed path equals $2\pi i$ times the sum of the residues of the poles that are inside the closed contour.*

The proof of this theorem involves taking the area enclosed by the contour, and filling it with small boxes that completely fill the space, as shown in Fig. 4.5(d). The integral over all of the small boxes equals the integral over the big contour. This statement is true since, for a box which is in the interior of the contour, the integral around its four sides are cancelled by the integrals around the sides of its neighboring boxes. A box on the edge has one edge that contributes to the larger integral

$$\oint dzf(z) = \sum_l \oint_{box_l} dzf(z) \tag{4.101}$$

Finally, those small boxes that enclose a pole will give the residue of the pole. The contour integral around small boxes, that have no poles, gives zero. The theorem is summarized by the equation

$$\oint dzf(z) = 2\pi i \sum_{j \in \bigcirc} R_j \tag{4.102}$$

This theorem is used to evaluate many integrals. One of the major uses of the theory of complex variables is to evaluate complicated integrals.

The integral along two different paths A and B that connect the same end points differ by the sum of the residues in the region enclosed by the two paths.

$$\int_B dzf(z) = \int_A dzf(z) + \int_{-A} dzf(z) + \int_B dzf(z)$$

$$= \int_A dzf(z) + \oint dzf(z) = \int_A dzf(z) + 2\pi i \sum_{j \in (AB)} R_j \tag{4.103}$$

An example of two paths A and B are shown in Fig. 4.4. The integral along the path $-A$ gives the opposite result as the integral along the path A. Therefore, $\int_A + \int_{-A} = 0$. However, the summation of the path B and the path $-A$ is a closed contour. By the prior theorem, it equals the sum of the residues of any poles in the area enclosed by the two paths (AB).

Some examples of integrals that contain only simple poles are presented. One should keep in mind that an integral is always real if it has a real integrand and the integral is taken over a real path. Complex answers require that either the integrand or the path involve complex numbers.

1.
$$J_1 = \int_{-\infty}^{\infty} \frac{dx}{x^2 + a^2} \tag{4.104}$$

The integral can be evaluated without contour methods. The indefinite integral is the arctangent that is $\pm \pi/2$ at $\pm \infty$

$$J_1 = \frac{1}{a} \tan^{-1}\left(\frac{x}{a}\right)\Big|_{-\infty}^{\infty} = \frac{\pi}{a} \tag{4.105}$$

A similar result is obtained using a contour integral that is closed in the upper half plane (UHP).

$$J_1 = \oint \frac{dz}{z^2 + a^2}$$

$$= \oint \frac{dz}{(z - ia)(z + ia)} = \frac{2\pi i}{2ia} = \frac{\pi}{a} \tag{4.106}$$

The integral over dz has two segments, the real axis from $(-\infty)$ to $(+\infty)$ and the semicircle of radius R in the upper half plane (UHP) as

$R \to \infty$ [see Fig. 4.5(e)]. The latter segment does not contribute, so the integral over dx is equal to the integral over dz. Using the theorem, the integral over dz equals the residues of the poles. There is one pole in the UHP at $z = ia$ and the residue there is $1/(2ia)$. It is important to prove that closing the contour by adding a semicircle at infinity does not change the value of the integral. We need to prove the semicircle gives zero. Along this segment, set $z = Re^{i\phi}$, $dz = id\phi z$. The segment is

$$\lim_{R = \infty} i \int_0^\pi \frac{d\phi Re^{i\phi}}{R^2 e^{2i\phi} + a^2} = \lim_{R = \infty} \frac{i}{R} \int_0^\pi d\phi e^{-i\phi} = 0 \qquad (4.107)$$

The contour at infinity will vanish as long as the denominator has a higher power of R than the numerator. The contour could also be closed in the lower half plane (LHP) and this semicircle also vanishes.

2.
$$J_2 = \int_{-\infty}^\infty \frac{dx}{(x - i)(x - 3i)} = \oint \frac{dz}{(z - i)(z - 3i)}$$

$$= 2\pi i \left[\frac{1}{i - 3i} + \frac{1}{3i - i} \right] = 0 \qquad (4.108)$$

The contour for dz is the same as the last problem: One closes the contour along the semicircle $z = Re^{i\theta}$ as $R \to \infty$ that makes no contribution. The integral equals the summation of the residues at the poles in the integrand. There are two poles, at $z = i$ and $z = 3i$. Their residues are equal and opposite, so that the net integral is zero. The zero result could be obtained another way by closing the contour by a semicircle in the LHP. There are no poles in the LHP so that the integral is zero. This observation leads to another theorem:

THEOREM. *If an integrand has all of its poles in the UHP, or in the LHP, the integral along the real axis is zero. There must be more than one pole.*

3. The exception to the above theorem when there is only one pole is done next. Let $\eta > 0$ and consider the integral

$$J_3 = \int_{-\infty}^\infty \frac{dx}{x - i\eta} \qquad (4.109)$$

The integral is evaluated by considering the same form for a complex integral. Two segments of the contour must be considered, and neither are zero

$$I' = \oint \frac{dz}{z - i\eta} = 2\pi i$$

$$= \int_{-\infty}^\infty \frac{dx}{x - i\eta} + \lim_{R \to \infty} \left\{ iR \int_0^\pi d\theta e^{i\theta} \frac{1}{Re^{i\theta} - i\eta} \right\}$$

$$= \int_{-\infty}^\infty \frac{dx}{x - i\eta} + i\pi \qquad (4.110)$$

In the limit of very large R, the factor of $i\eta$ is neglected in the denominator, so that the factor of R and factor of $\exp(i\theta)$ cancel. One is left with a simple result

$$\int_{-\infty}^{\infty} \frac{dx}{x - i\eta} = i\pi \, \text{sign}(\eta) \tag{4.111}$$

In this case the integral around the pole gives $i\pi$ rather than $2\pi i$. The difference is that the contribution from closing the contour at infinity could not be neglected. Closing the contour is required in order to use the theorem for evaluating contour integrals. If $\eta < 0$ then the integral is $-i\pi$.

4. Consider the integral

$$J_4 = \int_{-\infty}^{\infty} \frac{dx}{(x^2 + a^2)(x^2 + b^2)} \tag{4.112}$$

The integrand can be separated using partial fractions to get that

$$J_4 = \int_{-\infty}^{\infty} dx \left[\frac{1}{x^2 + a^2} - \frac{1}{x^2 + b^2} \right] \frac{1}{b^2 - a^2}$$

$$= \left[\frac{\pi}{a} - \frac{\pi}{b} \right] \frac{1}{b^2 - a^2} = \frac{\pi}{ab(a + b)} \tag{4.113}$$

which uses the result for J_1. A direct evaluation gives

$$J_4 = \int_{-\infty}^{\infty} \frac{dx}{(2ia)(2ib)} \left[\frac{1}{x - ia} - \frac{1}{x + ia} \right] \left[\frac{1}{x - ib} - \frac{1}{x + ib} \right] \tag{4.114}$$

The denominator has been separated using partial fractions. In the first bracket, the two terms have a pole in the UHP and LHP, respectively and the same applies to the second bracket. Multiplying these terms together gives these four terms: both poles in the UHP, both poles in the LHP and one poles in each plane. The first two integrals vanish since both poles are in the same plane—the contour can be closed in the other plane, which gives zero since no poles are encircled. The nonzero result for the present integral is found from the two terms that have one pole in each plane. These two terms give the same contribution, that is found by closing the contour at infinity in either plane

$$J_4 = -\frac{2\pi i}{4ab} \left[-\frac{2}{i(a + b)} \right] = \frac{\pi}{ab(a + b)} \tag{4.115}$$

5. The symbol P denotes the principal part of an integral. It is an instruction to omit the part of the integral over a singular point in the

path

$$J_5 = P \int_{-\infty}^{\infty} \frac{dx}{x - b} = 0$$

$$\equiv \lim_{\varepsilon \to 0} \left\{ \int_{-\infty}^{b-\varepsilon} + \int_{b+\varepsilon}^{\infty} \right\} \frac{dx}{x - b} \qquad (4.116)$$

There are several ways to show that this integral vanishes. The easiest is to change variables $x' = x - b$ and write it as

$$J_5 = P \int_{-\infty}^{\infty} \frac{dx'}{x'} \qquad (4.117)$$

The integrand has positive values for $x' > 0$ and negative values for $x' < 0$. These two segments are equal and opposite and cancel. That it vanishes can also be shown using complex variables. Consider the contour integral

$$\oint \frac{dz}{z} = 0 \qquad (4.118)$$

where the path is shown in Fig. 4.6(a). The contour goes along the real axis, except there is a semicircle above the origin of radius ε. The path is closed by a semicircle with a radius R as $R \to \infty$. The integral is zero since no poles are enclosed by this contour. The integral may be written as the sum of the segments along its path

$$0 = \oint \frac{dz}{z} = P \int_{-\infty}^{\infty} \frac{dx}{x} + \int_{\varepsilon} \frac{dz}{z} + \int_{R} \frac{dz}{z} \qquad (4.119)$$

$$\int_{\varepsilon} \frac{dz}{z} = i \int_{\pi}^{0} d\theta = -i\pi \qquad (z = \varepsilon e^{i\theta}) \qquad (4.120)$$

$$\int_{R} \frac{dz}{z} = i \int_{0}^{\pi} d\theta = +i\pi \qquad (z = R e^{i\theta}) \qquad (4.121)$$

The sum of the two segments, that are semicircles, exactly cancel and the principle part integral vanishes.

6. The same technique can be used to evaluate related integrals such as

$$J_6 = P \int_{-\infty}^{\infty} \frac{dx}{x^2 - b^2} = P \int_{-\infty}^{\infty} \frac{dx}{2b} \left[\frac{1}{x - b} - \frac{1}{x + b} \right]$$

$$= 0 \qquad (4.122)$$

Partial fractions have been used to split the denominator into two terms. Each has the form in Eq. (4.116) and equals zero.

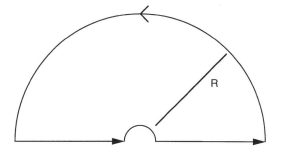

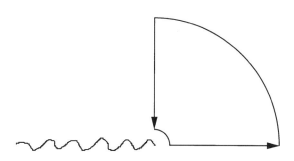

Figure 4.6 Contours of integration.

7. Partial fractions can be used to evaluate similar integrals by breaking them into simpler parts

$$J_7 = P \int_{-\infty}^{\infty} \frac{dx}{(x^2 + a^2)(x^2 - b^2)}$$

$$= \frac{1}{a^2 + b^2} P \int_{-\infty}^{\infty} dx \left[\frac{1}{x^2 - b^2} - \frac{1}{x^2 + a^2} \right]$$

$$= -\frac{\pi}{a(a^2 + b^2)} \tag{4.123}$$

The bracket has two terms and both have been done before. The first integral is zero and the second is π/a.

8. Consider the integral

$$J_8 = \int_0^{\infty} \frac{x \, dx \, \sin(ax)}{(x^2 + 1)(x^2 + 4)} = \frac{1}{2} \int_{-\infty}^{\infty} \frac{x \, dx \, \sin(ax)}{(x^2 + 1)(x^2 + 4)} \tag{4.124}$$

Since the integrand is an even function of the variable x, the same result is obtained by including the integral over $(-\infty, \infty)$ and then divide by 2.

Assume the constant a is positive, and express the sine function as the imaginary part of an exponential

$$J_8 = \tfrac{1}{2}\mathrm{Im}\{J_8'\} \tag{4.125}$$

$$J_8' = \oint \frac{z\,dze^{iaz}}{(z^2+1)(z^2+4)} \tag{4.126}$$

Close the contour as a semicircle of radius R, above the real axis, in the limit that $R \to \infty$. The contribution from this semicircle vanishes since the exponent has $\exp[-aR\sin(\theta) + iaR\cos(\theta)]$. The real part $-aR\sin(\theta)$ is always negative in the interval $0 < \theta < \pi$ where $\sin(\theta) > 0$. The integral along the real axis equals the sum of the poles in the integrand. There are two poles in the UHP, at $z = i$ and $z = 2i$. The residues give the result

$$J_8' = 2\pi i \left[\frac{e^{-a}}{6} - \frac{e^{-2a}}{6} \right] \tag{4.127}$$

$$J_8 = \frac{\pi}{6}[e^{-a} - e^{-2a}] \tag{4.128}$$

Note that we did not try to evaluate the contour integral for the function $\sin(az)$. It has two exponentials, $\exp(\pm iaz)$ and if one $[\exp(iaz)]$ converges in the UHP the other $[\exp(-iaz)]$ converges in the LHP. So a sine or cosine function can only be evaluated by treating differently the separate parts of $\exp(\pm iaz)$. In most cases it is easier to do one exponential and then to take the real or imaginary part of the answer to get the result for the sine or cosine.

9. Another integral that can be evaluated by contours is

$$J_9 = \int_0^\infty dx\,\frac{\sin(ax)}{x} = \frac{\pi}{2}\mathrm{sign}(a) \tag{4.129}$$

The factor of $\mathrm{sign}(a)$ is plus or minus one depending on the sign of the constant a. The integrand is well-behaved at the point $x = 0$ since the ratio $\sin(ax)/x \to a$ as $x \to 0$. There is no need for a principal part symbol, since there is no singularity at this point. The integrand is an even function of x, so it can be extended to negative infinity

$$J_9 = \frac{1}{2}\int_{-\infty}^\infty dx\,\frac{\sin(ax)}{x} \tag{4.130}$$

Write the factor of $\sin(ax)$ as the imaginary part of $\exp(iax)$. Unfortunately, this step makes the integrand of $\exp(iax)/x$ have a singularity at the point $x = 0$. This problem is resolved by introducing a semicircle that encircles and avoids the point $x = 0$. This semicircle should not change the final answer since the integrand does not actually diverge at this point. So consider the contour in Fig. 4.6(a)

$$J_9' = \oint \frac{dz}{z} e^{iaz} = 0$$

$$= P \int_{-\infty}^{\infty} \frac{dx}{x} e^{iaz} + \int_{\varepsilon} \frac{dz}{z} e^{iaz} + \int_{R} \frac{dz}{z} e^{iaz} \qquad (4.131)$$

$$\int_{\varepsilon} \frac{dz}{z} e^{iaz} = \lim_{\varepsilon \to 0} i \int_{\pi}^{0} d\theta e^{ia\varepsilon e^{i\theta}} = -i\pi \qquad (4.132)$$

$$\int_{R} \frac{dz}{z} e^{iaz} = \lim_{R \to \infty} i \int_{0}^{\pi} d\theta e^{iaR e^{i\theta}} = 0 \qquad (4.133)$$

$$P \int_{-\infty}^{\infty} \frac{dx}{x} e^{iaz} = i\pi \qquad (4.134)$$

$$J_9 = \frac{1}{2} \text{Im}\{J_9'\} = \frac{1}{2} \text{Im}\{i\pi\} = \frac{\pi}{2} \qquad (4.135)$$

The contour integral goes along the real axis. There is a semicircle above the origin and another semicircle at a radius R as $R \to \infty$. The semicircle at the origin gives a nonzero contribution. The semicircle at R gives zero since the exponent $\exp iaz = \exp(aR[-\sin(\theta) + i\cos(\theta)]) \to 0$. The real part of the exponent is always a large negative number in the UHP, so this contribution vanishes assuming that $a > 0$. If $a < 0$ close the contour in the LHP and get the same answer with the opposite sign.

10. An integral similar to the last one is

$$J_{10} = \int_{0}^{\infty} \frac{dx}{\sqrt{x}} \sin(ax) \qquad (4.136)$$

It is assumed that $a > 0$. If $a < 0$ the result has the opposite sign. There is a branch point at $x = 0$. Run the branch cut along the negative real axis. This integral cannot be done by extending the integral to negative values of x because the integrand is not a symmetric function. Instead, this important integral is done by another method, which also involves complex variables. Again express $\sin(ax)$ as the imaginary part of $\exp(iax)$, then rotate the contour up the imaginary axis. The rotation is done by considering the integral

$$J_{10}' = \oint \frac{dz}{\sqrt{z}} e^{iaz} \qquad (4.137)$$

The closed contour starts at the origin ($z = 0$) and runs along the real axis to R, as shown in Fig. 4.6(b). There is a quarter circle in the first quadrant, so set $z = R \exp(i\theta)$ while the values of $0 < \theta < \pi/2$. The rotation brings the contour to the imaginary axis at the point $z = iR$. Then return the contour down the imaginary axis. The fourth segment is a quarter circle around the origin. It has a radius ε and vanishes in the limit that $\varepsilon \to 0$.

$$\lim_{\varepsilon \to 0} i\sqrt{\varepsilon} \int_{\pi/2}^{0} d\theta e^{i\theta/2} \to 0 \tag{4.138}$$

This small segment is required to avoid the branch point at the origin. The closed contour encloses no poles, so gives zero. Eventually take the limit that $R \to \infty$. The quarter circle segment at large R gives zero since the exponent has the factor of $iaR\exp(i\theta) = -aR[\sin(\theta) - i\cos(\theta)]$. The real part is large and negative and causes the exponential to vanish. Therefore the integral along the real axis is equal to the integral along the imaginary axis. The integral along the imaginary axis is done by setting $z = iy/a, dz = i\,dy/a$, that gives

$$J_{10} = \text{Im}\left\{ \frac{\sqrt{i}}{\sqrt{a}} \int_{0}^{\infty} \frac{dy}{\sqrt{y}} e^{-y} \right\} \tag{4.139}$$

The integral is now totally real. The prefactor contains the term $\sqrt{i} = (1 + i)/\sqrt{2}$ and its imaginary part is $1/\sqrt{2}$.

$$J_{10} = \frac{1}{\sqrt{2a}} I_{10} \tag{4.140}$$

$$I_{10} = \int_{0}^{\infty} \frac{dy}{\sqrt{y}} e^{-y} \tag{4.141}$$

The integral $I_{10} = \sqrt{\pi}$. To prove this identity, first change variables to $x = \sqrt{y}$, $2\,dx = dy/\sqrt{y}$, so

$$I_{10} = 2 \int_{0}^{\infty} dx e^{-x^2} = \int_{-\infty}^{\infty} dx e^{-x^2} \tag{4.142}$$

$$I_{10}^2 = \int_{-\infty}^{\infty} dx \int_{-\infty}^{\infty} dx' e^{-x^2 - x'^2} \tag{4.143}$$

The integrand is now a symmetric function of x, so the integral can be extended to negative infinity. The integral was squared in the second line. The squared integral is two dimensional over all of two dimensional space. Change integration variables to polar coordinates: $\rho = \sqrt{x^2 + x'^2}$, $\tan(\theta) = x'/x$ and $dx\,dx' = \rho d\rho d\theta$. The integrand has no dependence on θ so it it is integrated around the circle to give 2π. The ρ integral is now simple

$$I_{10}^2 = 2\pi \int_{0}^{\infty} \rho\,d\rho e^{-\rho^2} = -\pi e^{-\rho^2}\big|_{0}^{\infty} = \pi \tag{4.144}$$

$$I_{10} = \sqrt{\pi} \tag{4.145}$$

$$J_{10} = \sqrt{\frac{\pi}{2a}} \tag{4.146}$$

The integral I_{10} is identified below as the gamma function of argument one half.

11. The *step function* is defined as

$$\Theta(x) = \frac{1}{2\pi i} \int_{-\infty}^{\infty} \frac{dt}{t - i\eta} e^{itx} \tag{4.147}$$

where the constant $\eta \to 0^+$. This integral is evaluated for three cases:

- For $x > 0$ close the t-integral as a semicircle in the UHP in the limit that $R \to \infty$. This part of the path gives zero because the real part of the exponent $ixRe^{i\theta}$ is always negative and large. So the integral along the real axis equals the contribution at the pole $t = i\eta$ that gives

$$\Theta(x) = \lim_{\eta \to 0} e^{-\eta x} = 1, \quad \text{for } x > 0 \tag{4.148}$$

- For $x < 0$ close the contour in the LHP as a semicircle of radius R in the limit as $R \to \infty$. The contribution to this semicircle always vanishes. The integral equals the contribution from the poles. There are none, so the integral is zero

$$\Theta(x) = 0, \quad \text{for } x < 0 \tag{4.149}$$

- For the case that $x = 0$ the result was evaluated in J_3 that gives $\Theta(0) = 1/2$. These results are summarized as

$$\Theta(x) = \begin{cases} 1 & x > 0 \\ 1/2 & x = 0 \\ 0 & x < 0 \end{cases} \tag{4.150}$$

The step function occurs often in physics.

12. The *gamma function* is denoted as $\Gamma(x)$ and is defined according to the integral

$$\Gamma(x) = \int_0^{\infty} dt\, t^{x-1} e^{-t} \tag{4.151}$$

It is related to the factorial function $n!$ for integers. The factorial function $n!$ is the product of all integers between one and n:

$$n! = n(n-1)(n-2) \cdots 3 \cdot 2 \cdot 1 \tag{4.152}$$

The factorial function has the property that $n! = n \cdot (n-1)!$. The gamma function has a similar property that $\Gamma(x + 1) = x\Gamma(x)$. In fact, for integers, the gamma function is just the factorial function, $\Gamma(n + 1) = n!$. The gamma function is an extension of the factorial function to noninteger values of the argument. First prove that Eq. (4.151) has the proper recursion relation that $\Gamma(x + 1) = x\Gamma(x)$. Integrate by parts

$$\Gamma(x + 1) = \int_0^\infty dt\, t^x e^{-t} \tag{4.153}$$

$$\int v\, du = uv - \int u\, dv, \quad v = t^x, \quad u = -e^{-t} \tag{4.154}$$

$$\Gamma(x + 1) = (-t^x e^{-t})_0^\infty + x \int_0^\infty dt\, t^{x-1} e^{-t} \tag{4.155}$$

$$= x\Gamma(x) \tag{4.156}$$

The integral has the correct recursion relation for the gamma function. One can also show that $\Gamma(1) = 1 = \Gamma(2)$. The integral I_{10} is identified as the gamma function of argument one half, so $\Gamma(1/2) = \sqrt{\pi}$. This result provides the method of evaluating the factorial function for half integer arguments. If asked to evaluate $\Gamma(5/2)$ the answer is

$$\Gamma(5/2) = (3/2)\Gamma(3/2) = (3/2)(1/2)\Gamma(1/2) = \frac{3\sqrt{\pi}}{4} \tag{4.157}$$

The factorial function is easily evaluated for all positive half integers.

13. A common type of integral has the integral over all angles θ of a function of $\sin(\theta)$ or $\cos(\theta)$. An example is

$$I_{11} = \int_0^{2\pi} \frac{d\theta}{2\pi} \frac{1}{a + \sin(\theta)} \tag{4.158}$$

Evaluate this integral for the case that $a > 1$. Change variables to $z = e^{i\theta}$, $id\theta = dz/z$. As the angle θ varies between 0 and 2π the new variable z goes around the unit circle. The variable change has turned the integral into a contour integral where the contour is the unit circle

$$I_{11} = \oint \frac{dz}{2\pi i z} \frac{1}{a + 1/2i(z - 1/z)} = \oint \frac{dz}{\pi} \frac{1}{z^2 + 2iaz - 1} \tag{4.159}$$

The denominator is a quadratic polynomial with two poles, $z_{1,2} = -ia \pm i\sqrt{a^2 - 1}$. For $a > 1$ the pole $z_1 = -i(a + \sqrt{a^2 - 1})$ is outside the circle while $z_2 = -i(a - \sqrt{a^2 - 1})$ is inside. The integral equals $2\pi i$ times the residue at the pole inside of the circle

$$I_{11} = \frac{2\pi i}{\pi(z_2 - z_1)} = \frac{1}{\sqrt{a^2 - 1}} \tag{4.160}$$

If $a < -1$ the integral equals $-1/\sqrt{a^2 - 1}$ and if $-1 < a < 1$ the integral is zero.

Now it is time for some more theorems.

THEOREM 1. *A meromorphic function $f(z)$ is analytic except for a finite number of simple poles.*

THEOREM 2. *A meromorphic function can always be represented by a series over its poles in the form*

$$f(z) = f(0) + \sum_j \left[\frac{R_j}{z - z_j} + \frac{R_j}{z_j} \right] \tag{4.161}$$

where z_j are the simple poles and R_j are the residues at those poles. This formula assumes that $f(0)$ is not divergent, i.e., there is no pole at the origin. It also assumes that as $|z| \to \infty$ the function $f(z)$ goes to a constant or to zero.

An example of a meromorphic function is

$$f(z) = \frac{1}{z^2 + a^2} = \frac{1}{(z - ia)(z + ia)} = \frac{1}{2ia} \left(\frac{1}{z - ia} - \frac{1}{z + ia} \right) \tag{4.162}$$

There are two simple poles at $z = ia$ and $z = -ia$ that have residues of $1/(2ia)$ and $-1/(2ia)$, respectively. The expression can be expanded in partial fractions to give an expression that is the summation of two poles and residues at those poles. This result is the same as obtained for the series expansion for meromorphic functions. The constant terms at the poles cancel $f(0)$.

The above theorem is proved using contour integrals. Assume the function $f(z)$ has simple poles inside a contour of integration. These poles occur at the points z_j with residues R_j. The function $f(z)$ is assumed to go to zero or a constant as $|z| \to \infty$. With these conditions, consider the contour integral over a circle of radius infinity

$$I(z) = \oint \frac{dw}{2\pi i} \frac{f(w)}{w(w - z)} \tag{4.163}$$

The integral $I(z)$ is zero since it vanishes on the circular path at infinity. The contour also equals the summation of the residues at the poles in the integrand. These poles are at the points, $w = 0$, $w = z$, and $w = z_j$, so that

$$0 = \frac{f(0)}{-z} + \frac{f(z)}{z} + \sum_j \frac{R_j}{z_j(z_j - z)} \tag{4.164}$$

$$f(z) = f(0) - z \sum_j \frac{R_j}{z_j(z_j - z)} = f(0) + \sum_j R_j \left[\frac{1}{z - z_j} + \frac{1}{z_j} \right] \tag{4.165}$$

As an example, consider the function

$$f(z) = 3 + \frac{1}{z + 1} \tag{4.166}$$

There is a simple pole at $z_1 = -1$ with residue $R_1 = +1$. Also $f(0) = 4$. Our theorem gives the correct result

$$f(z) = 4 + \left[\frac{1}{z+1} + \frac{1}{-1}\right] = 3 + \frac{1}{z+1} \tag{4.167}$$

It is necessary to have the term $f(0)$ to get the correct answer. When $f(\infty)$ is a nonzero constant, the formula needs to have the constant factors.

The other case is when $f(z) \to 0$ as $|z| \to \infty$. In this case the constant factors are not needed, in fact, they cancel: $f(0) + \Sigma_j R_j/z_j = 0$. The cancellation is proved by taking the limit of $|z| \to \infty$ in Eq. (4.161). If $f(z)$ vanishes, along with the series that depends on z, there remains only the terms

$$0 = f(0) + \sum_j \frac{R_j}{z_j} \tag{4.168}$$

that proves the constant terms cancel in this case, and are not needed.

As an example of meromorphic functions, it was shown earlier that the function $f(z) = 1/\sin(z)$ had poles along the real axis at the points $z_n = n\pi$ with residues $R_n = (-1)^n$. This function is given exactly by the series

$$\frac{1}{\sin(z)} = \sum_n \frac{(-1)^n}{z - n\pi} \tag{4.169}$$

Since this function has an infinite number of poles it is not meromorphic, but the series is valid nevertheless. Further examples are presented in Chapter 5.

4.6. Higher Poles

All of the singularities discussed up to this point have involved simple poles. Now consider the evaluation of contour integrals around poles of the form

$$I_n(z_0) = \oint dz \, \frac{f(z)}{(z - z_0)^n} = \frac{2\pi i}{(n-1)!} \frac{\partial^{(n-1)}}{\partial z_0^{(n-1)}} f(z_0) \tag{4.170}$$

$$f(z_0) = \oint \frac{dz}{2\pi i} \frac{f(z)}{z - z_0} \tag{4.171}$$

$$\frac{\partial}{\partial z_0} f(z_0) = \oint \frac{dz}{2\pi i} f(z) \frac{\partial}{\partial z_0} \frac{1}{z - z_0} = 1! \oint \frac{dz}{2\pi i} \frac{f(z)}{(z - z_0)^2} \tag{4.172}$$

$$\frac{\partial^2}{\partial z_0^2} f(z_0) = \oint \frac{dz}{2\pi i} f(z) \frac{\partial^2}{\partial z_0^2} \frac{1}{z - z_0} = 2! \oint \frac{dz}{2\pi i} \frac{f(z)}{(z - z_0)^3} \tag{4.173}$$

$$\frac{\partial^n}{\partial z_0^n} f(z_0) = n! \oint \frac{dz}{2\pi i} \frac{f(z)}{(z - z_0)^{(n+1)}} \tag{4.174}$$

These steps prove the theorem.

As an example, let the function be $f(z) = z^L$. Assume L is an integer greater than zero, so the function is analytic. Evaluate the above integral by setting $z = z_0 + w$, where the complex variable becomes w

$$I = \oint dz \frac{z^L}{(z - z_0)^n} = \oint \frac{dw}{w^n}(z_0 + w)^L$$

$$= \sum_{m=0}^{L} \binom{L}{m} z_0^{L-m} \oint dw \frac{w^m}{w^n} \tag{4.175}$$

The only term in the series that is nonzero is when $m = n - 1$. A nonzero answer requires that $L \geqslant n - 1$, that is now assumed. If $L < n - 1$ the integral is zero.

$$I = 2\pi i \binom{L}{n-1} z_0^{L+1-n} = \frac{2\pi i}{(n-1)!} \frac{\partial^{(n-1)}}{\partial z_0^{(n-1)}} z_0^L \tag{4.176}$$

These steps prove the theorem for higher poles for the case that $f(z) = z^L$. The same method can be used to prove the result if $f(z)$ is a summation of such polynomials.

As a nontrivial example of using this theorem, consider the evaluation of the integral

$$J_{11} = \int_{-\infty}^{\infty} \frac{dx}{(x^2 + a^2)^3} \tag{4.177}$$

A closed contour can be achieved by closing the contour as a semicircle in the UHP as in Fig. 4.5(e). The large power of z^6 in the denominator assures that this contribution gives zero as the semicircle radius R goes to infinity. The integral equals the summation of the poles which are encircled in the contour. There is only one pole in the UHP at $z_1 = ia$. In this case $n = 3$ and $f(z) = 1/(z + ia)^3$

$$J_{11} = \frac{2\pi i}{2!} \frac{\partial^2}{\partial z^2} \frac{1}{(z + ia)^3}\bigg|_{z=ia} = \frac{i\pi(-3)(-4)}{(2ia)^5}$$

$$= \frac{3\pi}{8a^5} \tag{4.178}$$

4.7. Integrals Involving Branch Cuts

There have been no branch points or branch cuts in the integrals discussed so far. In many cases, integrals must be done where branch cuts are involved. Often they are helpful or often a nuisance. The important rule is that *contour integrals cannot cross branch cuts*. If the path of integration wants to intersect a branch cut, the path must go around the cut. Some examples are given of evaluating integrals involving branch cuts.

1. Consider the integral

$$K_1 = \int_0^\infty dx \frac{\ln(x^2 + 1)}{x^2 + a^2} = \frac{1}{2} \int_{-\infty}^\infty dx \frac{\ln(x^2 + 1)}{x^2 + a^2} \qquad (4.179)$$

The first step is to note that the integrand is symmetric in x. Extend the integral to negative infinity and divide by two. The problem with this integral is that the function $\ln(z^2 + 1)$ has two branch points, at $z = i$ and $z = -i$. Both branch cuts must extend to infinity. The log function can be written as

$$\ln(x^2 + 1) = \ln[(x + i)(x - i)] = \ln(x + i) + \ln(x - i) \qquad (4.180)$$

The integral for each term is done separately. The second term gives a result that is the complex conjugate of the first. It is only necessary to do one term, and then take twice the real part. This observation simplifies the integral to the expression

$$K_1 = \text{Re}\left[\int_{-\infty}^\infty dx \frac{\ln(x + i)}{x^2 + a^2}\right] \qquad (4.181)$$

where Re denotes the real part of the integral. This change has a happy consequence because there is only one branch point, at $z = -i$. The branch cut for this branch point is run down the imaginary axis. The contour of integration is closed in the UHP, that completely avoids the branch cut. This contour is shown in Fig. 4.7(a). Therefore consider the contour integral

$$K_1' = \oint \frac{dz \ln(z + i)}{z^2 + a^2} = \frac{2\pi i}{2ia} \ln[i(a + 1)] = \frac{\pi}{a}[i\pi/2 + \ln(a + 1)] \qquad (4.182)$$

$$K_1 = \text{Re}[K_1'] = \frac{\pi}{a} \ln(a + 1) \qquad (4.183)$$

The closed contour of integration goes along the real axis, and then closes as a semicircle in the UHP. The semicircle of radius R has a contribution that vanishes in the limit that $R \to \infty$. The closed contour equals the sum of the residues of the poles in the integrand. There is only one pole in the UHP that is at the point $z_1 = ia$. Its pole gives the above evaluation. The contour along the real axis equals the contribution from the pole, which gives the final result.

2. Consider the integral

$$K_2 = \int_0^\infty \frac{x \, dx}{x^3 + 1} \qquad (4.184)$$

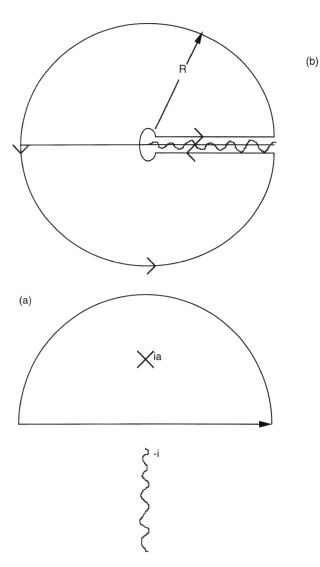

Figure 4.7   Contour integrals with branch cuts.

This integral cannot be extended to negative values of x since the integrand has a different value there. There is a special technique for treating these cases, that is now introduced. Examine the contour integral

$$K'_2 = \oint dz \, \ln(z) P(z) \tag{4.185}$$

where the original integral was $\int dx P(x)$. The function $P(z)$ must possess certain properties:

- It has no branch cuts or poles on the positive real axis.

- It vanishes at infinity. This condition is given as

$$\lim_{|z| \to \infty} P(z) \to \frac{C}{z^\alpha} \tag{4.186}$$

where α is a positive constant greater than one.

The function $\ln(z)$ was introduced. It has a branch point at the origin. The branch cut is put along the positive real axis. The contour integral has the path shown in Fig. 4.7(b). It is a circle of radius R, except that it must go around the branch cut. It also goes around the origin with a circle of small radius ε. Eventually take $R \to \infty$, $\varepsilon \to 0$. The path of the integral has four segments. The integral is evaluated on each of these segments:

- Upper real axis. Along the branch cut specify that $z = \rho \exp(i\theta)$ and $\theta = 0$ is on the upper side of the branch cut. The integral along the upper real axis is

$$I_1 = \int_\varepsilon^R dx \, \ln(x) P(x) \tag{4.187}$$

- Lower real axis. Here $z = x \exp(2\pi i)$ so that the integral is

$$I_2 = \int_R^\varepsilon dx P(x)[2i\pi + \ln(x)] \tag{4.188}$$

$$I_1 + I_2 = -2i\pi \int_\varepsilon^R dx P(x) \tag{4.189}$$

The contribution I_2 changes sign when reversing the direction of integration.

- The contribution from the semicircle at infinity is evaluated by setting $z = R \exp(i\theta)$, $dz = zi \, d\theta$

$$I_3 = \lim_{R \to \infty} iR \int_0^{2\pi} d\theta e^{i\theta} P(Re^{i\theta})[\ln(R) + i\theta]$$

$$\to \frac{iC}{R^{\alpha-1}} \int_0^{2\pi} d\theta e^{-i\theta(\alpha-1)}[\ln(R) + i\theta] = 0 \tag{4.190}$$

This term vanishes if $\alpha > 1$, according to the assumption made regarding the asymptotic properties of $P(z)$.

- The last segment is the circle around the origin of radius ε. Set $z = \varepsilon \exp(i\theta)$ and find

$$I_4 = \lim_{\varepsilon \to 0} i\varepsilon \int_{2\pi}^0 d\theta e^{i\theta} P(\varepsilon e^{i\theta})[\ln(\varepsilon) + i\theta] \to 0 \tag{4.191}$$

This term vanishes due to the vanishing prefactor ε. The function $P(z)$ is well-behaved at the origin, according to our assumptions.

The end result of these four integrals is that the only nonzero contributions are from $I_1 + I_2$. The above steps prove the identity

$$\int_0^\infty dx\, P(x) = -\frac{1}{2\pi i} \oint dz\, P(z)\ln(z) = -\sum_j R_j \ln(z_j) \qquad (4.192)$$

where z_j are the poles of $P(z)$ and R_j are the residues at those poles. The integral from zero to infinity equals the summation of the residues at these poles multiplied by $\ln(z_j)$. This formula is exact and now it is used to evaluate K_2 where $P(x) = x/(x^3 + 1)$. There are three poles and residues, $z_1 = \exp(i\pi/3)$, $z_2 = \exp(i\pi)$, and $z_3 = \exp(i5\pi/3)$. Note that the latter pole cannot be expressed as $\exp(-i\pi/3)$ due to the branch cut. All angles must be positive starting from the upper part of the positive real axis. Write the denominator as $(z - z_1)(z - z_2)(z - z_3)$. The result is

$$K_2 = -\left[\frac{e^{i\pi/3}(i\pi/3)}{(e^{i\pi/3} - e^{i\pi})(e^{i\pi/3} - e^{i5\pi/3})} + \frac{e^{i\pi}(i\pi)}{(e^{i\pi} - e^{i\pi/3})(e^{i\pi} - e^{i5\pi/3})}\right.$$

$$\left. + \frac{e^{i5\pi/3}(i5\pi/3)}{(e^{i5\pi/3} - e^{i\pi/3})(e^{i5\pi/3} - e^{i\pi})}\right]$$

$$= \frac{2\pi}{3\sqrt{3}} \qquad (4.193)$$

The latter result is obtained after some algebra. This procedure can be used to evaluate any integral from zero to infinity where the integrand $P(x)$ obeys the assumed conditions.

3. Consider the integral

$$K_3 = \int_0^\infty \frac{dx\, x^a}{x^2 + x + 1} \qquad (4.194)$$

where $-1 < a < 1$. The constant a is assumed to be irrational. The case $a = 1$ can be solved using the techniques of the prior example. In the present case, the $\ln(z)$ function is not needed since the integral has its own branch point and branch cut. The branch point is at $z = 0$. Choose the branch cut to run along the positive real axis to infinity. Then consider the integral over the closed contour

$$K'_3 = \oint \frac{dz\, z^a}{z^2 + z + 1} \qquad (4.195)$$

The contour is exactly the same as the last example [see Fig. 4.7(b)]. Again there are four segments to evaluate. The two circular segments vanish as long as $-1 < a < 1$. The answer is the sum of the two segments along the real axis—above and below the branch cut. Setting $z = x$ above the branch cut, and $z = x \exp(2\pi i)$ below, the two integrals are

$$I_1 = \int_\varepsilon^R \frac{dx\, x^a}{x^2 + x + 1} \tag{4.196}$$

$$I_2 = \int_R^\varepsilon \frac{dx\, x e^{2i\pi a}}{x^2 + x + 1} = -e^{2i\pi a} \int_\varepsilon^R \frac{dx\, x^a}{x^2 + x + 1} \tag{4.197}$$

$$K_3' = (1 - e^{2\pi i a}) \int_0^\infty \frac{dx\, x^a}{x^2 + x + 1} \tag{4.198}$$

The integral K_3' also equals to the summation of the residues at the poles inside of the contour. The equation $0 = z^2 + z + 1$ has two roots that are the two poles, at $z_1 = \exp(i2\pi/3)$ and $z_2 = \exp(i4\pi/3)$. The sum over the residues gives

$$K_3' = \frac{2\pi i}{z_1 - z_2} [z_1^a - z_2^a] = \frac{2\pi i e^{i2\pi a/3}}{e^{i2\pi/3} - e^{i4\pi/3}} [1 - e^{i2\pi a/3}]$$

$$= \frac{2\pi}{\sqrt{3}} t(1 - t) \tag{4.199}$$

$$t = e^{i2\pi a/3} \tag{4.200}$$

Equate the two expressions for K_3' and then derive

$$K_3 = \int_0^\infty \frac{dx\, x^a}{x^2 + x + 1} = \frac{2\pi}{\sqrt{3}} \frac{t(1 - t)}{1 - t^3}$$

$$= \frac{2\pi}{\sqrt{3}} \frac{t}{1 + t + t^2} = \frac{2\pi}{\sqrt{3}\,[1 + 2\cos(2\pi a/3)]} \tag{4.201}$$

4. Consider the integral

$$K_4 = \int_0^1 \frac{dx}{x - a} \left(\frac{1 - x}{x} \right)^\alpha \tag{4.202}$$

where we stipulate that $a > 1$ and $0 < \alpha < 1$. This integrand has branch points at $x = 0, 1$ that can be connected by a finite branch cut as shown in Fig. 4.8(a). There is also a simple pole at $x = a$ which is not on the branch cut. This integral is evaluated by considering the following contour integral

$$K_4' = \oint \frac{dz}{z - a} \left(\frac{z - 1}{z} \right)^\alpha \tag{4.203}$$

The contour is a counterclockwise circle of radius $R \to \infty$. Setting $dz = iz\, d\theta$ gives the integral

$$K_4' = \lim_{R \to \infty} i \int_0^{2\pi} \frac{d\theta}{1 - a e^{-i\theta}/R} (1 - e^{-i\theta}/R)^\alpha = 2\pi i \tag{4.204}$$

The contour integral also equals the contribution from encircling the poles (residues) plus the contribution from encircling the finite branch cut.

$$K'_4 = 2\pi i = \oint_{\text{pole}} dz\{\cdots\} + \oint_{\text{bc}} dz\{\cdots\} \qquad (4.205)$$

$$\oint_{\text{pole}} dz\{\cdots\} = 2\pi i \left(\frac{a-1}{a}\right)^{\alpha} \qquad (4.206)$$

$$\oint_{\text{bc}} dz\{\cdots\} = \sum_j \int_j dz\{\cdots\} \qquad (4.207)$$

The integral around the branch cut is shown in Fig. 4.8. There are four segments, two straight paths between zero and one and above and below the branch cut. There are also two circular paths around the two branch points. The two circular integrals are zero. Around $z = 0$ we set $z = \varepsilon \exp(i\theta)$ and take the limit as $\varepsilon \to 0$. A zero limit is obtained as long as $\alpha < 1$. Around the point $z = 1$ we set $z = 1 + \varepsilon \exp(i\theta)$ and this contribution also vanishes as $\varepsilon \to 0$. There remains the contributions of the two segments between zero and one. Define $z - 1 = \rho \exp(i\theta)$ where θ is zero along the x axis $x > 1$. The integral above (I_a) and below (I_b) are

$$I_a = \int_{1-\varepsilon}^{\varepsilon} \frac{dx}{x-a} \left(\frac{1-x}{x}\right)^{\alpha} e^{i\pi\alpha} \qquad (4.208)$$

$$I_b = \int_{\varepsilon}^{1-\varepsilon} \frac{dx}{x-a} \left(\frac{1-x}{x}\right)^{\alpha} e^{-i\pi\alpha} \qquad (4.209)$$

$$I_a + I_b = [e^{-i\pi\alpha} - e^{i\pi\alpha}] \int_0^1 \frac{dx}{x-a} \left(\frac{1-x}{x}\right)^{\alpha}$$

$$= -2i \sin(\pi\alpha) \int_0^1 \frac{dx}{x-a} \left(\frac{1-x}{x}\right)^{\alpha} \qquad (4.210)$$

Equate the two results for K'_4 and then divide by the complex factor in the above equation, that produces the final result

$$K_4 = \int_0^1 \frac{dx}{x-a} \left(\frac{1-x}{x}\right)^{\alpha} = \frac{\pi}{\sin(\alpha\pi)} \left[\left(\frac{a-1}{a}\right)^{\alpha} - 1\right] \qquad (4.211)$$

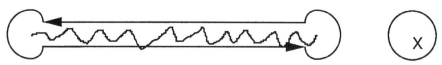

Figure 4.8   Contour integral for finite branch cut.

A similar technique is used when $0 < a < 1$ so the pole is on the branch cut.

$$K_5 = P \int_0^1 \frac{dx}{x-a} \left(\frac{1-x}{x} \right)^\alpha \qquad (4.212)$$

$$K_5' = \oint \frac{dz}{z-a} \left(\frac{z-1}{z} \right)^\alpha \qquad (4.213)$$

Then one starts out before showing $K_5' = 2\pi i$ when the contour is at infinity. Then $2\pi i$ equals the contour around the branch cut (Fig. 4.8). There are eight segments to integrate. The four straight segments give the desired integral times a complex factor

$$\int_\varepsilon^{a-\varepsilon} \frac{dx}{x-a} \left(\frac{1-x}{x} \right)^\alpha e^{-i\pi\alpha} + \int_{a+\varepsilon}^{1-\varepsilon} \frac{dx}{x-a} \left(\frac{1-x}{x} \right)^\alpha e^{-i\pi\alpha} + \int_{1-\varepsilon}^{a+\varepsilon} \frac{dx}{x-a} \left(\frac{1-x}{x} \right)^\alpha e^{i\pi\alpha}$$

$$+ \int_{a-\varepsilon}^\varepsilon \frac{dx}{x-a} \left(\frac{1-x}{x} \right)^\alpha e^{i\pi\alpha}$$

$$= (e^{-i\pi\alpha} - e^{i\pi\alpha}) P \int_0^1 \frac{dx}{x-a} \left(\frac{1-x}{x} \right)^\alpha \qquad (4.214)$$

The circular integrals around the two branch points gives zero as long as $0 < \alpha < 1$. The two semicircular integrals at $z = a + \varepsilon \exp(i\theta)$ give for the top and bottom, respectively,

$$\int_0^\pi id\theta \left(\frac{1-a}{a} \right)^\alpha e^{i\pi\alpha} = i\pi \left(\frac{1-a}{a} \right)^\alpha e^{i\pi\alpha} \qquad (4.215)$$

$$\int_{-\pi}^0 i\,d\theta \left(\frac{1-a}{a} \right)^\alpha e^{-i\pi\alpha} = i\pi \left(\frac{1-a}{a} \right)^\alpha e^{-i\pi\alpha} \qquad (4.216)$$

Adding these contributions gives

$$2\pi i = 2\pi i \left(\frac{1-a}{a} \right)^\alpha \cos(\pi\alpha) - 2i \sin(\pi\alpha) P \int_0^1 \frac{dx}{x-a} \left(\frac{1-x}{x} \right)^\alpha \qquad (4.217)$$

$$P \int_0^1 \frac{dx}{x-a} \left(\frac{1-x}{x} \right)^\alpha = \frac{\pi}{\sin(\pi\alpha)} \left[\left(\frac{1-a}{a} \right)^\alpha \cos(\pi\alpha) - 1 \right] \qquad (4.218)$$

which is slightly different from the result when the pole is off the cut.

4.8. Approximate Evaluation of Integrals

With modern computers any integral in one dimension can be evaluated numerically in a matter of seconds. Nevertheless, it is still useful to have approximate methods for evaluating integrals that cannot be done exactly analytically.

Steepest descent is a method used when the integral in one dimension has, along its path, one high maximum. Graph the integrand $P(x)$ as a function of x and find the place where the integrand $P(x)$ is the largest. The integral is approximated by a *Gaussian integral* around this maximum point. A Gaussian integral is one of the form

$$I(a, b) = \int_{-\infty}^{\infty} dx e^{-b(x-a)^2} = \sqrt{\frac{\pi}{b}} \qquad (4.219)$$

This integral was done before under I_{10}. The result does not depend on the parameter a since the change of variables $x' = (x - a)\sqrt{b}$ eliminates it and reduces the integral to exactly I_{10}. The steps in the evaluation using steepest descent are:

1. Write the integral as

$$I = \int dx P(x) = \int dx e^{f(x)} \qquad (4.220)$$

$$f(x) = \ln[P(x)] \qquad (4.221)$$

2. Find the maximum point x_0 of the function $f(x)$ by solving

$$\frac{df}{dx} = 0 = \frac{dP}{P\,dx} \qquad (4.222)$$

3. Expand the function $f(x)$ in a Taylor series around the point x_0 and retain the first two nonzero terms in the series

$$\delta x = x - x_0 \qquad (4.223)$$

$$f(x) = f(x_0) + \frac{(\delta x)}{1!}\frac{df(x_0)}{dx_0} + \frac{(\delta x)^2}{2!}\frac{d^2 f(x_0)}{dx_0^2} + \cdots + \qquad (4.224)$$

At the point x_0 then $f'(x_0) = 0$, $f''(x_0) < 0$. The first two nonzero terms are the constant $f(0)$ and the quadratic term $\propto (\delta x)^2$. These two terms produce the Gaussian integral

$$\int dx P(x) \approx P(x_0) \int dx e^{-(\delta x)^2 |f''|/2}$$

$$= \sqrt{\frac{2\pi}{|f''|}} P(x_0) \qquad (4.225)$$

where $|f''|$ denotes the absolute magnitude of the second derivative. The maximum point x_0 must be within the limits of integration.

As an example, consider the integral for the factorial function

$$n! = \int_0^\infty t^n \, dt \, e^{-t} \tag{4.226}$$

An approximate expression for this integral is obtained for the case that the integer n is large, say greater than 10. Write the integral in the above form of an exponential function $f(t) = n \ln(t) - t$. The maximum value of this function occurs at

$$f' = \frac{n}{t_0} - 1 = 0 \tag{4.227}$$

$$t_0 = n \tag{4.228}$$

$$f(t_0) = n \ln(n) - n \tag{4.229}$$

$$f'' = -\frac{n}{t_0^2} = -\frac{1}{n} \tag{4.230}$$

$$n! \approx \sqrt{2\pi n} \, n^n e^{-n} \tag{4.231}$$

The last formula is called *Stirlings approximation* to the factorial function. It is a famous and a very useful result.

4.8.2. Saddle Point Integrals

Saddle point integrals are contour integrals whose path takes them over a mountain pass. That is, the function $f(z)$ along the path of integration has its maximum value at some point z_0. The integral is evaluated by steepest descent using the point $f(z_0)$ as the top of the Gaussian. Usually the original integral is over a path that does not include the point z_0. It is usually quite simple to change the path of a contour integral to another path that does include z_0. The integrals over the two paths differ only by the contribution of poles or finite branch cut in the area defined by the two different paths. If there are no poles or branch cuts, the integral over the two paths are identical. The philosophy of saddle point integrals is to choose the path that goes over the point z_0. This point is always a saddle point.

If you are trekking over a mountain pass, most trails go over the low point in the mountain ridge. The low point is called a *saddle*. At the top of the pass, the path goes down in both directions of the trail. However, hiking perpendicular to the trail, one has to go up in either direction. A *saddle point* is defined as the top of the low point in the ridge. Write the contour integral as

$$I = \int dz f(z) = \int dz e^{F(z)} \tag{4.232}$$

$$F(z) = \ln[f(z)] \tag{4.233}$$

The saddle point is defined as the point where $dF(z)/dz = 0$.

THEOREM. *A saddle point is any place where the first derivative of F(z) vanishes while the second derivative of F(z) does not vanish.*

Say that the first derivative vanishes at z_0, so that $F'(z_0) = 0$. Expand the function $F(z)$ in a Taylor series about the point z_0 and retain only the first two terms. Keeping only two terms is the Gaussian approximation:

$$F(z) = F(z_0) + (z - z_0)F'(z_0) + \frac{1}{2!}(z - z_0)^2 F''(z_0) + \cdots +$$

$$\approx F(z_0) + (z - z_0)^2 A(z_0) \tag{4.234}$$

$$2A(z_0) = F''(z_0) \tag{4.235}$$

Define $A(z_0)$ as one half the second derivative of the function $F(z)$ at the point z_0. The next step is to show it is a saddle point. The complex number $A(z_0) = R\exp(i\theta_0)$ where R is the amplitude and θ_0 is the phase. The quantity $z - z_0 = \rho\exp(i\theta)$ is the displacement from the point z_0. The second derivative term in the Taylor series has the form

$$\Xi = A(z_0)(z - z_0)^2 = R\rho^2 e^{i(\theta_0 + 2\theta)} \tag{4.236}$$

The prefactor of $R\rho^2$ is strictly positive. The sign of Ξ depends on the total phase angle $\theta_0 + 2\theta$.

- At $\theta = -\theta_0/2$ then Ξ is real and positive.
- At $\theta = \pi/2 - \theta_0/2$ then Ξ is real and negative.
- At $\theta = \pi - \theta_0/2$ then Ξ is real and positive.
- At $\theta = 3\pi/2 - \theta_0/2$ then Ξ is real and negative.

Ξ changes sign every 90° from plus, to minus, to plus, to minus. That is the behavior at a saddle point. Along one path it is negative in both directions; that is the direction of the trail over the mountain pass. Along the perpendicular line, it is positive in both directions. This result does not change if higher order derivatives are included in the Taylor series. As long as the displacement ρ is small, the contribution of $O(\rho^2)$ will be larger than those of higher derivatives, $O(\rho^n)$ and $n > 2$. The second-order terms define the property of the saddle.

The path of the contour integral is selected to go along the same path as the trail. Choose $z = z_0 + i\rho\exp(-i\theta_0/2)$ so that Ξ is negative in both directions from the top of the path. Use steepest descent to evaluate the integral along this path

$$I \approx if(z_0)e^{-i\theta_0/2}\int d\rho\, e^{-R\rho^2} = i\sqrt{\frac{2\pi}{|F''(z_0)|}}\, f(z_0)e^{-i\theta_0/2} \tag{4.237}$$

This formula is quite general. The derivation of Stirling's approximation by steepest descent had $z_0 = n$, $\theta_0 = \pi$, and $\theta = 0$. The path over the saddle was along the real axis.

The Airy function is useful in quantum mechanics since it is the eigenfunction of a particle undergoing nonrelativistic motion in a linear potential, such as $V = -Fx$ where F is a constant force. This function is defined by an integral

$$Ai(u) = \frac{1}{\pi} \int_0^\infty dt \cos\left(\frac{t^3}{3} + ut\right) \tag{4.238}$$

The Airy function will be evaluated at very large values of $|u|$. Such expressions are called *asymptotic expansions* that are discussed in Chapter 5. First do the case that $u \gg 0$. The saddle point method will be used. The path of integration is extended to negative infinity since the cosine is an even function of its arguments.

$$Ai(u) = \text{Re}[J(u)] \tag{4.239}$$

$$J(u) = \frac{1}{2\pi} \int_{-\infty}^\infty dz e^{F(z)} \tag{4.240}$$

$$F(z) = i\frac{z^3}{3} + iuz \tag{4.241}$$

The first step is to find the saddle point. Set $F' = 0 = i(z_0^2 + u)$ that has the solution $z_0 = \pm i\sqrt{u}$. Only one of these two choices is needed and in order to find it, set $z = iy$ so that $F(iy) = y^3/3 - uy$. A graph of $F(iy)$ shows that $y = \sqrt{u}$ is a minimum point on the y axis, which is the saddle point. The other choice $y = -\sqrt{u}$ is a maximum point along the y axis, so set $z_0 = i\sqrt{u}$.

The next step is to find the path of integration over this saddle point. The second derivative is

$$F'' = 2iz \tag{4.242}$$

$$A(z_0) = \frac{1}{2}\frac{d^2F(z_0)}{dz_0^2} = iz_0 = -\sqrt{u} \tag{4.243}$$

This choice results in $R = \sqrt{u}$ and $\theta_0 = \pi$. The directions that make a downward descent from the saddle are along the $\pm x$ direction. The saddle point integral is evaluated by setting $z = x + i\sqrt{u}$ where x is the variable of integration, as shown in Fig. 4.9(a)

$$F(z) = \frac{i}{3}(i\sqrt{u} + x)^3 + iu(i\sqrt{u} + x) = -\frac{2}{3}u^{3/2} - \sqrt{u}\,x^2 + ix^3/3 \tag{4.244}$$

$$J(u) \approx \frac{e^{-2u^{3/2}/3}}{2\pi} \int dx e^{-\sqrt{u}\,x^2}$$

$$= \frac{1}{2\sqrt{\pi}\sqrt{u}} \exp(-2u^{3/2}/3)$$

$$\lim_{u \gg 1} Ai(u) \rightarrow \frac{1}{2\sqrt{\pi}\sqrt{u}} \exp(-2u^{3/2}/3) \tag{4.245}$$

Since $J(u)$ is real, it is the same as the Airy function.

At what point was u assumed to be large? The answer is that the term of $O(x^3)$ can only be neglected when u is large. In the Gaussian integral, the important values of x occur when $\sqrt{u}\,x^2 \sim 1$ which means that $x \sim 1/u^{1/4}$. Terms of $O(x^3)$ are then of $O(1/u^{3/4})$. Clearly u must be large if this term can be neglected. The derivation is completed for positive values of u.

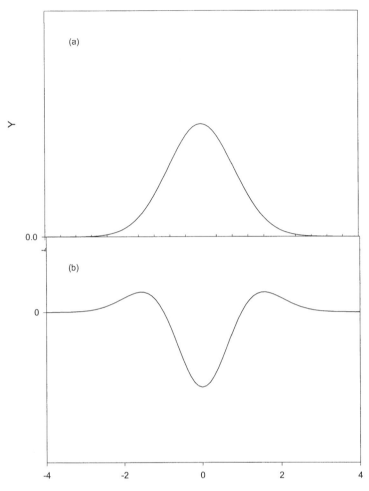

Figure 4.9 The path of saddle point integrals for Airy functions. (a) $u > 0$ and (b) $u < 0$.

Next consider the similar expression for large negative arguments. Treat u as a positive number and evaluate $Ai(-u)$. The steps are similar but the saddle point has moved. The equation for the exponent is

$$F(z) = \frac{iz^3}{3} - iuz \qquad (4.246)$$

$$F' = i[z^2 - u] \qquad (4.247)$$

$$F'' = 2iz \qquad (4.248)$$

$$z_0 = \pm\sqrt{u} \qquad (4.249)$$

Setting $F' = 0$ again gives two choices of saddle point. In this case, retain them both. Now the path of integration goes over two mountain ranges, that requires two Gaussian integrals. The path is shown in Fig. 4.9(b). Evaluate each in turn.

1. At $z_0 = \sqrt{u}$ then $A(z_0) = F''/2 = iz_0 = i\sqrt{u}$. In this case $R = \sqrt{u}$, $\theta_0 = \pi/2$. The path of integration over the saddle has an angle $\theta = \pi/2 - \theta_0/2 = \pi/4$. It bisects the real axis at 45°. Set $z = \sqrt{u} + \sqrt{i}\,\xi$ where ξ is the integration variable. The contribution from this saddle point is

$$F(z) = \frac{i}{3}(\sqrt{u} + \sqrt{i}\,\xi)^3 - iu(\sqrt{u} + \sqrt{i}\,\xi)$$

$$= -\frac{2i}{3}u^{3/2} - \sqrt{u}\,\xi^2 + O(\xi^3) \tag{4.250}$$

$$J^{(+)}(-u) = \frac{\exp(-2iu^{3/2}/3)}{2\pi}\sqrt{i}\int d\xi\, e^{-\sqrt{u}\xi^2}$$

$$= \frac{1}{2\sqrt{\pi}\sqrt{u}}\exp(i\pi/4 - 2iu^{3/2}/3) \tag{4.251}$$

2. At $z_0 = -\sqrt{u}$ then $A(z_0) = F''/2 = iz_0 = -i\sqrt{u}$. In this case $R = \sqrt{u}$, $\theta_0 = -\pi/2$. The path of integration over the saddle has an angle $\theta = \pi/2 - \theta_0/2 = 3\pi/4$. It bisects the real axis at $-45°$. Set $z = -\sqrt{u} + \sqrt{-i}\,\xi$ where ξ is the integration variable. The contribution from this saddle point is

$$F(z) = \frac{i}{3}(-\sqrt{u} + \sqrt{-i}\,\xi)^3 - iu(-\sqrt{u} + \sqrt{-i}\,\xi)$$

$$= \frac{2i}{3}u^{3/2} - \sqrt{u}\,\xi^2 + O(\xi^3) \tag{4.252}$$

$$J^{(-)}(-u) = \frac{\exp(2iu^{3/2}/3)}{2\pi}\sqrt{-i}\int d\xi\, e^{-\sqrt{u}\xi^2}$$

$$= \frac{1}{2\sqrt{\pi}\sqrt{u}}\exp(-i\pi/4 + 2iu^{3/2}/3) \tag{4.253}$$

$$J(-u) = J^{(+)} + J^{(-)} = \frac{1}{\sqrt{\pi}\sqrt{u}}\cos(-\pi/4 + 2u^{3/2}/3) \tag{4.254}$$

$$\lim_{u \gg 1} Ai(-u) \to \frac{1}{\sqrt{\pi}\sqrt{u}}\cos(-\pi/4 + 2u^{3/2}/3)$$

$$= \frac{1}{\sqrt{\pi}\sqrt{u}}\sin(\pi/4 + 2u^{3/2}/3) \tag{4.255}$$

The last two expressions, in terms of sines and cosines, are identical. Most books list the version using the sine function. Again the result is valid at large values of u that are required to make the omission of the term $O(\xi^3)$ an accurate approximation. The path of integration is shown in Fig. 4.9(b). The path goes from negative to positive infinity but drops below the real axis in order to go over both saddles. Another integral to

$$F_N(\omega) = \int_{-\infty}^{\infty} \frac{dt}{2\pi} e^{it\omega}[\cos(t)]^N \qquad (4.256)$$

The integral is simple for small values of N, and results in a summation of few delta functions.

- $N = 0$ has $F_0(\omega) = \delta(\omega)$.
- $N = 1$ has

$$F_1(\omega) = \tfrac{1}{2}[\delta(\omega + 1) + \delta(\omega - 1)] \qquad (4.257)$$

- $N = 2$ has $\cos^2(t) = [1 + \cos(2t)]/2$ and

$$F_2(\omega) = \tfrac{1}{4}[2\delta(\omega) + \delta(\omega + 2) + \delta(\omega - 2)] \qquad (4.258)$$

As N becomes large, there are many delta functions. The evaluation of this integral by a saddle point method makes $F_N(\omega)$ a continuous function of ω. In this case the result is actually the distribution of delta function amplitudes.

The exponent in the saddle point is $F(z) = iz\omega + N \ln[\cos(z)]$. This integral oscillates wildly when integrated along the real axis. The saddle point method moves it off the real axis where the integrand becomes smoother. Setting the derivative equal to zero gives

$$F' = i\omega - N \tan(z_0) = 0 \qquad (4.259)$$

$$\tan(z) = z + \frac{z^3}{3} + \cdots + \qquad (4.260)$$

For small values of z_0, the series expansion is shown for the tangent. Taking the first term in the series gives that $z_0 = i\omega/N$. In this approximation $F(z_0) = -\omega^2/(2N)$ and the second derivative gives $F'' = -N$. Here the angle of A is $\theta_0 = \pi$ so that $\theta = 0$ for the direction of the saddle point integral. The saddle is on the imaginary axis, and the saddle point goes over it along the x direction.

$$F_N \approx e^{-\omega^2/2N} \int \frac{dx}{2\pi} e^{-x^2 N/2} = \frac{e^{-\omega^2/2N}}{\sqrt{2\pi N}} \qquad (4.261)$$

Other terms in the series for the tangent are neglected since they are smaller. Since the above integral uses values of x in the range $x \sim 1/\sqrt{N}$ then the relevant values of $z \sim 1/\sqrt{N} + i\omega/N$ and terms that are $Nz^3 \sim O(1/\sqrt{N})$ and become small for large N. In Chapter 5 it is shown how these terms cause asymptotic corrections to the saddle point results.

The tangent is a periodic function of its arguments. However, that is not important here since the saddle is on the imaginary axis. Set $z = iy$ and the equation for y_0 is

$$F' = 0 = iN \left[\frac{\omega}{N} - \tanh(y_0) \right] \qquad (4.262)$$

which only has one solution $y_0 = \tanh^{-1}(\omega/N)$.

Problems

1. Use complex variables to prove Poincarés theorem, where (a, b, c) are all real

$$P\left(\frac{1}{(a-b)(a-c)}\right) = P\left[\frac{1}{b-c}\left(\frac{1}{a-b} - \frac{1}{a-c}\right)\right] + \pi^2\delta(a-b)\delta(a-c) \quad (4.263)$$

2. Reduce the following analytic functions to the form $f = u(x, y) + iv(x, y)$.

   (a) $f = \ln(z)$
   (b) $f = \ln(z^2 + a^2)$
   (c) $f = \sqrt{z - 1}$.

3. Use the Cauchy–Riemann conditions to see whether the following functions are analytic. If they are, can you guess the functional form $f(z)$?

   (a) $u = -x(3y^2 - x^2)$, $v = y(3x^2 - y^2)$

   (b) $u = \dfrac{\sin(x)}{\cos(x) + \cosh(y)}$, $v = \dfrac{\sinh(y)}{\cos(x) + \cosh(y)}$

   (c) $u = [\sqrt{x^2 + y^2} + x]^{1/2}$, $v = [\sqrt{x^2 + y^2} - x]^{1/2}$

4. Find the branch points of the following functions. Discuss whether branch cuts are finite or infinite.

   (a) $f(z) = [z(z^2 - 4)]^{1/3}$

   (b) $f(z) = \ln(z^2 - b^2)$

   (c) $f(z) = \ln\left(\dfrac{z+b}{z-b}\right)$

   (d) $f(z) = \dfrac{z+3}{2 + \sqrt{z^2 - 1}}$

5. Evaluate the following integrals that have a circular contour at the indicated radius R.

$$I_a = \oint_{R=2} dz\, \frac{\cosh(\pi z)}{z(z^2 - 1)}$$

$$I_b = \oint_{R=3} \frac{dz}{2\pi i}\, \frac{\sin(az)}{(z^2 + 1)}$$

$$I_c = \oint_{R=2} dz \ln\left(\frac{z+1}{z-1}\right)$$

$$I_c = \oint_{R=2} dz \ln\left(\frac{z+1}{z-1}\right)$$

6. Evaluate the following integrals using contour integrals

   (a) $\displaystyle\int_{-\infty}^{\infty} \frac{dx}{(x-i)(x+3i)}$

   (b) $\displaystyle\int_{-\infty}^{\infty} dx\, \frac{\cos(ax)}{x^2 + b^2}$

(c) $P \int_{-\infty}^{\infty} dx \, \frac{\cos(ax)}{x^2 - b^2}$

(d) $\int_{-\infty}^{\infty} \frac{dx}{x^4 + a^4}$

(e) $P \int_{-\infty}^{\infty} \frac{dx}{x^4 - a^4}$

(f) $\int_{-\infty}^{\infty} dx \, \frac{\cos(ax)}{(x - b)^2 + c^2}$

(g) $\int_{0}^{2\pi} \frac{d\phi}{2\pi} \frac{1}{1 + ia \cos(\phi)}$

7. Evaluate the following that involve branch cuts

(a) $\int_{0}^{\infty} \frac{dx}{\sqrt{x}} \cos(bx)$

(b) $\int_{0}^{\infty} \frac{dx}{x^2 + x + 5/2}$

(c) $\int_{-1}^{+1} \frac{dx}{\sqrt{1 - x^2}} \frac{1}{x - y}, \, |y| > 1$

(d) $P \int_{-1}^{+1} \frac{dx}{\sqrt{1 - x^2}} \frac{1}{x - y}, \, |y| < 1$

(e) $\int_{-1}^{1} dx \, \frac{\sqrt{1 - x^2}}{x^2 - y^2}, \, |y| > 1$

(f) $\int_{0}^{\infty} \frac{\sqrt{x} \, dx}{x^2 + a^2}$

8. Evaluate the contour integral of radius R as $R \to \infty$. Also, evaluate it by encircling the finite branch cut between $-1 < z < 1$, and also the pole at $y > 1$. Do these two methods give the same answer?

$$\oint \frac{dz}{z - y} \ln\left(\frac{z + 1}{z - 1}\right)$$

9. For the following meromorphic function, give the expansion in terms of its poles and residues

(a) $f(z) = \dfrac{z}{z^4 + 1}$

(b) $f(z) = \dfrac{1}{z^4 - b^4}$

(c) $f(z) = \dfrac{z^2}{z^2 + a^2}$

10. Evaluate the following integrals using steepest descent

$$I_a = \int_{-1}^{1} dx [1 - x^2]^n$$

$$I_b = \int_{0}^{\infty} dx \exp[-ax^2 - b/x^2]$$

$$I_c = \int_{0}^{\infty} dx x^a e^{-x^4}$$

11. Evaluate the following integrals for large $u \gg 0$ using saddle point methods

(a) $J_0(u) = \dfrac{2}{\pi} \int_{0}^{\infty} dt \sin[u \cosh(t)]$

(b) $I_0(u) = \int_{-\infty}^{\infty} \dfrac{dx}{2\pi} \sin(x^2 + xu)$

(c) $K_0(u) = \int_{-\infty}^{\infty} dx \cos(x^4 + 2ux^2)$

The first is a Bessels function.

Series

Many functions $f(z)$ can be represented by a series of terms, where each term depends on the variable z. One type of series has the form

$$f(z) = \sum_n a_n(z - z_0)^n \tag{5.1}$$

The coefficients a_n are constants, as is z_0. We continue to use the language of complex variables, and will treat the variable z as complex. However, for many applications the variable z is actually real. The above form of the series is not the only possible one. Others are discussed below. The exponent n is not limited to positive integers. When the exponent n includes negative and positive integers, the function is called a *Laurent series* (see Section 5.3).

5.1. Taylor Series

Taylor series are the most common and the most useful series. They have the above form, but the exponents n are limited to nonnegative integers.

$$f(z) = \sum_{n=0}^{\infty} a_n(z - z_0)^n \tag{5.2}$$

$$a_n = \frac{1}{n!}\left(\frac{d^n f(z_0)}{dz_0^n}\right) \tag{5.3}$$

The coefficients a_n are given by the nth order derivative of the function $f(z)$, divided by n! This result is proved easily using the theorems in Chapter 4. Take the series Eq. (5.2) and divide by $(z - z_0)^m$ and then perform a contour integral around the point $z = z_0$

$$\oint dz \frac{f(z)}{(z - z_0)^m} = \sum_n a_n \oint dz(z - z_0)^{(n-m)} \tag{5.4}$$

The integral on the right is nonzero only for the term with $m = n + 1$, and then it equals $2\pi i$. These steps show that

$$a_n = \oint \frac{dz}{2\pi i} \frac{f(z)}{(z - z_0)^{(n+1)}} = \frac{1}{n!} \frac{d^n}{dz_0^n} f(z_0) \tag{5.5}$$

where the last result was proven in Chapter 4.

There are several comments that need to be made regarding this derivation. First, the function $f(z_0)$ has to be an analytic function in the neighborhood of the point z_0. Each term in the series is an analytic function. However, the summation of terms, which is the function $f(z)$, could be nonanalytic although each term in the series is analytic. Examples of this behavior are given below. Another assumption is that it is acceptable to exchange the order of the two steps, integration and summation. As remarked above, this assumption is not always valid. These exceptional cases are treated below.

Some examples of Taylor series are:

1. $f(z) = e^z$. This function has the property that all derivatives are the same

$$\frac{d^n}{dz^n} e^z = e^z, \qquad a_n = \frac{e^{z_0}}{n!} \tag{5.6}$$

$$e^z = e^{z_0} \sum_{n=0}^{\infty} \frac{(z - z_0)^n}{n!} \tag{5.7}$$

$$e^z = \sum_{n=0}^{\infty} \frac{z^n}{n!} \tag{5.8}$$

The last equation is the way the series is usually presented, where the expansion is about the point $z_0 = 0$. However, the first series shows that the exponent can be expanded around any point.

2. $f(z) = \sin(z)$. Denote the nth derivative as ∂^n. For the sine function one has $\partial f = \cos(z)$, $\partial^2 f = -\sin(z)$, $\partial^3 f = -\cos(z)$, and $\partial^4 f = \sin(z)$. Every fourth derivative brings $\sin(z)$ back to the original function. Setting $z_0 = 0$, $\sin(z_0) = 0$, $\cos(z_0) = 1$ gives the series

$$\sin(z) = \frac{z}{1!} - \frac{z^3}{3!} + \frac{z^5}{5!} - \cdots - \tag{5.9}$$

3. $f(z) = \ln(z)$. Here the derivatives are $\partial f = 1/z$, $\partial^n f = -(-1)^n (n - 1)!/z^n$ that gives the series

$$\ln(z) = \ln(z_0) + \frac{(z - z_0)}{z_0} - \frac{(z - z_0)^2}{2z_0^2} + \frac{(z - z_0)^3}{3z_0^3} + \cdots + \tag{5.10}$$

Note that the denominator contains n rather than $n!$. This series is usually presented for cases that $z_0 = 1$ and $z = 1 + x$ or $z = 1 - x$

$$\ln(1 + x) = x - \frac{x^2}{2} + \frac{x^3}{3} - \frac{x^4}{4} + \cdots + \tag{5.11}$$

$$\ln(1 - x) = -\left[x + \frac{x^2}{2} + \frac{x^3}{3} + \frac{x^4}{4} + \cdots + \right] \tag{5.12}$$

4. $f(z) = 1/z$, $\partial^n f = (-1)^n n!/z^{(n+1)}$ so that the series for $z = 1 \pm x$ are

$$f(1 - x) = \frac{1}{1 - x} = 1 + x + x^2 + x^3 + \cdots + \tag{5.13}$$

$$f(1 + x) = \frac{1}{1 + x} = 1 - x + x^2 - x^3 + \cdots + \tag{5.14}$$

$$\frac{1}{(1 - x)^2} = 1 + 2x + 3x^2 + 4x^3 + \cdots + \tag{5.15}$$

The last series is derived from $f(z) = 1/z^2$. Its derivation is left as an exercise. The function $f(z) = 1/(1 - z)$ is not analytic at the point $z = 1$, although every term in its series expansion is analytic at this point. This case is an example where each term in a series is analytic but the function is not analytic at a point.

5. $f(z) = \tan(z)$. The expansion about the point $z_0 = 0$ is

$$\tan(z) = z + \frac{z^3}{3} + \cdots + \tag{5.16}$$

Since $\tan(z)$ is an odd function of z, only odd powers of z appear in the series. The function $\tan(z)$ has poles at the points where $\cos(z) = 0$ that are at $z = \pi(n + 1/2)$ where n is any integer. The radius of convergence of the above series is for $|z| < \pi/2$. However, the function can be evaluated about other points. The function also vanishes where $\sin(z) = 0$ which is at the points $z = n\pi$. For $-\pi/2 < z - n\pi < \pi/2$ expand

$$\tan(z) = (z - n\pi) + \frac{(z - n\pi)^2}{3} + \cdots + \tag{5.17}$$

A series expansion exists for any value of z for this function.

5.2. Convergence

So far nothing has been said regarding whether the series converge to a sensible answer. Of the above series, some converge absolutely (for any value of z) while others only have regions of convergence.

The function

$$f(z) = \frac{1}{1-z} \tag{5.18}$$

has a simple pole at $z = 1$. Its Taylor series fails to converge at this point. For $z = 1$ the series is

$$f = 1 + z + z^2 + z^3 + \cdots + = 1 + 1 + 1 + \cdots + \tag{5.19}$$

that gives infinity after enough terms! However, the series fails to converge for any real value of z larger than one. For example, take $z = 2$ and then

$$f(2) = \frac{1}{1-2} = -1 \neq 1 + 2 + 2^2 + 2^3 + 2^4 + \cdots + \tag{5.20}$$

Minus one cannot equal a sum of positive numbers. Similar problems happen along the negative axis. Set $z = -2$ then get

$$f(-2) = \frac{1}{1-(-2)} = \frac{1}{3} \neq 1 - 2 + 4 - 8 + 16 \ldots \tag{5.21}$$

A careful investigation of the function $f = 1/(1-z)$ finds that the Taylor series around the point $z_0 = 0$ fails to converge for any value of $|z| \geq 1$. The domain of convergence of this function is a circle of radius $|z| < 1$ about the origin as shown in Fig. 5.1. The radius of convergence is determined by the pole at $z = 1$.

The radius of convergence is determined by the terms with large values of n. Suppose at large n the constant $a_n \rightarrow \mu^n$ and the terms in the series become $(\mu z)^n$. In this case the radius of convergence is $|z| < 1/|\mu|$. If for some large value of $n > N$, a_n can be represented by μ^n, then write

$$f(z) = \sum_{n=0}^{N-1} a_n z^n + \sum_{n=N}^{\infty} (\mu z)^n \tag{5.22}$$

$$= \sum_{n=0}^{N-1} a_n z^n + \frac{(\mu z)^N}{1 - \mu z} \tag{5.23}$$

The only singular part is the last term. Its radius of convergence ($|z| < 1/\mu$) is determined by the parameter μ. It governs the behavior of the entire series. These remarks can be summarized by the theorem.

Define the parameter μ as

$$\mu = \lim_{n \rightarrow \infty} |a_n|^{1/n} \tag{5.24}$$

and the convergence of the series $f(z) = \Sigma a_n z^n$ is determined by three possible limits:

- $\mu = 0$. In this case the radius of convergence is infinity, meaning that the series converges for all possible values of z.

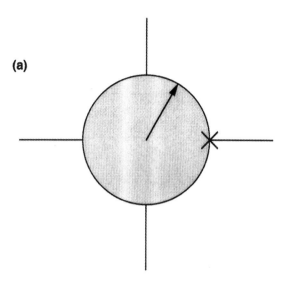

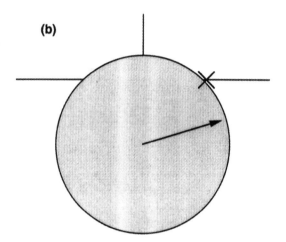

Figure 5.1 Radius of convergence of the function $f(z) = 1/(1 - z)$. (a) For expansion about $z_0 = 0$ and (b) for expansion about $z_0 = -i$.

- μ = constant that is not zero. In this case the series has a finite radius of convergence $1/|\mu|$.
- $\mu \to \infty$. In this case the series never converges for any nonzero value of z.

The series

$$f(z) = \sum_n a_n (z - z_0)^n \tag{5.25}$$

has a radius of convergence around the point z_0, where $|z - z_0| < 1/|\mu|$.

If the function $f(z)$ has a pole or branch point the radius of convergence must avoid these singular points. The function $f = 1/(1 - z)$ has a radius of convergence of unity in order to avoid the pole at $z = 1$. This phenomena is general: The radius of convergence avoids poles and branch cuts.

Examples of finding the parameter μ are presented for various series.

- The series

$$-\ln(1 - z) = z + \frac{z^2}{2} + \frac{z^3}{3} + \cdots + = \sum_{n=1}^{\infty} \frac{z^n}{n} \tag{5.26}$$

In this case $a_n = 1/n$ and

$$\mu = \lim_{n \to \infty} \left(\frac{1}{n}\right)^{1/n} = \lim_{n \to \infty} \exp[(1/n)\ln(1/n)] = e^0 = 1 \tag{5.27}$$

$$\mu = 1 \tag{5.28}$$

As $\varepsilon \to 0$ the function $\varepsilon \ln(\varepsilon) \to 0$. The exponent has this form by setting $\varepsilon = 1/n$. The radius of convergence is the unit circle. This result is obvious since the branch point is at $z = 1$.

- The series for $\exp(z)$ has $a_n = 1/n!$. At large values of n use Stirling's approximation for the factorial

$$\mu = \lim_{n \to \infty} \left(\frac{1}{n!}\right)^{1/n}$$

$$= \lim_{n \to \infty} \exp[-(1/n)\ln(n!)] \tag{5.29}$$

$$\exp[-(1/n)\ln n!] = \exp\left\{-\frac{1}{n}\left[\frac{1}{2}\ln(2\pi n) - n + n\ln(n)\right]\right\} \to \frac{e}{n} \to 0 \tag{5.30}$$

$$\mu = 0 \tag{5.31}$$

Here the series converges for all values of z.

- Consider the series

$$f(z) = \sum_n n! z^n \tag{5.32}$$

In this case $a_n = n!$ that is the inverse of the prior case. Doing exactly the same steps show that $\mu \to \infty$ as n increases. This case is an example of a series whose radius of convergence is zero. There is no value of z, aside from $z = 0$, for which the series gives a finite value.

- Consider the series:

$$f(z) = 1 - 2z^2 + 3z^4 - 4z^6 + \cdots + \tag{5.33}$$

that has $a_{2n} = (-1)^n(n + 1)$, $a_{2n+1} = 0$. It is an example where the negative signs are ignored and the absolute magnitude of a_{2n} is used to find μ

$$\mu = \lim_{n \to \infty} (n + 1)^{1/2n}$$

$$= \lim_{n \to \infty} \exp\left[\frac{1}{2n} \ln(n + 1)\right] = 1 \qquad (5.34)$$

The exponent vanishes at large n so that $\mu = 1$. The radius of convergence is the unit circle. In fact, the function is actually $f(z) = 1/(1 + z^2)^2$. It has two poles, at $z = \pm i$. The radius of convergence must avoid them and is unity.

- The function $f(z) = 1/(1 - z)$ can be expanded about points other than zero. Take $z_0 = -i$. Then the expansion parameter is $(z + i)^n$. The series is

$$f(z) = \frac{1}{1 + i - (z + i)} = \frac{1}{1 + i}\frac{1}{1 - (z + i)/(1 + i)}$$

$$= \frac{1}{1 + i}\left[1 + \left(\frac{z + i}{1 + i}\right) + \left(\frac{z + i}{1 + i}\right)^2 + \left(\frac{z + i}{1 + i}\right)^3 + \cdots + \right] \qquad (5.35)$$

The radius of convergence is determined by $|1 + i| = \sqrt{2} = 1/\mu$. This radius is exactly the distance from the center of expansion point $z_0 = -i$ to the singular point $z = 1$, as shown in Fig. 5.1(b).

- What is the radius of convergence of the derivative of a function?

$$\frac{df}{dz} = \sum_n n a_n z^{n-1} = \sum_n (n + 1)a_{n+1}z^n \qquad (5.36)$$

$$\mu = \lim_{n \to \infty} [(n + 1)a_{n+1}]^{1/n} = \lim_{n \to \infty} (a_n)^{1/n} \exp\left[\frac{1}{n}\ln(n + 1)\right]$$

$$= \lim_{n \to \infty} (a_n)^{1/n} \qquad (5.37)$$

The radius of convergence of the derivative f' is identical to the radius of the function f.

Listed below are several theorems.

1. *A series represents an analytic function within its radius of convergence.*

    This theorem restates a fact mentioned earlier that there can be no poles or branch points within the radius of convergence. To prove this theorem, assume that the function can be represented by its Taylor series, that is, rigorously valid within the radius of convergence. Each term in the Taylor series is an analytic function. Note that z^n obeys the Cauchy–Riemann conditions since it is a function only of z. The sum of such terms, that is, the function $f(z)$, is therefore analytic.

2. *Cauchy's inequality* states that if M is the upperbound of $|f(z)|$ on the circle of radius $|z| = R$ then $|a_n R^n| < M$. The radius R must be within the radius of convergence.

The proof is to start with the integral definition of a_n and to evaluate the contour using the bound on $f(z)$. One takes the absolute magnitude of every factor in the integrand

$$a_n = \frac{1}{2\pi i} \oint_{|z|=R} dz \frac{f(z)}{z^{n+1}} \leqslant \frac{M}{2\pi i R^n} i \int_0^{2\pi} d\theta |e^{-in\theta}| \qquad (5.38)$$

$$|a_n| \leqslant \frac{M}{R^n} \qquad (5.39)$$

that proves the inequality.

3. *Liouville theorem.* A function $f(z)$ that is bounded for all values of z is a constant.

This theorem is really a lemma on Cauchy's inequality. If $f(z)$ is bounded by M for all z then extend the radius R in the above proof to $R \to \infty$ that gives

$$|a_n| \leqslant \lim_{R \to \infty} \frac{M}{R^n} = 0 \qquad (5.40)$$

Therefore, $a_n = 0$ for all values of $n > 0$. The only nonzero value of a_n is a_0 that gives $f = a_0$ as a constant.

A consequence of Liouville's theorem is that every function $f(z)$ diverges for some value of z. For example, $\sin(z) = \sin(x + iy)$ does not diverge as $x \to \infty$ but does diverge as $y \to \infty$.

5.3. Laurent Series

Taylor series always have terms $(z - z_0)^n$ with exponents n that are positive. Laurent series can have any power of the exponent, including negative values. Some examples follow.

1. Consider the familiar function $f(z) = 1/(1 - z)$. For $|z| < 1$ this function could be expanded in a Taylor series in powers of z^n. For the other case of $|z| > 1$, another series can be derived

$$f(z) = -\frac{1}{z}\frac{1}{1 - 1/z} = -\frac{1}{z}\left[1 + \frac{1}{z} + \frac{1}{z^2} + \frac{1}{z^3} + \cdots + \right] \qquad (5.41)$$

Now the expansion is a Taylor series in inverse powers of z. It is an example of a Laurent series. Actually, the function $f(z)$ is a one term Laurent series itself, with $z_0 = 1$ and $n = -1$.

2. The following function has a well-behaved series expansion except at the point $z = 0$

$$e^{1/z} = \sum_{n=0}^{\infty} \frac{1}{z^n n!} \qquad (5.42)$$

3. The function $f(z) = \ln(1-z)$ has a Taylor series for $|z| < 1$ (see above). The other case is $|z| > 1$. This function has a branch point at $z = 1$. The Taylor series for small z did not have to deal with this branch point. In the case $z > 1$ the branch cut has to be specified. Write the function as

$$\ln(1-z) = \ln\left[(-z)\left(1 - \frac{1}{z}\right)\right] = \ln(-z) - \left[\frac{1}{z} + \frac{1}{2z^2} + \frac{1}{3z^3} + \cdots + \right] \tag{5.43}$$

The function $\ln(-z)$ now has a branch point at the origin. The series converges for $|z| > 1$. The Laurent series has one term that is a logarithm. This function is an acceptable term in a series. The exponent $n = 0$ can be either a constant or a logarithm function.

4. The function $f(z) = \sqrt{z^2 - 1}$ has a branch point at $z = \pm 1$. However, for value of $|z| > 1$ it has a Laurent series of the form

$$\sqrt{z^2 - 1} = z\sqrt{1 - \frac{1}{z^2}} = z\left[1 - \frac{1}{2z^2} + \frac{1}{8z^4} + \cdots + \right] \tag{5.44}$$

The branch points do not appear in the series expression. They disappear because the expansion is in a region that does not contain the branch points.

5. The final example is

$$f(z) = \frac{1}{(z-2)(z-i)} \tag{5.45}$$

This function has several regions of convergence. There is a different Laurent series for each region. There are two poles, at $z = i$ and $z = 2$. One region is inside both poles, $|z| < 1$. The second region is an annulus between the two poles $1 < |z| < 2$ and the third region is beyond both poles $|z| > 2$.

- For $|z| < 1$ expand both denominators in a Taylor series and then multiply them together

$$f = \frac{1}{(2i)(1 - z/2)(1 + iz)}$$

$$= \frac{1}{2i}\left[1 + \frac{z}{2} + \frac{z^2}{2^2} + \cdots + \right][1 - iz + (iz)^2 + \cdots +]$$

$$= \frac{1}{2i}\left[1 + z\left(\frac{1}{2} - i\right) + z^2\left(\frac{1}{4} - \frac{i}{2} + i^2\right) + \cdots + \right] \tag{5.46}$$

• The next easiest region is outside of both poles, that has $|z| > 2$ where the expansion is done in inverse powers of z

$$f(z) = \frac{1}{z^2(1 - 2/z)(1 - i/z)}$$

$$= \frac{1}{z^2}\left[1 + \frac{2}{z} + \left(\frac{2}{z}\right)^2 + \cdots + \right]\left[1 + \frac{i}{z} + \frac{i^2}{z^2} + \cdots + \right]$$

$$= \frac{1}{z^2}\left[1 + \frac{1}{z}(2 + i) + \frac{1}{z^2}(2^2 + i^2 + 2i) + \cdots + \right] \qquad (5.47)$$

• The annular region $1 < |z| < 2$ has a Laurent series with both positive and negative values of exponent n.

$$f = \frac{1}{2 - i}\left(\frac{1}{z - 2} - \frac{1}{z - i}\right)$$

$$= -\frac{1}{2 - i}\left(\frac{1}{2(1 - z/2)} + \frac{1}{z(1 - i/z)}\right)$$

$$= -\frac{1}{2 - i}\left(\frac{1}{2}[1 + z/2 + (z/2)^2 + \cdots +] + \frac{1}{z}[1 + i/z + (i/z)^2 + \cdots +]\right)$$

$$(5.48)$$

These examples show the different kinds of Laurent series. For any function $f(z)$, first find the poles and branch points. The series expansion must avoid the region of a pole or branch point. One series is valid inside a circle that bisects such a point, while another series is valid outside such points.

5.4. Meromorphic Functions

Meromorphic functions were defined in Chapter 4. There is also a theorem that states that any meromorphic function can be expanded as a power series of the form $R_j/(z - z_j)$, where the simple poles are at z_j and the residues are R_j. This theorem can also be used to generate some interesting infinite series. The trick is to find a function with an infinite number of poles. A meromorphic function is defined as one with a finite number of poles. However, the theorem for expressing the function as a series is valid even for functions with an infinite number of simple poles.

As an example, consider the function $f(z) = 1/\cos(z)$. It has an infinite number of poles. The cosine function vanishes on the real axis at the points $z_n = \pi(n + 1/2)$ with a residue equal to $-(-1)^n$. The integer n can have any value, including negative numbers, zero, and positive numbers. Obviously $f(0) = 1$. Using the form of the series that includes the constant term $f(0)$ gives the series

$$\frac{1}{\cos(z)} = 1 - \sum_n(-1)^n\left[\frac{1}{z - \pi(n + 1/2)} + \frac{1}{\pi(n + 1/2)}\right]$$

$$= -\sum_n\frac{(-1)^n}{z - \pi(n + 1/2)} \qquad (5.49)$$

$$\frac{\pi}{4} = \sum_{n=0}^{\infty} \frac{(-1)^n}{2n+1} = 1 - \frac{1}{3} + \frac{1}{5} - \cdots - \qquad (5.50)$$

To prove the constant terms cancel, start with Eq. (5.49) and let $z = iy$, $\cos(z) = \cosh(y)$, and $y \to \infty$. The terms in Eq. (5.49) which depend on y, vanish, leaving only the constant terms to must cancel.

Another example is $f(z) = 1/\sin(z)$. It is not a meromorphic function since it has an infinite number of poles. Nevertheless, the same function can be used to expand this function. Eliminate the simple pole at the origin by, instead, defining the function as

$$f(z) = \frac{1}{\sin(z)} - \frac{1}{z} = \sum_{l \neq 0} (-1)^l \left[\frac{1}{z - l\pi} + \frac{1}{l\pi} \right] = \sum_{l \neq 0} \frac{(-1)^l}{z - l\pi}$$

$$\frac{1}{\sin(z)} = \frac{1}{z} + \sum_{l \neq 0} \frac{(-1)^l}{z - l\pi} = \sum_{l} \frac{(-1)^l}{z - l\pi} \qquad (5.52)$$

The constant term in the summation $[(-1)^l/(l\pi)]$ gives zero when we take positive and negative values of l. The end result is an expression for the inverse sine function as an infinite series over its poles and its residues at these poles.

5.5. Asymptotic Series

Asymptotic series are those that appear to converge but do not. If we take a few terms in the series, the correct answer is obtained to several significant figures. If more terms in the series are evaluated, the answer diverges. This strange behavior is best illustrated by a simple example. Consider the evaluation of the simple integral

$$I(x) = \int_0^{\infty} \frac{dt}{1 + t^2} e^{-xt} \qquad (5.53)$$

It cannot be evaluated analytically. At large values of the parameter x we might consider using a series expansion. Change the integration variable to $y = xt$ and find

$$I(x) = \frac{1}{x} \int_0^{\infty} \frac{dy}{1 + y^2/x^2} e^{-y} \qquad (5.54)$$

Expand the denominator in a series and evaluate the integral term by term

$$I(x) = \frac{1}{x} \int_0^{\infty} dy\, e^{-y} \left[1 - \frac{y^2}{x^2} + \frac{y^4}{x^4} - \cdots - \right]$$

$$= \frac{1}{x} \left[1 - \frac{2!}{x^2} + \frac{4!}{x^4} - \frac{6!}{x^6} + \cdots + \right] \qquad (5.55)$$

Table 5.1 Asymptotic Series for Integral in Eq. (5.55)

x	$I(x)$	1	$-2!/x^2$	$+4!/x^4$	$-6!/x^6$	$+8!/x^8$
5	0.18814	0.20000	0.18400	0.19168	0.18246	0.20311
10	0.09819	0.10000	0.09800	0.09824	0.09817	0.09821
15	0.06610	0.06667	0.06607	0.06611	0.06610	0.06610
20	0.04976	0.05000	0.04975	0.04976	0.04976	0.04976

The exact result is the second column. The columns to the right are the results obtained by adding successive terms to the series.

The series $n!z^n$ diverges for all values of z, so the above series does not converge. However, it is an asymptotic series. After a few terms it gives an answer that is numerically accurate. The accuracy increases with larger values of x. This behavior is shown in Table 5.1. The first row shows the value for $x = 5$, that is not a large number. The exact numerical result is 0.18814. The asymptotic series has nearly this value after three or four terms. Adding one more term to the series $+8!/x^8$ makes the series diverge away from the correct answer. This behavior is more apparent for $x = 10$ where the result after four terms (0.09817) is quite close to the exact result of 0.09819. The next term is also causing the series to move away from the exact result. For $x = 15, 20$ the series homes in on the exact result after a few terms and then stays at those values. If we take more terms in the series, eventually it will diverge in these cases also.

Evaluating an asymptotic series is knowing when to quit. It gives a pretty good answer after a few terms. Keeping too many terms causes the series to diverge. A rule of thumb is that the term in the series with the minimum absolute value is the last one to keep.

Asymptotic series are often used in the evaluation of functions at large values of their argument. Stirling's approximation to the factorial function was derived in the previous chapter. Usually this result is expressed as an asymptotic series

$$\lim_{z \gg 1} \Gamma(z) = \sqrt{\frac{2\pi}{z}} z^z e^{-z} \left[1 + \frac{1}{12z} + \frac{1}{288z^2} - \frac{139}{5140z^3} - \cdots - \right] \qquad (5.56)$$

The part in brackets is an asymptotic series. Using the first couple of terms in this series increases the accuracy of Stirling's approximation. Rather than derive this result, instead the similar series is derived for the factorial function. The basic steps are similar to the evaluation of the steepest descent integration which went according to

$$n! = \int_0^\infty dt\, e^{f_n(t)} \qquad (5.57)$$

$$f_n(t) = -t + n \ln(t) \qquad (5.58)$$

Recall that the function $f_n(t)$ had its minimum value at $f' = 0$ which is the point $t_0 = n$. The function $f_n(t)$ is expanded in a Taylor series about the point $t_0 = n$

$$f_n(t) = f_n(n) + a_1(t - n) + a_2(t - n)^2 + a_3(t - n)^3 + \cdots + \qquad (5.59)$$

$$a_l = \frac{1}{l!} \frac{d^l f}{dt^l} \qquad (5.60)$$

$$f_n(t) = [-n + n \ln(n)] - \frac{(t - n)^2}{2n} + \frac{(t - n)^3}{3n^2} - \frac{(t - n)^4}{4n^3} + \cdots + \qquad (5.61)$$

$$n! = n^n e^{-n} \int_0^\infty dt \exp\left[-\frac{(t - n)^2}{2n} + \frac{(t - n)^3}{3n^2} - \frac{(t - n)^4}{4n^3} + \cdots + \right] \qquad (5.62)$$

Change variables of integration to $s = (t - n)/\sqrt{2n}$.

$$n! = \sqrt{2n}\, n^n e^{-n} \int_{-\sqrt{n/2}}^\infty ds \exp\left[-s^2 + \frac{s^3}{3}\sqrt{\frac{8}{n}} - \frac{s^4}{n} + \cdots + \right] \qquad (5.63)$$

Keep the Gaussian factor of $-s^2$ in the exponent. The remaining terms are expanded in a Taylor series using $\exp(f) = 1 + f + f^2/2! + \cdots +$

$$n! \approx \sqrt{2n}\, n^n e^{-n} \int_{-\sqrt{n/2}}^\infty ds\, e^{-s^2}\left[1 + \left(\frac{s^3}{3}\sqrt{\frac{8}{n}} - \frac{s^4}{n} \right) + \frac{1}{2!}\left(\frac{s^3}{3}\sqrt{\frac{8}{n}} - \frac{s^4}{n} \right)^2 + \cdots + \right] \qquad (5.64)$$

For n large the lower limit of integration can be extended to negative infinity. This step introduces an error of $O(e^{-n/2})$ that is negligible. The term in the brackets proportional to s^3 integrates to zero and is an odd function of s. The term $O(1/n)$ comes from two terms, $O(s^4/n)$ and the square of $O(s^3)$

$$\int_{-\infty}^\infty ds\, s^n e^{-s^2} = \Gamma\left(\frac{n + 1}{2} \right) \qquad (5.65)$$

$$n! = \sqrt{2\pi n}\, n^n e^{-n}\left[1 - \frac{3}{4n} + \frac{1}{2}\frac{8}{9n}\frac{15}{8} - \cdots - \right]$$

$$= \sqrt{2\pi n}\, n^n e^{-n}\left[1 + \frac{1}{12n} + \cdots + \right] \qquad (5.66)$$

The series is only asymptotic since several steps in the derivation are of dubious rigor. The Taylor series expansion for $f_n(t)$ as a power series in $(t - n)^l$ has a finite radius of convergence. The integral between $(-\infty, \infty)$ cannot be correctly evaluated using this series. The series obviously does not converge for a value, say $t = 3n$. In spite of these dubious steps in the derivation, the resulting asymptotic series gives an accurate numerical result using a few terms in the series.

Another way to derive an asymptotic series is by successive integration by parts. For example, consider the evaluation of the integral

$$I(x) = \int_0^\infty dt \frac{\sin(t)}{t + x} = \frac{1}{x}\left[1 - \frac{2!}{x^2} + \frac{4!}{x^4} - \frac{6!}{x^6} + \cdots + \right] \tag{5.67}$$

The asymptotic series has been given on the right. Notice that it is identical to Eq. (5.55). The two integrals are equal to each other. The series is derived by repeated integration by parts. The first is

$$\int u \, dv = uv - \int v \, du \tag{5.68}$$

$$u = \frac{1}{x + t}, \qquad du = -\frac{dt}{(x + t)^2} \tag{5.69}$$

$$dv = dt \sin(t), \qquad v = -\cos(t) \tag{5.70}$$

$$I(x) = -\left.\frac{\cos(t)}{(x + t)}\right|_0^\infty - \int_0^\infty dt \frac{\cos(t)}{(x + t)^2} \tag{5.71}$$

$$= \frac{1}{x} - \int_0^\infty dt \frac{\cos(t)}{(x + t)^2} \tag{5.72}$$

Do another integration by parts on the integral with $u = 1/(x + t)^2$, $v = \sin(t)$. In this case the constant of integration is zero:

$$I(x) = \frac{1}{x} - 2! \int_0^\infty dt \frac{\sin(t)}{(x + t)^3} \tag{5.73}$$

Another integration by parts with $u = 1/(x + t)^3$, $v = -\cos(t)$ gives a constant term

$$I(x) = \frac{1}{x}\left[1 - \frac{2!}{x^2}\right] - 3! \int_0^\infty dt \frac{\cos(t)}{(x + t)^4} \tag{5.74}$$

Repeating this process many times produces the series shown in Eq. (5.67).

Asymptotic series are very useful for the accurate evaluation of functions at large values of their arguments. These series become more accurate as the constant x increases in value.

5.6. Summing Series

Until now this chapter has been devoted to expanding functions in various types of series. This section treats the reverse problem—how to sum a series. Some type of series can be summed analytically into simple expressions. The process is the reverse of the series generated by meromorphic functions. The only type of series discussed here are infinite series. The meromorphic functions with an infinite number of terms in the series are function such as $1/\cos(z)$ or $1/\sin(z)$. Finding an analytic expression for an

Table 5.2 Functions that Generate Infinite Series

Function	z_n	R_n
$\dfrac{\pi}{\sin(\pi z)}$	n	$(-1)^n$
$\pi \cot(\pi z)$	n	1
$\dfrac{(\pi/2)}{\cos(\pi z/2)}$	$2n+1$	$(-1)^n$
$(\pi/2)\tan(\pi z/2)$	$(2n+1)$	1
$\dfrac{1}{e^z - 1}$	$2\pi i n$	1
$\dfrac{1}{e^z + 1}$	$i\pi(2n+1)$	-1

infinite series is achievable only if it can be identified with such a function. Table 5.2 gives the different functions with an infinite number of poles. The residues at these poles are also listed, which is important information. The various factors of π and $\pi/2$ are included to give poles at either all integers, odd integers, or complex integers. Note that terms such as $1/\sin(\pi z)$ have residues that can alternate sign. However, multiplying it by $\cos(\pi z)$ makes $\cot(\pi z)$ and gives residue that are all the same sign. This trick is useful since the residues could either be all of the same sign, or alternate in sign, and one should use the appropriate function for each case.

Next, several examples are presented and all of them involve constructing a complex function that is evaluated by taking a circular contour at infinity.

1. Consider the series

$$S_1(a) = \sum_{n=-\infty}^{\infty} \frac{1}{n^2 + a^2} \tag{5.75}$$

The denominator of this series can be written as

$$(a^2 + n^2) = (a - in)(a + in) \tag{5.76}$$

It has poles at the imaginary integers. The function that generates poles at these points (Table 5.2), is $1/(e^z - 1)$. The residues at these poles all have the same sign, that agree with the series: It has no sign alternation. Construct the contour integral

$$0 = \oint \frac{dz}{2\pi i} \frac{1}{e^z - 1} \frac{1}{z^2 - b^2} \tag{5.77}$$

The value for b is derived below. The integral is zero when the contour is a circle of radius R in the limit that $R \to \infty$. The integral also equals the summation of the residues of the poles of the integrand. The function $1/(e^z - 1)$ has poles at $z_n = 2\pi in$. The other function has poles at $z = \pm b$. The residues give

$$0 = -\sum_{n=-\infty}^{\infty} \frac{1}{(2\pi n)^2 + b^2} + \frac{1}{2b}\left[\frac{1}{e^b - 1} - \frac{1}{e^{-b} - 1}\right] \qquad (5.78)$$

The bracket is equal to $\coth(b/2)$. Divide the denominator of the first term by $(2\pi)^2$, which shows that $b = 2\pi a$. Rearranging the above formula gives

$$S_1(a) = \frac{\pi}{a}\coth(\pi a) \qquad (5.79)$$

The series $S(a)$ is generated by the function $\coth(\pi a)$ times the factor of π/a. The key step in the derivation is to construct a contour integral. Its integrand has two factors, one is the function that is being summed, the other is a function from Table 5.2 that generates poles with the correct residues.

2. Another example is

$$S_2(a) = \sum_{n=-\infty}^{\infty} \frac{1}{(n + 1/2)^2 + a^2} \qquad (5.80)$$

Here the poles occur at $a = \pm i(n + 1/2) = i(2n + 1)/2$. One must use the function in Table 5.2 that generates poles at odd integers along the imaginary axis, that is $1/(e^z + 1)$. Therefore, consider the contour integral at infinity

$$0 = \oint \frac{dz}{2\pi i} \frac{1}{e^z + 1} \frac{1}{z^2 - b^2} \qquad (5.81)$$

$$0 = \sum_n \frac{1}{\pi^2(2n + 1)^2 + b^2} + \frac{1}{2b}\left[\frac{1}{e^b + 1} - \frac{1}{e^{-b} + 1}\right] \qquad (5.82)$$

$$S_2(a) = \frac{\pi}{a}\tanh(\pi a) \qquad (5.83)$$

where $b = 2\pi a$. The series is converted into S_2 by dividing the denominator by $(2\pi)^2$.

3. Another series is

$$S_3(a) = \sum_{n-\infty}^{\infty} \frac{(-1)^n}{(n^2 + a^2)^2} \qquad (5.84)$$

Now the terms alternate in sign. Use as the pole generator $\pi/\sin(\pi z)$ that has poles at the points n with a residue that goes as $(-1)^n$. The denominator of the sum can be factored as $(n + ia)^2(n - ia)^2$ that shows it has double poles. The residues there are evaluated using the first derivative

$$0 = \oint \frac{dz}{2\pi i} \frac{\pi}{\sin(\pi z)} \frac{1}{(z^2 + a^2)^2}$$

$$= \sum_n \frac{(-1)^n}{(n^2 + a^2)^2} + \pi \left[\frac{\partial}{\partial z} \frac{\csc(\pi z)}{(z - ia)^2} \right]_{z = -ia} + \pi \left[\frac{\partial}{\partial z} \frac{\csc(\pi z)}{(z + ia)^2} \right]_{z = ia} \tag{5.85}$$

$$S_3(a) = \frac{\pi}{2a^3 \sinh(\pi a)} [1 + \pi a \coth(\pi a)] \tag{5.86}$$

Other examples are given in the problems.

5.7. Padé Approximants

All the series under discussion so far had coefficients a_n of x^n that are known for all n. In many cases we can sum the series analytically and deduce the original generator function. However, there are many examples where we know only the first few terms of an infinite series. Given that one has a few terms, and that the general form of a_n is not transparent, Padé approximants are a good way to approximate the series. For example, if asked to sum the series

$$f(x) = 1 + \frac{x}{2} - \frac{5}{8}x^2 + \frac{13}{16}x^3 - \frac{141}{128}x^4 + \cdots + \tag{5.87}$$

the constant a_n is not obvious. The generator for this series is given below.

The general procedure for using Padé approximants[1,2] is the following. Given a polynomial of degree N of the form

$$f_N(x) = \sum_{j=0}^{N} a_j x^j \tag{5.88}$$

the Padé approximants is to represent this series by the ratio of two polynomials whose degrees add to N

$$f_N(x) = \frac{P_L(x)}{Q_M(x)} \equiv f_N^{(L,M)}(x), \qquad N = L + M \tag{5.89}$$

$$P_L(x) = \sum_{l=0}^{L} p_l x^l \tag{5.90}$$

$$Q_M(x) = \sum_{i=0}^{M} q_i x^i \tag{5.91}$$

There are $N + 1$ pieces of information (a_j) and $(L + 1) + (M + 1) = N + 2$ unknowns (p_l, q_i). If P_L and Q_M are multiplied by the same constant the ratio is unchanged. It is conventional to choose the constant such that $q_0 = 1 = Q_M(0)$. Then only $N + 1$ unknown coefficients are found from $N + 1$ values of

a_j. The choice of $L + M = N$ uses all the information available. If $L + M > N$ there are not enough values of a_j to find the unknown values of (p_l, q_i). If $L + M < N$ one is not using all of the available information to construct the Padé approximants.

If the Padé approximants is expanded in a Taylor series the first N terms are identical to the original series. The easy way to find the coefficients (q_i, p_l) is to multiply Eq. (5.89) by Q_M and find

$$Q_M(x) f_N(x) = P_L(x) \tag{5.92}$$

$$[1 + q_1 x + q_2 x^2 + \cdots +][a_0 + a_1 x + a_2 x^2 + \cdots +] = p_0 + p_1 x + p_2 x^2 \tag{5.93}$$

Multiply together the two series on the left and then equate the terms on both sides of the equal sign with the same power of x^n, starting with x^0. The first three equations are

$$a_0 = p_0$$

$$a_1 + a_0 q_1 = p_1 \tag{5.94}$$

$$a_2 + a_1 q_1 + a_0 q_2 = p_2$$

Keep in mind that the coefficients a_j are known for $0 \leqslant j \leqslant N$. Assume that $a_j = 0$ for $j > N$, $q_i = 0$ for $i > M$, and $p_l = 0$ for $l > L$. The number of unknown values of (q_i, p_l) are $M + L + 1 = N + 1$. That is the number of equations that must be kept in the above series. The last one is the coefficient of x^{L+M} where $N = L + M$

$$a_N + a_{N-1} q_1 + \cdots + a_L q_M = 0 \tag{5.95}$$

The set of $N + 1$ equations are solved for the unknown coefficients.

As an example construct the Padé approximants for Eq. (5.87). Here $N = 4$ and choose $L = M = 2$. The coefficients a_j are $a_0 = 1$, $a_1 = 1/2$, $a_2 = -5/8$, $a_3 = 13/16$, $a_4 = -141/128$.

$$a_0 = p_0 \Rightarrow p_0 = 1 \tag{5.96}$$

$$a_1 + a_0 q_1 = p_1 \Rightarrow \frac{1}{2} + q_1 = p_1 \tag{5.97}$$

$$a_2 + a_1 q_1 + a_0 q_2 = p_2 \Rightarrow -\frac{5}{8} + \frac{q_1}{2} + q_2 = p_2 \tag{5.98}$$

$$a_3 + a_2 q_1 + a_1 q_2 = 0 \Rightarrow \frac{13}{16} - \frac{5}{8} q_1 + \frac{q_2}{2} = 0 \tag{5.99}$$

$$a_4 + a_3 q_1 + a_2 q_2 = 0 \Rightarrow -\frac{141}{128} + \frac{13}{16} q_1 - \frac{5}{8} q_2 = 0 \tag{5.100}$$

The last two equations are a pair of linear equations for the two unknowns (q_1, q_2). Solving them gives

$$\begin{pmatrix} a_2 & a_1 \\ a_3 & a_2 \end{pmatrix} \begin{pmatrix} q_1 \\ q_2 \end{pmatrix} = -\begin{pmatrix} a_3 \\ a_4 \end{pmatrix} \tag{5.101}$$

$$q_1 = \frac{11}{4}, \quad q_2 = \frac{29}{16} \tag{5.102}$$

The first three equations are evaluated to give (p_0, p_1, p_2)

$$\begin{pmatrix} p_0 \\ p_1 \\ p_2 \end{pmatrix} = \begin{pmatrix} a_0 & 0 & 0 \\ a_1 & a_0 & 0 \\ a_2 & a_1 & a_0 \end{pmatrix} \begin{pmatrix} 1 \\ q_1 \\ q_2 \end{pmatrix} \tag{5.103}$$

$$p_1 = \frac{13}{4}, \quad p_2 = \frac{41}{16} \tag{5.104}$$

$$f^{(2,2)}(x) = \frac{1 + \frac{13}{4}x + \frac{41}{16}x^2}{1 + \frac{11}{4}x + \frac{29}{16}x^2} \tag{5.105}$$

The choice $L = 2$, $M = 2$ is not unique. We may try other choices such as $(L = 1, M = 3)$, or $(L = 3, M = 1)$. The most accurate choices generally have $L \sim M$. These latter two choices are

$$f^{(1,3)} = \frac{1 + \frac{363}{200}x}{1 + \frac{263}{200}x - \frac{13}{400}x^2 + \frac{41}{1600}x^3} \tag{5.106}$$

$$f^{(3,1)} = \frac{1 + \frac{193}{104}x + \frac{11}{208}x^2 - \frac{29}{832}x^3}{1 + \frac{141}{104}x} \tag{5.107}$$

The next step is to compare these Padé approximants with the original series Eq. (5.87) and with the function that generated the series

$$f(x) = \sqrt{\frac{1 + 2x}{1 + x}} \tag{5.108}$$

These results are shown in Table 5.3. The first column is x, the second column is the actual function $f(x)$, the third column is the fourth-order Taylor series in Eq. (5.87), and the last three columns are Padé approximants. The most obvious result is that series is a poor approximation for $x > 0.5$. Note that the function has branch points at $x = -0.5$, -1.0. The series has a radius of convergence of $|x| \leqslant 0.5$. Naturally it gives a lousy numerical value for larger values of x. The three Padé approximants' are much better at approximating the function for larger values of x. In fact, the function $f^{(2,2)}(x)$ gives an answer with high accuracy at all positive values of x. Even as $x \to \infty$ one has

$$\lim_{x \to \infty} \begin{cases} f(x) \\ f^{(2,2)}(x) \end{cases} = \begin{matrix} \sqrt{2} = 1.4142 \\ 41/29 = 1.4138 \end{matrix} \tag{5.109}$$

It is interesting that the Padé approximants $(2, 2)$ gives a much more accurate representation to the function $f(x)$ than does the fourth-order polynomial from

Table 5.3 Comparison of Padé Approximants with the Series f_4 and the Original Function f

x	$f(x)$	$f_4(x)$	$f^{(1,3)}$	$f^{(2,2)}$	$f^{(3,1)}$
0.0	1.0000	1.0000	1.0000	1.0000	1.0000
0.5	1.1547	1.1265	1.1543	1.1547	1.1543
1.0	1.2247	0.5859	1.2196	1.2247	1.2199
2.0	1.2910	−11.625	1.2497	1.2909	1.2513
5.0	1.3540	−599.0	1.0110	1.3538	0.9312

which it was derived. Padé approximants are able to overcome the finite radius of convergence of the series. That Padé approximants are better than the original series is a common feature. It is based on the assumption that the original function $f(x)$ was smoothly varying.

A further comment is useful regarding Eqs. (5.101) and (5.103). In the series of equations indicated in Eq. (5.94) there are $L + 1$ nonzero values of p_l on the right. The last M equations have zero on the right since it is assumed that $p_l = 0$ for $l > L$. Since there are M unknown values of q_i, and since the last M equations have zero on the right, they are a set of linear equations for the unknown (q_i)

$$\begin{pmatrix} a_L & a_{L-1} & \cdots & a_{L-M+1} \\ s_{L+1} & a_L & \cdots & a_{L-M+2} \\ \vdots & \vdots & \ddots & \vdots \\ a_N & a_{N-1} & \cdots & a_L \end{pmatrix} \begin{pmatrix} q_1 \\ q_2 \\ \vdots \\ q_M \end{pmatrix} = - \begin{pmatrix} a_{L+1} \\ a_{L+2} \\ \vdots \\ a_N \end{pmatrix} \tag{5.110}$$

where $a_J = 0$ if $J > 0$. This matrix equation is the generalization of Eq. (5.101). It is solved easily for the unknown values of (q_i). The generalization of Eq. (5.103) is given for the case $L = M$

$$\begin{pmatrix} p_0 \\ p_1 \\ \vdots \\ p_L \end{pmatrix} = \begin{pmatrix} a_0 & 0 & \cdots & 0 \\ a_1 & a_0 & \cdots & 0 \\ \vdots & \vdots & \ddots & \vdots \\ a_L & a_{L-1} & \cdots & a_0 \end{pmatrix} \begin{pmatrix} 1 \\ q_1 \\ \vdots \\ q_L \end{pmatrix} \tag{5.111}$$

that determines the unknowns (p_l). Similar results can be given for $L \neq M$. In this case the above matrix is not square. The evaluation of the coefficients in the Padé approximants is straightforward.

The present section is only a brief introduction to the properties of Padé approximants.

References

1. G. A. Baker, Jr., *Essentials of Padé Approximants.* Academic, 1975.

2. G. A. Baker, Jr. and P. Graves-Morris, *Padé Approximants*, 2nd ed. Cambridge University Press, 1996.

1. Find the Taylor series expansion about $z_0 = 0$ for the following functions. In each case give the first three nonzero terms on the right of the equal sign.

(a) $\cos(z)$ (5.112)

(b) $\tan(z)$ (5.113)

(c) $\arctan(z)$ (5.114)

(d) $\dfrac{1}{(1-x)^{3/2}}$ (5.115)

2. For the Taylor series $f(z) = \Sigma_n a_n z^n$, find the radius of convergence parameter μ for the following examples

(a) $a_n = \dfrac{(-1)^n}{n^2}$ (5.116)

(b) $a_n = nb^n$ (5.117)

(c) $a_n = \dfrac{2^n (n!)^2}{(2n)!}$ (5.118)

3. Give the expansion in poles and residues of the following functions
   (a) $f(z) = 1/\cos(z)$
   (b) $f(z) = \tan(z)$
   (c) $f(z) = 1/(e^z + 1)$

4. Expand the function $f(z)$ in a Laurent series about the point $z_0 = 2$. Discuss the convergence of the series $\Sigma_n a_n (z-2)^n$ in various regions of z.

$$f(z) = \frac{1}{z^2 - 1} \tag{5.119}$$

5. Derive the first three terms in the asymptotic series for the following functions

$$I_a = \int_0^\infty \frac{dt\, e^{-t}}{t + x} \tag{5.120}$$

$$I_b = \int_0^\infty \frac{dt\, e^{-t^2}}{t^2 + x^2} \tag{5.121}$$

$$I_c = \int_x^\infty \frac{dt}{t^n} e^{-t} \tag{5.122}$$

6. Derive the first three terms in the asymptotic expansion for the following integral by successive integration by parts or by other means

$$I_d = \int_0^\infty dt\, e^{-t} \ln(x + t) \tag{5.123}$$

7. Find the first term in the asymptotic series for large values of u using saddle point methods

(a) $J_0(u) = \dfrac{2}{\pi} \displaystyle\int_0^\infty dt\, \sin[u \cosh(t)]$ (5.124)

(b) $Ai(u) = \dfrac{1}{\pi} \displaystyle\int_0^\infty dt \cos\left(\dfrac{t^3}{3} + ut\right)$ (5.125)

8. Sum the series

$$S_a = \sum_n \frac{(-1)^n}{(n + 1/4)^2 + a^2}$$ (5.126)

$$S_b = \sum_n \frac{1}{n^4 - a^4}$$ (5.127)

9. Use the seventh-order Taylor series for the arctan(x) to construct the two Padé approximants' $f^{(3,4)}$ and $f^{(5,2)}$. Compare them with the original functions and with the series for values of $x = 0.5, 1.0, 1.5$.

$$\tan^{-1}(x) = x - \frac{x^3}{3} + \frac{x^5}{5} - \frac{x^7}{7} + \cdots +$$ (5.128)

Hint: The series can be written as $f = x\tilde{f}(x^2)$ and the Padé approximants for $\tilde{f}(x^2)$ contains only even powers of x.

6

Conformal Mapping

Conformal mapping is the name given to the process of using complex variables to map one figure onto another. The general procedure is to construct a figure or a region of space in the $z = x + iy$ two dimensional space, then change to another space by defining a new complex variable $w = u + iv = G(z)$. In the new two dimensional space with coordinates (u, v) the figure will usually change its shape. This procedure can be used to solve a variety of interesting problems in many branches of engineering and physics. Many applications of mapping are for solutions to Laplace's equation in two dimensions and this feature is explained first.

6.1. Laplace's Equation

Laplace's equation appears in a variety of physics problems and several examples are provided below. The relevance of Laplace's equation to complex variables is provided by the following important theorem.

THEOREM. *Any analytic function of z is a solution of the two dimensional Laplace equation.*

Proof. In two dimensions the form of Laplace's equation is

$$0 = \nabla^2\phi = \left[\frac{\partial^2}{\partial x^2} + \frac{\partial^2}{\partial y^2}\right]\phi(x, y) \tag{6.1}$$

An analytic function depends on the variables x and y only in the combination of $\phi(x + iy)$. Therefore

$$\frac{\partial\phi}{\partial x} = \frac{\partial\phi}{\partial z}\frac{\partial z}{\partial x} = \frac{\partial\phi}{\partial z} \tag{6.2}$$

$$\frac{\partial^2\phi}{\partial x^2} = \frac{\partial^2\phi}{\partial z^2} \tag{6.3}$$

$$\frac{\partial \phi}{\partial y} = \frac{\partial \phi}{\partial z}\frac{\partial z}{\partial y} = i\frac{\partial \phi}{\partial z} \tag{6.4}$$

$$\frac{\partial^2 \phi}{\partial y^2} = i^2 \frac{\partial^2 \phi}{\partial z^2} \tag{6.5}$$

$$\left[\frac{\partial^2}{\partial x^2} + \frac{\partial^2}{\partial y^2}\right]\phi(x + iy) = (1 + i^2)\frac{\partial^2 \phi}{\partial z^2} = 0 \tag{6.6}$$

since $i^2 = -1$ and $(1 + i^2) = 0$. This simple but important theorem is why conformal mapping is useful to physicists and engineers. It allows the solution to Laplaces' equation for a wide variety of problems in two dimensions. It is unfortunate that no similar technique is available in three dimensions.

Some examples of Laplace's equation are:

- Electrostatic potential problems have

$$\nabla^2 \phi(\mathbf{r}) = \frac{e}{\varepsilon_0}n(\mathbf{r}) \tag{6.7}$$

where $n(\mathbf{r})$ is the density of free charge and $\phi(\mathbf{r})$ is the potential. If there are no free charges in the space of concern, then $\nabla^2\phi(x, y) = 0$. Generally the potential field is created by boundary conditions. The interior space of our system is taken to be free of charges, so that the potential can be described by an analytical function.

- The diffusion equation has the general form

$$\frac{\partial}{\partial t}n(\mathbf{r}, t) - D\nabla^2 n(\mathbf{r}, t) = \text{sources} \tag{6.8}$$

where the variable $n(\mathbf{r}, t)$ is the density of particles and D is the diffusion coefficient. The steady state solution has no dependence on time, so the first term is absent. If all of the sources are at the boundaries, then the diffusion equation is just Laplace's equation ($\nabla^2 n = 0$). It can be readily solved in two dimensions using analytic functions.

- Heat flow problems are governed by the diffusion equation for the temperature $T(\mathbf{r}, t)$

$$\frac{\partial}{\partial t}T(\mathbf{r}, t) - D\nabla^2 T(\mathbf{r}, t) = \text{sources} \tag{6.9}$$

The steady state heat flow is governed by Laplace's equation ($\nabla^2 T = 0$) if the sources are confined to the boundaries.

Start with a complex function of z such as $F(z) = U(x, y) + iV(x, y)$, where the symbol U denotes the potential function. It is the electrostatic potential, the temperature, or the density of particles when solving $\nabla^2 U = 0$. The complex part of this function $V(x, y)$ is called the *stream function*. The stream function is so named because it is the direction of currents. Solutions to Laplace's equation usually have a current associated with them.

- For electrostatic problems where $U(x, y)$ or $U(u, v)$ is a potential energy, then $\nabla^2 U = 0$. There may be an electrical current $\mathbf{J} = -\sigma\nabla U$, where σ is the electrical conductivity. In a purely two dimensional problem, the current density $\mathbf{J}$ has units of amperes per meter, U has units of volts, and the conductivity σ has units of Siemens.

- For temperature problems we solve $\nabla^2 T = 0$. There is a heat current given by

$$\dot{Q} = -K\nabla T \tag{6.10}$$

where K is the thermal conductivity. In purely two dimensional problems the heat flow is Watts/meter and the thermal conductivity is Watts/degree.

- For the static diffusion of particles, $\nabla^2 n = 0$. There is a diffusion current given by

$$\mathbf{J} = -D\nabla n \tag{6.11}$$

where D is the diffusion coefficient in m^2/s and $\mathbf{J}$ is particles/(m s).

In all of these cases, the flow of a current of particles or heat is determined by the gradient of the solution to Laplace's equation.

THEOREM. *The current flows along the lines of constant values of the stream function.*

In Chapter 4 it was shown that the lines of constant U are always perpendicular to the lines of constant V. The proof of the present theorem is to show that ∇U is always perpendicular to the line of constant U. Then the line of constant V and ∇U are both perpendicular to the lines of constant U, and must be parallel.

Let the unit vector $\hat{n}(x, y)$ point along the direction of constant U. If $U = f(x, y)$ then variations in U give

$$\delta U = \delta x \left(\frac{\partial f}{\partial x}\right)_y + \delta y \left(\frac{\partial f}{\partial y}\right)_x \tag{6.12}$$

Along the line of constant U, $\delta U = 0$ which gives

$$\delta y = -\delta x \frac{f_x}{f_y} \tag{6.13}$$

$$\hat{n} = \frac{(\delta x, \delta y)}{\sqrt{(\delta x)^2 + (\delta y)^2}} \tag{6.14}$$

$$= \frac{(f_y, -f_x)}{\sqrt{(f_x)^2 + (f_y)^2}} \tag{6.15}$$

$$\nabla U = \nabla f = (f_x, f_y) \tag{6.16}$$

$$\nabla U \cdot \hat{n} = \frac{f_x f_y (1 - 1)}{\sqrt{(f_x)^2 + (f_y)^2}} = 0 \tag{6.17}$$

The vector ∇U is perpendicular to the direction $\hat{n}$ along the line of constant U. The currents are parallel with the lines of constant V. This feature makes the stream function useful.

An important aspect of conformal mapping is that the equations continue to obey Laplace's equation even after a change of variable.

- Given a function $F(z) = G(x, y) + iH(x, y)$ where $F(z)$ is analytic and $\nabla^2 G = 0$, after the conformal mapping $w = f(z) = u + iv$, the new function also obeys Laplace's equation

$$\left[\frac{\partial^2}{\partial u^2} + \frac{\partial^2}{\partial v^2}\right] G(u, v) = 0 \qquad (6.18)$$

The proof takes a lot of derivatives. A shorthand notation is useful where $\partial G/\partial x \equiv G_x$.

$$G_x = G_u u_x + G_v v_x \qquad (6.19)$$

$$G_y = G_u u_y + G_v v_y \qquad (6.20)$$

$$G_{xx} = G_u u_{xx} + G_v v_{xx} + u_x[G_{uu} u_x + G_{uv} v_x] + v_x[G_{vu} u_x + G_{vv} v_x] \qquad (6.21)$$

$$G_{yy} = G_u u_{yy} + G_v v_{yy} + u_y[G_{uu} u_y + G_{uv} v_y] + v_y[G_{uv} u_y + G_{vv} v_y] \qquad (6.22)$$

$$0 = G_{xx} + G_{yy}$$
$$= G_u(u_{xx} + u_{yy}) + G_v(v_{xx} + v_{yy}) + G_{uu}(u_x^2 + u_y^2)$$
$$+ G_{vv}(v_x^2 + v_y^2) + 2G_{uv}(u_x v_x + u_y v_y) \qquad (6.23)$$

where $G_{uv} = G_{vu}$. The Cauchy–Riemann conditions are $u_x = v_y$, $u_y = -v_x$ and they give the following relationships

$$u_x v_x + u_y v_y = v_y v_x - v_x v_y = 0 \qquad (6.24)$$

$$u_{xx} + u_{yy} = \frac{\partial}{\partial x} v_y - \frac{\partial}{\partial y} v_x = 0 = v_{xx} + v_{yy} \qquad (6.25)$$

$$v_x^2 + v_y^2 = u_x^2 + u_y^2 \qquad (6.26)$$

$$0 = G_{xx} + G_{yy} = [G_{uu} + G_{vv}][u_x^2 + u_y^2] \qquad (6.27)$$

Generally, since $[u_x^2 + u_y^2] \neq 0$ then $G_{uu} + G_{vv} = 0$. A more trivial proof is to note that $G(u, v)$ is the real part of an analytic function of w, and therefore has to be a solution of Laplace's equation.

6.2. Mapping

Mapping can be understood using some simple examples. Figure 6.1(a) shows a shaded box twice as high as it is wide and is one unit wide and two units high. The corners in (x, y) space are at $(0, 0)$, $(1, 0)$, $(1, 2)$, and $(0, 2)$, going around the box in a counterclockwise direction. The mapping $w = 2z$ in Fig. 6.1(b) makes the box appear twice as large in the space of $w = u + iv$. In the coordinate system of (u, v), the four corners are at $(0, 0)$, $(2, 0)$, $(2, 4)$, and $(0, 4)$. In this example the box just doubled in size.

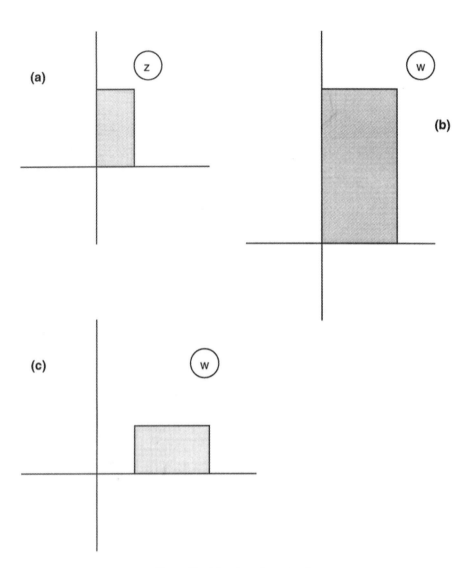

Figure 6.1 Mapping of a rectangle.

The second example is to map the box using $w = iz + 3$. Now the box appears on its side in Fig. 6.1(c). The constant "3" moved the box over three places. The factor of i rotated the box by an angle of $\pi/2 = 90°$.

The next example is more challenging. Take the figure as all of the upper-half plane of the space labeled by $z = x + iy$. The UHP is the region where $y \geqslant 0$. Consider how the UHP maps using the transformation $w = \ln(z)$. This mapping will use polar coordinates, so write

$$z = \rho e^{i\theta} \tag{6.28}$$

$$w = i\theta + \ln(\rho) = u + iv \tag{6.29}$$

$$u = \ln(\rho), \quad v = \theta \tag{6.30}$$

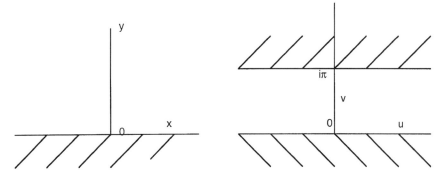

Figure 6.2 Mapping of $w = \ln(z)$ for UHP of z space.

When confining the space to the UHP the axis $y = 0$ is treated as a branch cut. All multivalued functions are defined from branch cuts by rotating angles in the UHP. This convention dictates the choice of using positive angles for θ. The angle θ is limited in value to $0 \leqslant \theta \leqslant \pi$, in order to stay in the UHP. The largest value of v is π. The length ρ can be small or large, so that $u = \ln(\rho)$ can be negative or positive. The function w is confined to a strip bounded by $0 \leqslant v \leqslant \pi$ but only for all positive or negative values of u. The half space of z is now a narrow strip in the space $w = \ln(z)$. This mapping is shown in Fig. 6.2. In the figures of these mappings, should the shaded region be the area mapped or should the mapping area be clear and the other regions shaded? Both options are utilized. In Fig. 6.2, the regions that are mapped are shown as clear.

THEOREM. *All angles are preserved in a conformal mapping except at singular points.*

In Chapter 4 it was proved that lines of constant u are always perpendicular to lines of constant v. This orthogonality is preserved in a conformal mapping. Indeed, all angles are the same. Figure 6.3 shows the neighborhood of a point z_0 in the space z which is mapped onto the point w_0 in the space w defined by the conformal mapping $w = f(z)$ so that $w_0 = f(z_0)$. Two short line segments emanate from z_0, that are labeled $z_{1,2}$ and the angle between them is θ_{12}.

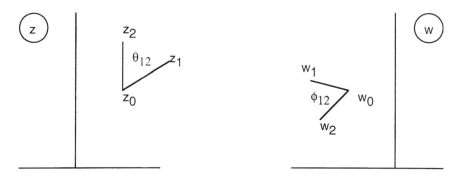

Figure 6.3 Angles are preserved in a mapping.

Choose the segments to be the same length ρ

$$z_1 - z_0 = \rho e^{i\theta_1} \tag{6.31}$$

$$z_2 - z_0 = \rho e^{i\theta_2} \tag{6.32}$$

$$\frac{z_1 - z_0}{z_2 - z_0} = e^{i(\theta_1 - \theta_2)} \tag{6.33}$$

$$\theta_{12} = \theta_1 - \theta_2 \tag{6.34}$$

The angle between the segments is the phase angle we obtain by dividing the two complex numbers for the two segments. The same procedure is used to find the phase angle ϕ_{12} in the space w

$$\frac{w_1 - w_0}{w_2 - w_0} = e^{i(\phi_1 - \phi_2)} \tag{6.35}$$

$$\phi_{12} = \phi_1 - \phi_2 \tag{6.36}$$

The theorem is that $\theta_{12} = \phi_{12}$. The proof uses Taylor series to expand the functions w_1, w_2 around the point w_0

$$w_1 = f(z_1) = f(z_0) + (z_1 - z_0)f' + O[(z_1 - z_0)^2] \approx w_0 + (z_1 - z_0)f'$$
$$\tag{6.37}$$

$$w_2 = f(z_2) = f(z_0) + (z_2 - z_0)f' + O[(z_2 - z_0)^2] \approx w_0 + (z_2 - z_0)f'$$
$$\tag{6.38}$$

$$\frac{w_1 - w_0}{w_2 - w_0} = \frac{(z_1 - z_0)f'}{(z_2 - z_0)f'} = \frac{z_1 - z_0}{z_2 - z_0} = e^{i(\theta_1 - \theta_2)} \tag{6.39}$$

Both first deriatives f' are taken at the point z_0 and are alike. They cancel, which shows that $\phi_{12} = \theta_{12}$, this proves the theorem. The proof requires that $f' \neq 0, \infty$. The case $f' = \infty$ is a singular point, and the angles are not preserved at singular points. The case $f' = 0$ also does not preserve angles.

Next prove a useful theorem that gives the potential $U(x, y)$ in the UHP if it is known on the real axis.

THEOREM. *Given the potential $U(x, 0)$ on the real axis, the potential in the UHP is*

$$U(x, y) = \frac{y}{\pi} \int_{-\infty}^{\infty} dx' \frac{U(x', 0)}{(x - x')^2 + y^2} \tag{6.40}$$

The proof proceeds in two steps. The first is to show that the proposed function satisfies Laplace's equation in two dimension. The proof is by differentiating twice on the kernel

$$K(x - x', y) = \frac{y}{(x - x')^2 + y^2} \tag{6.41}$$

$$\frac{d^2}{dx^2} K(x - x', y) = \frac{2y}{[(x - x')^2 + y^2]^3} [3(x - x')^2 - y^2] \tag{6.42}$$

$$\frac{d^2}{dy^2} K(x - x', y) = \frac{2y}{[(x - x')^2 + y^2]^3} [y^2 - 3(x - x')^2] \tag{6.43}$$

$$\nabla^2 K = 0 \tag{6.44}$$

The second step in the proof is to show that this function gives the correct potential on the real axis

$$\lim_{y \to 0} K(x - x', y) = \pi \delta(x - x') \tag{6.45}$$

$$\lim_{y \to 0} U(x, y) = \int_{-\infty}^{\infty} \frac{dx'}{\pi} U(x', 0) \pi \delta(x - x') = U(x, 0) \tag{6.46}$$

The limit of $y \to 0$ on the kernel $K(x - x', y)$ gives a delta function. The function satisfies Laplace's equation and gives the correct boundary condition. These two requirements are sufficient to prove that Eq. (6.40) is the correct potential. The solution is unique, in that there is only one function with these features.

A simple example is the case where $U(x, y) = U_0 \theta / \pi$ in the UHP. On the real axis $U = 0$ for $x > 0$ and $U = U_0$ for $x < 0$. The above integral is

$$U(x, y) = \frac{U_0 y}{\pi} \int_{-\infty}^{0} \frac{dx'}{(x - x')^2 + y^2} = \frac{U_0}{\pi} \tan^{-1} \left(\frac{x' - x}{y} \right)_{-\infty}^{0} \tag{6.47}$$

$$= \frac{U_0}{\pi} \left[\frac{\pi}{2} - \tan^{-1} \left(\frac{x}{y} \right) \right] = \frac{U_0}{\pi} \tan^{-1} \left(\frac{y}{x} \right) \tag{6.48}$$

The last equal sign gives an identity for the arctangent. The integral gives the same result that $U = U_0 \theta / \pi$.

As another example, consider that $U(x, 0) = U_0$ during the interval $-a \leqslant x \leqslant a$ and is zero elsewhere on the real axis. Then the potential in the UHP is

$$U(x, y) = \frac{U_0 y}{\pi} \int_{-a}^{a} \frac{dx'}{(x - x')^2 + y^2}$$

$$= \frac{U_0}{\pi} \tan^{-1} \left(\frac{x' - x}{y} \right)_{-a}^{a}$$

$$= \frac{U_0}{\pi} \left[\tan^{-1} \left(\frac{a - x}{y} \right) - \tan^{-1} \left(\frac{-a - x}{y} \right) \right] \tag{6.49}$$

Note that in the case that $(-a \leqslant x \leqslant a, y \to 0)$ then the first arctangent is $\pi/2$ and the second one is $-\pi/2$ so the bracket is $\pi/2 - (-\pi/2) = \pi$ and $U = U_0$ as required. Elsewhere on the real axis, the two arctangents cancel, and $U = 0$. This solution will be obtained in a later problem by mapping.

6.3. Examples

Examples of conformal mapping are given below.

Example 1. $w = \ln(z)$: Earlier it was discussed how $w = \ln(z)$ mapped the UHP of z space into a strip $0 \leqslant v \leqslant \pi$ in the w space. Consider this example again, but with a potential function in z space of

$$U + iV = -i\frac{U_0}{\pi}\ln(z) = \frac{U_0}{\pi}[\theta - i\ln(\rho)] \tag{6.50}$$

$$U(x, y) = \frac{U_0}{\pi}\theta = \frac{U_0}{\pi}\tan^{-1}\left(\frac{y}{x}\right) \tag{6.51}$$

$$V(x, y) = -\frac{U_0}{\pi}\ln(\rho) = -\frac{U_0}{2\pi}\ln(x^2 + y^2) \tag{6.52}$$

Lines of constant U are those with constant angle. If U is the potential, then V is the stream function. Lines of constant V are semicircles. The lines of constant U are always perpendicular to the lines of constant V. This behavior is shown in Fig. 6.4(a). Now consider the mapping $w = \ln(z)$ which gives

$$U + iV = -i\frac{U_0}{\pi}w = \frac{U_0}{\pi}[v - iu] \tag{6.f53}$$

In this space, lines of constant U have a constant value of v. Since $0 \leqslant v \leqslant \pi$, then $U = 0$ on the u-axis ($v = 0$), and $U = U_0$ at $v = \pi$. Lines of constant V are vertical, since they have constant u. This behavior is shown in Fig. 6.4(b).

Example 2. *Two fins:* Consider the region in the space z bounded by $y \geqslant 0$ and $-\pi/2 \leqslant x \leqslant +\pi/2$. This region is shown in Fig. 6.5. Let the boundaries at $x = \pm\pi/2$ represent the two plates of a capacitor. The left plate ($x = -\pi/2$) is at a potential of $U = -U_0$ while the right-hand plate ($x = +\pi/2$) is at the potential $U = +U_0$. Anywhere between them the potential is given by

$$U(x, y) = \frac{2U_0}{\pi}x \tag{6.54}$$

This potential satisfies Laplace's equation. Lines of constant U are vertical. If the general function is $F(z) = 2U_0z/\pi \equiv U + iV$, the stream function is $V = 2U_0y/\pi$. Lines of constant V are horizontal. This problem is quite simple and easy to understand.

Now produce a conformal mapping using $w = \sin(z)$. What happens to the figure? The best way to proceed is to take specific points on one graph

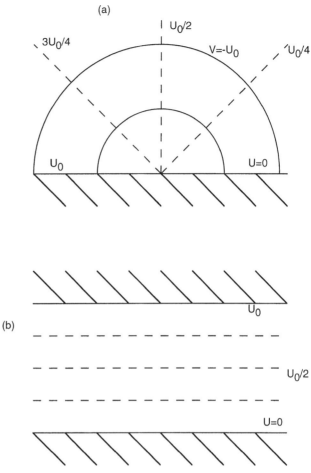

Figure 6.4 (a) For the function $\ln(z)$ the lines of constant U are at constant angle, while those of constant V are at constant radius. (b) The lines of constant U are horizontal and the lines of constant V are vertical.

to see where that point is on the new graph. The end result is shown in the figure. The points A, B, C, D, E on the map go to A′, B′, C′, D′, E′ on the other. There are two fins, that extend along the real axis for $|u| > 1$.

- Points along the $U = -U_0$ wall of the z space have $x = -\pi/2$, $y > 0$. This maps onto $w = \sin(-\pi/2 + iy) = -\cosh(y)$. This function is real and negative so it is along the negative u axis. The left-hand fin is at potential $-U_0$.

- Points along the right electrode in the z space have $x = \pi/2$, $y > 0$ that maps onto $w = \sin(\pi/2 + iy) = \cosh(y)$. This function is real and positive, so it is along the positive u axis. The right-hand fin is at the potential U_0

The resulting mapping is to a problem of the potential field between two fins that are pointing at each other, with a gap between their ends. The potential

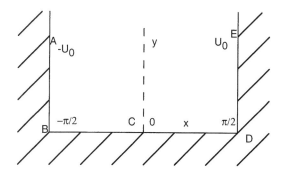

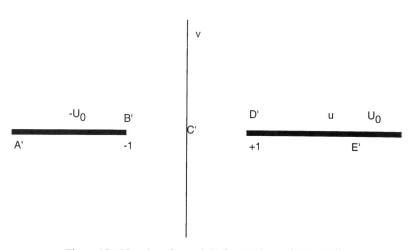

Figure 6.5 Mapping of $w = \sin(z)$ for $y \geqslant 0$, $-\pi/2 \leqslant x \leqslant \pi/2$.

field everywhere is given by

$$U(u, v) = \frac{2U_0}{\pi} \operatorname{Re}\{\sin^{-1}(u + iv)\} \tag{6.55}$$

which is the concise way to write the answer. Since the arcsine of a complex
number is rather obscure, the answer is derived by traditional methods

$$w = \sin(x + iy) = \sin(x)\cosh(y) + i\cos(x)\sinh(y) \tag{6.56}$$

$$u = \sin(x)\cosh(y), \quad v = \cos(x)\sinh(y) \tag{6.57}$$

$$\cosh(y) = \frac{u}{\sin(x)} \tag{6.58}$$

$$\sinh(y) = \frac{v}{\cos(x)} \tag{6.59}$$

$$1 = \cosh^2 y - \sinh^2 y = \frac{u^2}{\sin^2(x)} - \frac{v^2}{\cos^2(x)} \tag{6.60}$$

Solving the latter equation, with $\cos^2 x = 1 - \sin^2 x$ gives the quadratic equation

$$0 = \sin^4(x) - \sin^2(x)[1 + u^2 + v^2] + u^2 \tag{6.61}$$

$$\sin^2(x) = \frac{1}{2}[1 + u^2 + v^2 - \sqrt{(1 + u^2 + v^2)^2 - 4u^2}]$$

$$= \frac{1}{4}\left[\sqrt{(u+1)^2 + v^2} - \sqrt{(u-1)^2 + v^2}\right]^2 \tag{6.62}$$

$$U(u, v) = \frac{2U_0}{\pi}\sin^{-1}\left[\frac{1}{2}\left(\sqrt{(u+1)^2 + v^2} - \sqrt{(u-1)^2 + v^2}\right)\right] \tag{6.63}$$

The negative sign is chosen in front of the root since otherwise $\sin^2(x)$ would be larger than one. This formula can be used to graph lines of constant U.

Another type of result is to calculate the charge density on the electrodes. The electric field is the gradient of the potential, $\boldsymbol{E} = -\nabla U$. On the surface of an electrode the field is directly proportional to the surface charge $E = -\sigma/\varepsilon_0$. For example, in the z space where the capacitor had parallel electrodes the electric field is a constant

$$\frac{\sigma}{\varepsilon_0} = \frac{dU}{dx} = \frac{2U_0}{\pi} \tag{6.64}$$

The surface charge is the same value σ at all points along the parallel plates.

The more interesting result is to calculate the surface charge along fins in w space. The fins are equipotential surfaces and the electric field must be perpendicular to the fins, so that

$$\frac{\sigma(u)}{\varepsilon_0} = \left[\frac{\partial U(u, v)}{\partial v}\right]_{v=0, u>1} \tag{6.65}$$

where the right-hand fin has $u > 1$. Since the final mathematical operation is to take $v = 0$, it simplifies the algebra if the value of v is taken to be small before taking the derivative. Various terms are expanded in powers of v^2 using Taylor series. The factors in Eq. (6.62) are

$$\sqrt{(u+1)^2 + v^2} \approx u + 1 + \frac{v^2}{2(u+1)} \tag{6.66}$$

$$\sqrt{(u-1)^2 + v^2} \approx u - 1 + \frac{v^2}{2(u-1)} \tag{6.67}$$

$$\frac{1}{2}\left[\sqrt{(u+1)^2 + v^2} - \sqrt{(u+1)^2 - v^2}\right] \approx 1 - \frac{v^2}{2(u^2 - 1)} \equiv \xi \tag{6.68}$$

$$\frac{\partial U(u, v)}{\partial v} = \frac{2U_0}{\pi} \frac{\partial}{\partial v} \sin^{-1}[\xi(u, v)] = \frac{2U_0/\pi}{\sqrt{1 - \xi^2}} \frac{\partial \xi}{\partial v} \qquad (6.69)$$

$$1 - \xi^2 = (1 - \xi)(1 + \xi) = \frac{v^2}{u^2 - 1} \qquad (6.70)$$

$$\frac{\sigma(u)}{\varepsilon_0} = -\frac{2U_0}{\pi\sqrt{u^2 - 1}} \qquad (6.71)$$

The exact answer is simple and elegant. Note that it diverges near the tip of the electrode ($u = \pm 1$).

The problem of two fins can be applied to several physical problems. One is the flow of current between a source and drain of a semiconductor field effect transistor (FET). If there is no gate potential the current in the upper-half plane has exactly the geometry of an FET, where the left electrode is the source and the right electrode is the drain.

Example 3. *Fin and grounded plate:* A problem which is closely related to that of two fins is the problem of a fin of potential U_0 above a grounded plate. This geometry is shown in Fig. 6.6. In the problem of two fins, the v axis is an equipotential surface at $U = 0$. This assertion is obvious from the geometry, but can be proved from Eq. (6.62) by setting $u = 0$. Then the argument of the arcsine vanishes, as does the arcsine, and one is left with $U = 0$. The solution to two fins is also a solution to the problem of a grounded plate (a line in two dimensions) at potential $U = 0$ with a fin above it at a potential U_0. The result for $U(u, v)$ is exactly the same as in Eq. (6.62) with the restriction that $u > 0$. The vertical axis $u = 0$ has the potential of $U(0, v) = 0$. The fin ($v = 0, u > 1$)

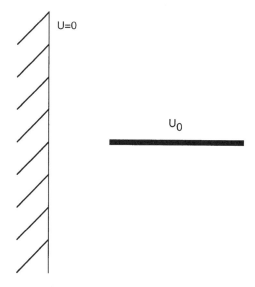

Figure 6.6   Fin above grounded plate.

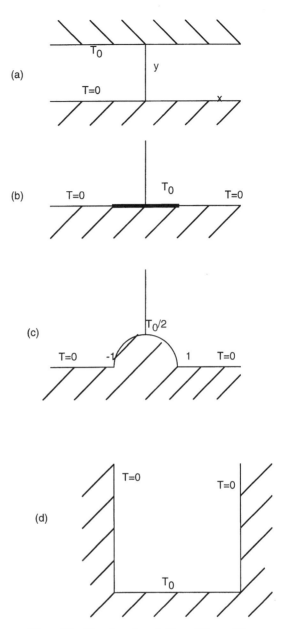

Figure 6.7 Mapping of $z = \ln[(w - 1)/(w + 1)]$.

still has potential U_0 since the argument of the arcsine is one, and the arcsine is $\pi/2$.

Example 4. $z = \ln[(w - 1)/(w + 1)]$: The next example is shown in Fig. 6.7. In z space there is one dimensional heat flow along the vertical axis. The temperature is

$$T(y) = \frac{T_0}{\pi} y = \frac{T_0}{\pi} \text{Im}[z] \tag{6.72}$$

The x axis ($y = 0$) has a temperature $T = 0$. At $y = \pi$ the temperature is T_0.
Now consider the mapping of

$$z = \ln\left(\frac{w-1}{w+1}\right), \qquad w = -\coth(z/2) \tag{6.73}$$

$$T(u, v) = \frac{T_0}{\pi} \operatorname{Im}\left[\ln\left(\frac{w-1}{w+1}\right)\right]$$

$$= \frac{T_0}{\pi}\left[\tan^{-1}\left(\frac{v}{u-1}\right) - \tan^{-1}\left(\frac{v}{u+1}\right)\right]$$

$$= \frac{T_0}{\pi}\left[\tan^{-1}\left(\frac{1-u}{v}\right) + \tan^{-1}\left(\frac{u+1}{v}\right)\right] \tag{6.74}$$

The last equality again uses the identity for the arctangent, $\tan^{-1}(1/a) = \pi/2 - \tan^{-1}(a)$. The above formula is identical to Eq. (6.49) where the solution was in (x, y) space, while here it is in (u, v) space. The form is the same, since both have the same boundary condition of a nonzero potential only along the real axis from $-1 \leqslant u \leqslant 1$.

In order to discern which figure results from this mapping, again it is useful to take selected points.

- At $y = 0$ along the x axis, then $w = -\coth(x/2) = u$ which is a real number with a magnitude greater than one. It is the u axis in w space as shown in Fig. 6.7(b). The axis has $T = 0$ for $u < -1$ and $u > 1$.
- At $y = \pi$, $w = -\coth[(x + i\pi)/2] = -\tanh(x/2) = u$. This function is real and has a magnitude $-1 \leqslant u \leqslant 1$. The u axis is at temperature T_0 for this interval and the allowed region in w space is the UHP.

Equation (6.74) is the solution to the temperature profile of a system which is confined to the upper-half plane and has a temperature $T = 0$ along the real axis except for the interval where $-1 < u < 1$ where the temperature is T_0. The resulting temperature for all values of $(u, v > 0)$ is given by Eq. (6.74).

Example 5. *Semicircle on grounded plate:* In an earlier example after doing one mapping, an equipotential was found which leads to a solution of a related problem. The same procedure is used here. Another way to express the above answer is

$$T(u, v) = \frac{T_0}{\pi} \operatorname{Im}\left[\ln\left(\frac{(w-1)(w^* + 1)}{|w+1|^2}\right)\right]$$

$$= \frac{T_0}{\pi} \operatorname{Im}\left[\ln\left(\frac{u^2 + v^2 - 1 + 2vi}{|w+1|^2}\right)\right]$$

$$= \frac{T_0}{\pi} \tan^{-1}\left(\frac{2v}{u^2 + v^2 - 1}\right) \tag{6.75}$$

In Eq. (6.64) set $u^2 + v^2 = 1$ making the argument of the artangent diverge to infinity. Then the angle is $\pi/2$ and $T = T_0/2$. So Eq. (6.74) is also the solution to the problem of the temperature profile in the UHP where the temperature is zero everywhere along the real axis, except for a semicircle of unit radius at the origin, where the temperature is $T_0/2$. This problem is shown in Fig. 6.7(c).

The flow of heat in two dimensions has the units of Watts/meter and is given by $\dot{\mathbf{Q}} = -K\nabla T$, where K is the thermal conductivity. For the above two problems the derivative of the arctangent in Eq. (6.74) gives

$$\dot{Q}_u = -\frac{KT_0}{\pi}\left[\frac{v}{(1+u)^2 + v^2} - \frac{v}{(1-u)^2 + v^2}\right] \tag{6.76}$$

$$\dot{Q}_v = \frac{KT_0}{\pi}\left[\frac{1+u}{(1+u)^2 + v^2} + \frac{1-u}{(1-u)^2 + v^2}\right] \tag{6.77}$$

Mapping has been used to find the flow of heat and the distribution of temperature of a reasonably complicated problem.

Example 6. *Map it again!:* Return to the mapping $z = \ln[(w-1)/(w+1)]$. It has T_0 on the real axis between $-1 < u < 1$ and $T = 0$ elsewhere on the real axis. Introduce a third complex space and call it $z' = x' + iy'$. Consider the mapping $w = \sin(z')$. It was considered in an earlier problem and z' space is bounded between $-\pi/2 < x' < \pi/2$, $y' > 0$. Set

$$z = \ln\left(\frac{\sin(z') - 1}{\sin(z') + 1}\right) \tag{6.78}$$

$$T(x', y') = \frac{T_0}{\pi}\text{Im}[z] \tag{6.79}$$

The mapping is shown in Fig. 6.7(d). The real axis has T_0 while the vertical boundaries at $x' = \pm\pi/2$ are at $T = 0$. The above result can be simplified. Write

$$\frac{\sin(z') - 1}{\sin(z') + 1} = \frac{[\sin(z') - 1][\sin(z')^* + 1]}{|\sin(z') + 1|^2} \tag{6.80}$$

$$[\sin(z') - 1][\sin(z')^* + 1] = [\sin(x')\cosh(y') - 1 + i\cos(x')\sinh(y')]$$
$$\times [\sin(x')\cosh(y') + 1 - i\cos(x')\sinh(y')]$$
$$= \sin^2(x')\cosh^2(y') - 1 + \cos^2(x')\sinh^2(y')$$
$$+ 2i\cos(x')\sinh(y')$$
$$= \sinh^2(y') - \cos^2(x') + 2i\cos(x')\sinh(y')$$
$$= [\sinh(y') + i\cos(x')]^2 \tag{6.81}$$

$$T(x', y') = \frac{2T_0}{\pi}\tan^{-1}\left(\frac{\cos(x')}{\sinh(y')}\right) \tag{6.82}$$

It is a simple way to give the temperature of a semifinite bar with one end at T_0 and the sides at $T = 0$. In Chapter 11 the same problem is solved using series which gives a less elegant result.

So far the procedure for mapping has been to write down a function such as $w = \sin(z)$ and then examine the kind of mapping it produces. A more systematic procedure is to have a final figure in mind. The method of starting with a figure and deriving the final formula is called a *Schwartz–Christoffel transformation*.

Figure 6.8 shows the z plane on top. There is a line along the real axis that goes over some points labeled x_1, x_2, x_3, x_4, x_5. The goal is to map the upper-half plane of z space into the figure below in w space. The points x_j on top become w_j below. In the space of z, the line runs from left to right. In the space of w, the line runs around the figure in a counterclockwise direction. These conventions are standard.

The points w_j occur at the corners of the figure. Recall the theorem that angles are preserved in mapping except at singular points. The angles are not preserved at the points $x_j \Leftrightarrow w_j$ and these points must be singular. If there are n singular points in general (our example has five), the mathematical formula must be of the form

$$\frac{dw}{dz} = \frac{A}{(z - x_1)^{\alpha_1}(z - x_2)^{\alpha_2} \cdots (z - x_n)^{\alpha_n}} \tag{6.83}$$

This functional form is required by the need to have dw/dz nonanalytic at each point $z = x_j$.

In Fig. 6.8 points b and c on the z axis corresponds to points B, C on the w figure. As $b \to -\infty$, $c \to +\infty$ the two points B and C become the same point. Then dw/dz must have the same value at B and C. If one goes from b to c the factor of $(z - x_j)$ changes sign and the factor of $(z - x_j)^{\alpha_j} \to (x_j - z)^{\alpha_j}e^{i\pi\alpha_j}$. In order that B and C have the same value of dw/dz the phase factors must add

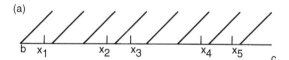

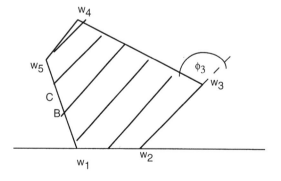

Figure 6.8 Mapping of the points $x_1, x_2, \ldots, x_5$ into the points $w_1, w_2, \ldots, w_5$.

to 2π or

$$\sum_j \alpha_j = 2 \tag{6.84}$$

Another feature of the figure in w space is that each corner turns an angle ϕ_j. The angle is called positive if it turns to the left and negative if it turns to the right. In order for B and C to be the same point then the sum of these angles must be 360°

$$\sum_j \phi_j = 2\pi \tag{6.85}$$

$$\alpha_j = \frac{\phi_j}{\pi} \tag{6.86}$$

The two constraints are satisfied if $\alpha_j = \phi_j/\pi$. This last step completes the derivation. The generalized Schwartz–Christoffel transformation is

$$w = w_0 + A \int_{-\infty}^{\infty} \frac{dz}{\Pi_j (z - x_j)^{\phi_j/\pi}} \tag{6.87}$$

The complex constants of integration w_0 and A are determined by matching the x_j points to the w_j points. The only remaining information is to state the method of treating points at infinity. They are omitted unless they are needed to change from one surface to another.

An interesting feature of the Schwartz–Christoffel transformation is that the shape of the figure in w space is determined entirely by the points along the real axis in z space. Given a transformation for which the figure is unknown it can be traced out in w space by choosing points along the real axis in z space.

The method is demonstrated by the following examples.

Example 1. $z = \sin(w)$: Figure 6.9(a) shows a figure in w space that occupies $-\pi/2 < u < \pi/2$, $v > 0$. There are three points, $w_1 = (-\pi/2, 0)$, $x_1 = -1$, $w_2 = (\pi/2, 0)$, $x_2 = +1$, and w_3 at infinity along the imaginary axis. The infinite points are ignored and the angles are $\phi_{1,2} = \pi/2$ so that $\alpha_{1,2} = 1/2$. The transformation is

$$\frac{dw}{dz} = \frac{A}{(z + 1)^{1/2}(z - 1)^{1/2}} = \frac{A}{\sqrt{z^2 - 1}} \tag{6.88}$$

$$w = w_0 + A \sin^{-1}(z) \tag{6.89}$$

The constants are $A = 1$ and $w_0 = 0$ in order that the correct values of $w_{1,2} = \pm\pi/2$ are found when $x_{1,2} = \pm 1$. The final result $w = \sin^{-1} z$, $z = \sin(w)$ agrees with the earlier discussion which mapped $w = \sin(z)$ rather than $z = \sin(w)$. The switch of variables just switches the labels on the maps.

Example 2. *Vertical fin:* Figure 6.9(b) shows the figure in w space as having a fin that projects along the imaginary axis to the point $w = ip$. The fin has an infinitesimal thickness but is given a finite width to aid the eye. The w space

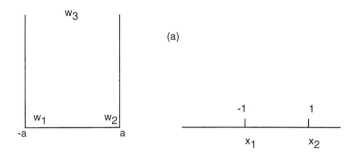

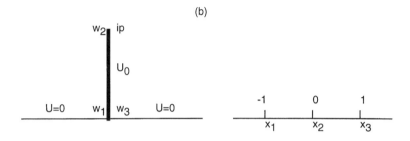

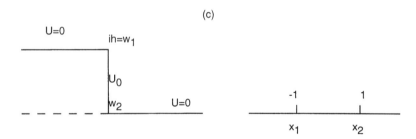

Figure 6.9 Mappings using the Schwartz–Christoffel transformation.

has three singular points where the line changes direction: w_1 and w_3 are both in the vicinity of $w = 0$ and are on opposite sides of the base of the fin. The point w_2 is at the top of the fin. The three angles are $\phi_{1,3} = \pi/2$, $\phi_2 = -\pi$ so that the three exponents are $\alpha_{1,3} = 1/2$, $\alpha_2 = -1$. These three points are assigned to the x axis as $-1, 0$, and $+1$. Only three points can be assigned arbitrarily. If there are more points they are given unknown values such as x_j that are determined later. The transformation is

$$\frac{dw}{dz} = \frac{Az}{\sqrt{z^2 - 1}} \tag{6.90}$$

$$w = w_0 + A\sqrt{z^2 - 1} \tag{6.91}$$

The integral is quite easy and gives a simple function. The constant $w_0 = 0$ since $w = 0$ at $z = \pm 1$. Also, at $z = 0$ then $A = p$, so the final transformation is $w = p\sqrt{z^2 - 1}$. The inverse is $z = \sqrt{1 + (w/p)^2}$.

An earlier theorem proved that if the potential $U(x, 0)$ is known on the real axis, then a simple integral gives it everywhere in the UHP. This technique adds power to the Schwartz–Christoffel transformation. By specifying the potential along the real axis, it can be determined as a function of z. If we can do the inverse transform to get $z = f(w)$, then the potential $U(u, v)$ is known in the transformed space. An example of this technique is now presented.

Assign the fin a potential U_0, while the rest of the real axis in w space has $U = 0$. In the z space the potential U_0 is found along the real axis from $-1 < x < 1$, and $U = 0$ elsewhere on the real axis. In this case the potential in the UHP is

$$
\begin{aligned}
U(x, y) &= \frac{U_0 y}{\pi} \int_{-1}^{1} \frac{dx'}{(x' - x)^2 + y^2} \\
&= \frac{U_0}{\pi} \left[\tan^{-1} \left(\frac{1 - x}{y} \right) + \tan^{-1} \left(\frac{1 + x}{y} \right) \right] \\
&= \frac{U_0}{\pi} \operatorname{Im} \left[\ln \left(\frac{z - 1}{z + 1} \right) \right]
\end{aligned}
\tag{6.92}
$$

The next step is to transform to the w space which is done by replacing z with $z(w)$

$$
U(u, v) = \frac{U_0}{\pi} \operatorname{Im} \left[\ln \left(\frac{\sqrt{1 + (w/p)^2} - 1}{\sqrt{1 + (w/p)^2} + 1} \right) \right]
\tag{6.93}
$$

This rather formidable result is the potential in the upper-half plane of the space with the vertical fin of height p along the imaginary axis. The formula is complicated since the problem is complicated. However, it has been solved analytically.

Consider the points along the imaginary axis. If $u = 0$ then

$$
\sqrt{1 + (w/p)^2} = \sqrt{1 - v^2/p^2}
\tag{6.94}
$$

- For the case that $0 < v < p$ then $\sqrt{1 - v^2/p^2}$ is real and less than one. The above expression is written as

$$
\begin{aligned}
U(0, v) &= \frac{U_0}{\pi} \operatorname{Im} \left[\ln \left(e^{i\pi} \frac{1 - \sqrt{1 - (v/p)^2}}{1 + \sqrt{1 - (v/p)^2}} \right) \right] \\
&= U_0
\end{aligned}
\tag{6.95}
$$

The only imaginary part of the bracket is $\ln[\exp(i\pi)] = i\pi$. The formula gives the correct answer that the fin has a potential of U_0.

- For $y > p$ the root is $i\sqrt{v^2/p^2 - 1}$ and the argument of the logarithm has several imaginary components

$$U(0, v) = \frac{U_0}{\pi} \text{Im}\left[\ln\left(e^{i\pi}\frac{1 - i\sqrt{(v/p)^2 - 1}}{1 + i\sqrt{(v/p)^2 - 1}}\right)\right]$$

$$= U_0\left[1 - \frac{2}{\pi}\tan^{-1}\sqrt{(v/p)^2 - 1}\right] = \frac{2U_0}{\pi}\tan^{-1}\left(\frac{1}{\sqrt{(v/p)^2 - 1}}\right) \tag{6.96}$$

The last result is obtained by the identity on the arctangent. As the value of v increases to infinity the value of the arctangent decreases to zero.

Example 3. *Step:* Figure 6.9(c) shows the w space with step of height h. There are only two singular points in the figure at $w_1 = (0, ih)$, $w_2 = (0, 0)$ and they are assigned to the z space as $x_1 = -1$ and $x_1 = +1$. The angles are $\phi_j = \pm\pi/2$ so that $\alpha_j = \pm1/2$. The formula is

$$\frac{dw}{dz} = A\left[\frac{z + 1}{z - 1}\right]^{1/2} \tag{6.97}$$

$$w = w_0 + A\left[\sqrt{z^2 - 1} + \ln\left(\frac{\sqrt{z + 1} + \sqrt{z - 1}}{\sqrt{z + 1} - \sqrt{z - 1}}\right)\right]$$

$$= w_0 + A[\sqrt{z^2 - 1} + \coth^{-1}(z)] \tag{6.98}$$

The answer has been written in two alternate but equivalent forms. The constant has to be evaluated carefully. At point $z = (1, 0)$, $w = (0, 0)$ which gives the equation $0 = w_0$. At point $z = (-1, 0)$, $w = (0, ih)$ and the equation gives

$$ih = A\ln(-1) = Ai\pi \tag{6.99}$$

$$A = \frac{h}{\pi} \tag{6.100}$$

$$w = \frac{h}{\pi}\left[\sqrt{z^2 - 1} + \ln\left(\frac{\sqrt{z + 1} + \sqrt{z - 1}}{\sqrt{z + 1} - \sqrt{z - 1}}\right)\right] \tag{6.101}$$

The transformation gives $w = f(z)$. There is no simple analytic inverse in the form of $z = g(w)$, due to all of the square roots and the logarithm. This absence means that even if the potential function $F(z) = U + iV$ is known, it cannot be generalized to w space in an analytical fashion of the form $F(w)$. However, usually the converse is achievable and it is possible to obtain $z(F)$ and therefore $w(F)$. In this case it is easy to plot lines of constant U, or lines of constant V.

As an example, in the present problem consider that along the real axis there is a potential U_0 between $-1 \leqslant x \leqslant 1$, that gives a potential in the UHP

$$F = -i\frac{U_0}{\pi}\ln\left(\frac{z-1}{z+1}\right) = U + iV \tag{6.102}$$

$$z = i\cot\left(\frac{\pi F}{2U_0}\right) = i\cot\left[\frac{\pi(U + iV)}{2U_0}\right] \tag{6.103}$$

$$\theta \equiv \frac{\pi F}{2U_0}, \quad \sqrt{z+1} = \sqrt{\frac{i}{\sin(\theta)}}\sqrt{\cos(\theta) - i\sin(\theta)} = \sqrt{\frac{i}{\sin(\theta)}}e^{-i\theta/2}$$

$$\tag{6.104}$$

$$\sqrt{z-1} = \sqrt{\frac{i}{\sin(\theta)}}e^{i\theta/2}, \quad \sqrt{z^2-1} = \frac{i}{\sin(\theta)}$$

$$w = \frac{h}{\pi}\left\{\frac{i}{\sin(\pi F/2U_0)} + \ln[i\cot(\pi F/4U_0)]\right\} \tag{6.105}$$

Equation (6.102) gives $F(z)$ which has been solved to give $z(F)$ in Eq. (6.103). The latter is used to give $w(F)$. This equation can be used to solve for lines of constant U or V. Just fix, say U, then vary V and calculate the different points $w = u + iv$ which gives the shape of the curve. Some examples are:

- Set $U = 0$ for a line of constant U. Then

$$\cot(\pi F/4U_0) = \cot(i\pi V/4U_0) = -i\coth(\pi V/4U_0) \tag{6.106}$$

$$w = \frac{h}{\pi}\left\{\frac{1}{\sinh(\pi V/2U_0)} + \ln[\coth(\pi V/4U_0)]\right\} \tag{6.107}$$

If $V > 0$ the right-hand side is real, so $v = 0$ and $u > 0$. If $V < 0$ the argument of the logarithm is negative, which gives a factor of $\ln(-1) = i\pi$ making $v = h$. In this case $u < 0$. These two segments are the boundaries with $U = 0$.

- Set $V = 0$ for a line of constant V, then define $\theta = \pi U/(2U_0)$ and

$$w = \frac{h}{\pi}\left\{\frac{i}{\sin(\theta)} + \frac{i\pi}{2} + \ln[\cot(\theta/2)]\right\} \tag{6.108}$$

When $U = U_0$ then

$$\theta = \pi/2, \ \sin(\theta) = 1, \ \cot(\theta/2) = 1, \ \text{and} \ u = 0, \ v = h(1/2 + 1/\pi) \tag{6.109}$$

This point is on the vertical step. As U and θ decrease in value, v increases in value rapidly and u increases in value slowly.

Example 4. *Fringing field:* As a comprehensive example, consider the geometry shown in Fig. 6.10(a). There is a film of thickness b and a metal electrode covers the entire bottom of the film. On top of the film the electrode covers only the right half of the film. The problem is to calculate the flow of current from one plate to another. Let us say the bottom electrode is at $U = 0$ and the

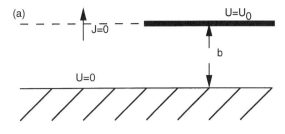

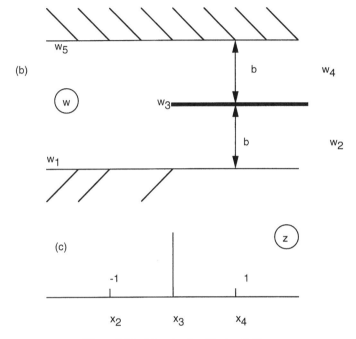

Figure 6.10 Mapping for fringing fields.

top electrode is at U_0. The potential satisfies Laplace's equation $\nabla^2 U = 0$ in the film, with these boundary conditions on the electrodes. The current is given by $\mathbf{J} = -\sigma \nabla U$. Another boundary condition is $J_v = 0$ along the top of the film where there is no electrode. This latter condition that $\partial U/\partial v = 0$ on the top of the film to the left of the electrode, makes this problem different from the case where the potential is allowed to have any value for ($u < 0$, $v = 0$).

Image theory is used to force the boundary condition that $J_v = 0$ where $u < 0$, $v = 0$. As shown in Fig. 6.10(b), the Schwartz–Christoffel transformation is found for the symmetric problem with two electrodes at $U = 0$, and a fin in the middle with U_0. Put the point of the fin at $w = 0$. The mapping has five singular points, in w space two are at infinity on the left, one is at zero, and two are infinity on the right. The two on the left can be omitted but the two on the right change sheets and must be included. The three points included are put at x_j points of, $-a$, 0, and $+a$. The angles ϕ_j are π, $-\pi$, and π so that the formula is

$$\frac{dw}{dz} = 2A\frac{z}{z^2 - a^2} \tag{6.110}$$

$$w = w_0 + A\ln\left(\frac{a^2 - z^2}{a^2}\right) \tag{6.111}$$

The constant was made $2A$ rather than A to eliminate the factor of $1/2$ in front of the logarithm, which is a nuisance. Since $w = 0$ at $z = 0$ then $w_0 = 0$. The sign of the argument of the logarithm [rather than $(z^2 - a^2)$] was chosen with this result in mind. These choices differ by a phase factor that is absorbed into w_0. Next, consider values of $x < -a$ where $z + a = \rho\exp(i\pi)$

$$w = A\left[i\pi + \ln\left(\frac{|a + z|(a - z)}{a^2}\right)\right] \tag{6.112}$$

$$A = -\frac{b}{\pi} \tag{6.113}$$

For $x < -a$ the w boundary is the bottom electrode that is along the imaginary line of $v = -ib$. This constraint sets $A = -b/\pi$. Similarly, for $x > a$ we are along the top electrode. In this case set $a - z = \rho\exp(-i\theta)$ where the angle θ is zero if it points toward $x = 0$ and is π if it points along $x > a$. In this case

$$\ln\left(\frac{a^2 - z^2}{a^2}\right) = -i\pi + \ln\left(\frac{z^2 - a^2}{a^2}\right) \tag{6.114}$$

$$w = ib \tag{6.115}$$

These arguments complete the derivation

$$w = -\frac{b}{\pi}\ln\left(\frac{a^2 - z^2}{a^2}\right) \tag{6.116}$$

The center electrode is at the potential U_0. In z space this electrode maps onto the real axis between $-a < x < a$, this gives the familiar potential

$$U(x, y) = \frac{U_0}{\pi}\text{Im}\left[\ln\left(\frac{z - a}{z + a}\right)\right] \tag{6.117}$$

This equation is the solution to Laplace's equation when $U(x, y = 0) = 0$ for $|x| > a$ and $U(x, y = 0) = U_0$ for $|x| < a$. The Schwartz–Christoffel transformation is solved for z

$$w = -\frac{b}{\pi}\ln\left(\frac{a^2 - z^2}{a^2}\right) \tag{6.118}$$

$$z = a\sqrt{1 - e^{-\pi w/b}} \tag{6.119}$$

$$U(u, v) = \frac{U_0}{\pi} \text{Im} \left\{ \ln \left(\frac{\sqrt{1 - e^{-\pi w/b}} - 1}{\sqrt{1 - e^{-\pi w/b}} + 1} \right) \right\} \quad (6.120)$$

The above expression is the complete solution of Laplace's equation for the potential everywhere in the film. For example, the current leaving the lower plate is

$$J_v(u, -ib) = -\sigma \left(\frac{\partial U}{\partial v} \right)_{v = -b} \quad (6.121)$$

In doing the above derivative, it is useful to set $w = u - ib + i\delta v$ where δv is small. Equation (6.120) is expanded in powers of δv and then the derivative is taken with respect to δv.

$$\exp(-\pi w/b) = \exp(i\pi - \pi u/b - i\pi\delta v/b) \approx -\lambda \left(1 - \frac{i\pi\delta v}{b} \right) \quad (6.122)$$

$$\lambda = e^{-\pi u/b}, \qquad \phi = \frac{\pi\delta v}{b} \quad (6.123)$$

$$\sqrt{1 - e^{-\pi w/b}} \approx \sqrt{1 + \lambda(1 - i\phi)} \approx \sqrt{\lambda + 1} - \frac{i\phi\lambda}{2\sqrt{1 + \lambda}} \quad (6.124)$$

Put this expression inside of the logarithm,

$$\ln \left(\frac{\sqrt{1 - e^{-\pi w/b}} - 1}{\sqrt{1 - e^{-\pi w/b}} + 1} \right) = \ln \left(\frac{\sqrt{\lambda + 1} - 1}{\sqrt{\lambda + 1} + 1} \right)$$
$$- \frac{i\phi\lambda}{2\sqrt{1 + \lambda}} \left[\frac{1}{\sqrt{\lambda + 1} - 1} - \frac{1}{\sqrt{\lambda + 1} + 1} \right] \quad (6.125)$$

$$\text{Im} \left\{ \ln \left(\frac{\sqrt{1 - e^{-\pi w/b}} - 1}{\sqrt{1 - e^{-\pi w/b}} + 1} \right) \right\} \approx -\frac{i\phi}{\sqrt{1 + \lambda}} = -\frac{i\pi\delta v}{b\sqrt{1 + \lambda}} \quad (6.126)$$

$$J_v(u, -ib) = \frac{\sigma U_0}{b} \frac{1}{\sqrt{1 + e^{-\pi u/b}}} \quad (6.127)$$

As $u \gg b$ then the exponent goes to zero, and we have the simple result $J_v = \sigma U_0/b$ which is the result for two parallel electrodes. The other limit of $u \ll -b$ makes the exponential grow to large values and the current becomes zero. Similar expressions for the current can be found for any point in the film.

Next, calculate the current entering the film with $U = U_0$ that is at $v = 0$ for $u > 0$. Here set $w = u + i\delta v$ where again δv is small,

$$\exp(-\pi w/b) = \exp(-\pi u/b - i\pi\delta v/b) = \lambda \left(1 - \frac{i\pi\delta v}{b} \right) \quad (6.128)$$

The steps are similar to those at $v = -ib$ except the sign of λ has changed. For the case that $u > 0$ then $\lambda < 1$. Following the same steps as above, with the opposite sign on λ, gives

$$J_v(u, 0^-) = \frac{\sigma U_0}{b} \frac{1}{\sqrt{1 - e^{-\pi u/b}}} \Theta(u) \tag{6.129}$$

The other possibility is that $u < 0$, $\lambda > 1$. In this case the equivalent of Eq. (6.124) is

$$\sqrt{1 - e^{-\pi w/b}} \approx \sqrt{1 - \lambda(1 - i\phi)} = i\sqrt{\lambda - 1} + \frac{\phi\lambda}{2\sqrt{\lambda - 1}} \tag{6.130}$$

The first term on the right is imaginary, while the term in $\phi \propto \delta v$ is real. When taking the imaginary part of the logarithm in Eq. (6.120), the part that is retained contains $\sqrt{\lambda - 1}$ and the part with δv drops out. Taking a derivative with respect to δv gives zero. These steps prove that the current $J_v = 0$ for $(u < 0, v = 0)$.

The convenient way to graph lines of constant U or constant V is to solve $z(F)$ and then $w(F)$, where $F = U + iV$. These steps are similar to those of prior examples,

$$F = -i\frac{U_0}{\pi} \ln\left(\frac{z - a}{z + a}\right) \tag{6.131}$$

$$z = ia \cot\left(\frac{\pi F}{2U_0}\right) \tag{6.132}$$

$$w = -\frac{b}{\pi} \ln\left(\frac{a^2 - z^2}{a^2}\right) = \frac{2b}{\pi} \ln\left[\sin\left(\frac{\pi F}{2U_0}\right)\right] \tag{6.133}$$

that is a simple formula. Some examples are:

- Set $U = U_0$ that give

$$\sin\left(\frac{\pi F}{2U_0}\right) = \sin\left(\frac{\pi}{2} + i\frac{\pi V}{2U_0}\right) = \cosh\left(\frac{\pi V}{2U_0}\right) \tag{6.134}$$

$$w(U_0, V) = u = \frac{2b}{\pi} \ln\left[\cosh\left(\frac{\pi V}{2U_0}\right)\right] \tag{6.135}$$

Since the hyperbolic cosine is always larger than one, the logarithm is positive and $u > 0$. So the equipotential is the central fin, in agreement with the boundary conditions.

- Set $U = 0$ that gives

$$\sin\left(\frac{\pi F}{2U_0}\right) = \sin\left(i\frac{\pi V}{2U_0}\right) = i \sinh\left(\frac{\pi V}{2U_0}\right) \tag{6.136}$$

$$\ln\left[\sin\left(\frac{\pi F}{2U_0}\right)\right] = \pm\frac{i\pi}{2} + \ln\left[\sinh\left(\frac{\pi|V|}{2U_0}\right)\right] \tag{6.137}$$

$$w(U_0, V) = u = \pm ib + \frac{2b}{\pi} \ln \left[\sinh \left(\frac{\pi |V|}{2U_0} \right) \right] \qquad (6.138)$$

When $V > 0$ then the argument of the logarithm is i so the phase is $i\pi/2$. However, when $V < 0$ then the argument is $(-i|V|)$ and the phase is $-i\pi/2$. The values $V > 0$ are the upper plate where $U = 0$ at $v = ib$. The values $V < 0$ are the lower plate where $U = 0$ at $v = -ib$.

- Set $U = U_0/2$ and

$$\sin \left(\frac{\pi F}{2U_0} \right) = \sin \left(\frac{\pi}{4} + i \frac{\pi V}{2U_0} \right)$$

$$= \frac{1}{\sqrt{2}} \left[\cosh \left(\frac{\pi V}{2U_0} \right) + i \sinh \left(\frac{\pi V}{2U_0} \right) \right] \qquad (6.139)$$

$$\ln \left[\sin \left(\frac{\pi F}{2U_0} \right) \right] = \frac{1}{2} \left\{ \ln \left[\cosh \left(\frac{\pi V}{U_0} \right) \right] - \ln(2) + 2i \tan^{-1} \left[\tanh \left(\frac{\pi V}{2U_0} \right) \right] \right\}$$

$$(6.140)$$

$$w(U_0/2, V) = u + iv$$

$$= \frac{b}{\pi} \left\{ \ln \left[\cosh \left(\frac{\pi V}{U_0} \right) \right] - \ln(2) + 2i \tan^{-1} \left[\tanh \left(\frac{\pi V}{2U_0} \right) \right] \right\}$$

$$(6.141)$$

When $\pi V/U_0 \gg 1$ the argument of the arctangent is unity, so its value is $\pi/4$. Then $v = b/2$. When $\pi V/U_0 \ll -1$ the argument of the arctangent is minus one, so its value is $-\pi/4$ and $v = -b/2$. These two cases are the points when $u \gg b$ and the system behaves as a parallel plate capacitor, with the potential $U_0/2$ midway between the plates. The other interesting point is where the line $U = U_0/2$ crosses the axis of $v = 0$, $V = 0$, that is at $u = -(b/\pi) \ln(2)$.

The Schwartz–Christoffel transformation is most useful when the integral over dw/dz can be done in a simple analytical formula. Generally this step is easy if there are only one or two singular points. If there are three, the integral is usually possible, although the result must often be looked up in a table of integrals. We have never succeeded in integrating a case with four or more singular points not at infinity.

6.5. van der Pauw

In material science the resistivity ρ of a solid is an important experimental parameter, that is often measured as a function of temperature, magnetic field, or impurity content. The usual case is to measure the resistance R of a bar of length L and cross-sectional area A, in which case $R = \rho L/A$. The resistance R has the units of Ohms while the resistivity ρ has the units of Ohm-meter. In the measurement on the bar, the largest experimental uncertainty is in the length and area of the bar. When characterizing new materials, the bars are often only millimeters or less in dimensions. Sometimes the experimentalist is able to grow the crystal in the shape of a thin platelet or as a thin film of uniform thickness d. In this case an accurate measurement of the resistivity ρ can be found using the method of van der Pauw[3]. The proof utilizes the concept of conformal mapping.

6.5.1. Currents

The understanding of the van der Pauw method requires the use of current density. In a three dimensional solid, the current density $\mathbf{J}$ has the units of A/m^2. It is the density of current at any point and a vector quantity. In two dimensions $\mathbf{J} = (J_x, J_y)$ has the units of A/m. It is convenient to express the two vector components (J_x, J_y) as the real and imaginary parts of a current function $J(z)$. First recall that

$$\mathbf{J} = -\sigma \nabla U(x, y) \tag{6.142}$$

$$J_x = -\sigma \frac{\partial U}{\partial x} \tag{6.143}$$

$$J_y = -\sigma \frac{\partial U}{\partial y} = \sigma \frac{\partial V}{\partial x} \tag{6.144}$$

The last equality uses the Cauchy–Riemann conditions. Since derivatives with respect to x are identical to those with respect to z, then

$$J(z) \equiv J_x - iJ_y = -\sigma \frac{d}{dz}[U + iV] = -\sigma \frac{dF}{dz}$$

The conductivity σ in two dimensions has the units of Siemen $= 1/\Omega$. The resistivity ρ is the inverse of the conductivity ($\sigma = 1/\rho$). In three dimensions the conductivity has the units of Siemens/m. The formula $\mathbf{J} = -\sigma \nabla U$ is dimensionally correct in both two and three dimensions. The procedure is to solve the problems in two dimensions and to convert to three dimensions at the end by adding the factor d for the film thickness.

The current $J(z)$ is an analytic function of z. Its real and imaginary parts are $(J_x, -J_y)$ and is distinct from the vector function $\mathbf{J}(x, y) = (J_x, J_y)$. The real part of $J(z)$ is J_x and the imaginary part is $-J_y$, rather than J_y. This latter convention follows directly from the definition of the current as the gradient of the potential. The above equation can be integrated

$$-\rho \int_{z_a}^{z_b} dz J(z) = F(z_b) - F(z_a) \tag{6.145}$$

$$U(z_b) - U(z_a) = -\rho \, \text{Re}\left[\int_{z_a}^{z_b} dz J(z) \right] \tag{6.146}$$

The difference of the potential between two points can be found as a contour integral over the current between these two points. The result is independent of the actual path as long as no poles or branch cuts are traversed during the contour.

Next, consider how a wire lead injects current into a two dimensional region. Figure 6.11(a) shows a current I coming up a wire lead which is

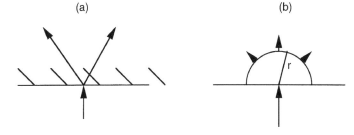

(a)   (b)

Figure 6.11   Current entering a sheet from a point along the edge.

connected to the edge of the two dimensional region. How does the current I (in amperes) get converted to the current density $J(z)$ (A/m)? Figure 6.11(b) shows in detail the region near to the contact. The semicircle about the contact point has a radius r that is small. The current density J is integrated along this semicircle and the total should be the injected current I. Of course, only the component J_n that is normal to the semicircle contributes to the net current along each segment $r\, d\theta$,

$$I = r \int_0^\pi d\theta J_n(\rho, \theta) \tag{6.147}$$

This formula accounts for all the electrons entering the two dimensional region. In the region of the contact, define a local coordinate system $z' = x' + iy'$

$$re^{i\theta} = x' + iy', \qquad r^2 = (x')^2 + (y')^2 \tag{6.148}$$

$$\hat{n} = \frac{(x', y')}{r} \tag{6.149}$$

The unit vector $\hat{n}$ is perpendicular to the semicircle at each point. Therefore

$$J_n = \hat{n} \cdot \mathbf{J} = \frac{1}{r}[x'J_x + y'J_y] = \frac{1}{r}\operatorname{Re}[z'J] \tag{6.150}$$

$$z'J = (x' + iy')(J_x - iJ_y) = x'J_x + y'J_y + i(y'J_x - x'J_y) \tag{6.151}$$

$$I = \int_0^\pi d\theta \operatorname{Re}[z'J] \tag{6.152}$$

The last formula is an identity that must be true in all circumstances. The obvious solution is

$$J(z') = \frac{I}{\pi z'} \tag{6.153}$$

Next, consider that the current is injected along the real axis at the point x_1. Then $z' = z - x_1$ and

$$J(z) = \frac{I}{\pi}\frac{1}{z - x_1} \tag{6.154}$$

The above formula gives the current $J(z)$ in the UHP where the current I is injected at the point x_1. Another option is to have the current injected at point x_1 and be removed by a current lead at point x_2. Removing a current only changes the sign of I. The current everywhere in the UHP is given by

$$J(z) = \frac{I}{\pi} \left[\frac{1}{z - x_1} - \frac{1}{z - x_2} \right] \tag{6.155}$$

This formula plays an essential role in proving the method of van der Pauw. The current must be given by an analytic function and this requirement comes from the definition that $\rho \mathbf{J} = -\nabla U$. The potential U must be an analytical function since it satisfies Laplace's equation.

6.5.2. Resistance

Figure 6.12 shows a two dimensional figure of a thin platelet. It has four current leads connected to it at the points A, B, C, D. Let a current I enter one of the four leads ($i = A$) and exit another ($j = B$). Measure the voltage difference between the two leads not carrying current. If (ij) are the current leads and (kl) are the voltage leads, the resistance measured by this process is defined as

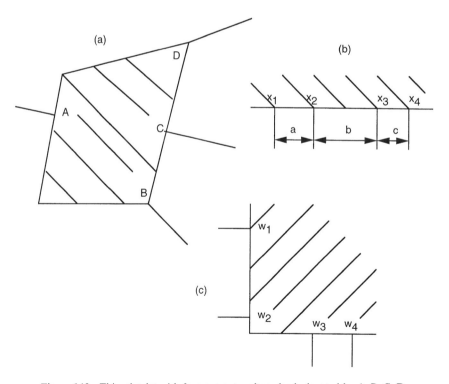

Figure 6.12 Thin platelet with four current–voltage leads denoted by A, B, C, D.

$$R_{ij:kl} = \frac{U_l - U_k}{I_{ij}} \tag{6.156}$$

The order of indices is important since

$$R_{ij:kl} = -R_{ji:kl} = -R_{ij:lk} \tag{6.157}$$

van der Pauw proved a theorem

$$1 = \exp\left[-\frac{\pi d}{\rho} R_{AB:CD}\right] + \exp\left[-\frac{\pi d}{\rho} R_{BC:DA}\right] \tag{6.158}$$

Two different measurements, such as $R_{AB:CD}$ and $R_{BC:DA}$ are sufficient for finding the ratio of $\pi d/\rho$ with a simple computer code, from which it is easy to find ρ. The proof of this theorem requires complex variables and mapping. Note the result is independent of the shape of the platelet and independent of the distance between the leads along the edge.

Figure 6.12(b) shows a possible arrangement for the measurement. The platelet is the entire UHP. The four wire connections are at the points x_1, x_2, x_3, x_4 along the x axis. They are separated by the distances a, b, c as shown in the figure. In the first measurement assume the current enters at x_1 and exits at x_2.

$$J = \frac{I}{\pi}\left[\frac{1}{(z - x_1)} - \frac{1}{(z - x_2)}\right] \tag{6.159}$$

The voltage drop between the other two points x_3, x_4 is the line integral of the current between these two points. For simplicity, take the integral along the x axis,

$$\begin{aligned}
U_4 - U_3 &= -\rho \int_{x_3}^{x_4} dx J_x \\
&= -\frac{I\rho}{\pi} \int_{x_3}^{x_4} dx \left[\frac{1}{x - x_1} - \frac{1}{x - x_2}\right] \\
&= \frac{I\rho}{\pi} \ln\left[\frac{(x_4 - x_2)(x_3 - x_1)}{(x_4 - x_1)(x_3 - x_2)}\right] \\
&= \frac{I\rho}{\pi} \ln\left[\frac{(a + b)(b + c)}{(a + b + c)b}\right]
\end{aligned} \tag{6.160}$$

Convert the two dimensional resistivity above into a three dimensional resistivity: $\rho_{2D} = \rho_{3D}/d$ where d is the film thickness. Here the current is still two dimensional, so the integral has the units of amperes, ρ has the units of ohm/meter, and V is in volts. The resistance is

$$R_{21:34} = \frac{\rho}{\pi d} \ln\left[\frac{(a + b)(b + c)}{(a + b + c)b}\right] \tag{6.161}$$

$$\exp\left[-\frac{\pi d}{\rho}R_{12:34}\right] = \frac{(a+b+c)b}{(a+b)(b+c)} \tag{6.162}$$

In the second measurement the current goes in at x_2 and comes out at x_3. The voltage is measured from x_4 to x_1.

$$\begin{aligned}
U_1 - U_4 &= \frac{I\rho}{\pi d}\int_{x_4}^{x_1} dx \left[\frac{1}{x-x_2} - \frac{1}{x-x_3}\right] \\
&= \frac{I\rho}{\pi}\ln\left[\frac{(x_4-x_2)(x_3-x_1)}{(x_4-x_3)(x_2-x_1)}\right] \\
&= \frac{I\rho}{\pi}\ln\left[\frac{(a+b)(b+c)}{ac}\right] \tag{6.163}
\end{aligned}$$

$$R_{23:41} = \frac{\rho}{\pi d}\ln\left[\frac{(a+b)(b+c)}{ac}\right] \tag{6.164}$$

The final step in the derivation is to add these formulas in a way suggested by van der Pauw,

$$\begin{aligned}
\exp\left[-\frac{\pi d}{\rho}R_{12:34}\right] + \exp\left[-\frac{\pi d}{\rho}R_{23:41}\right] &= \frac{b(a+b+c)}{(a+b)(b+c)} + \frac{ac}{(a+b)(b+c)} \\
&= \frac{(a+b)(b+c)}{(a+b)(b+c)} = 1 \tag{6.165}
\end{aligned}$$

The distance factors combine to give one. This result proves the van der Pauw theorem for the case that all of the currents and voltages are connected at points along the x axis and the material is in the UHP.

6.5.3. Mapping

Under a Schwartz–Christoffel transformation the points along the real axis get mapped onto the border of the figure in w space. The current and voltage leads are still connected at edge points for any general figure. The second part of the proof is that conformal mapping does not change the result, since the potential is an analytic function. The current is integrated to get the potential $U(z)$ in the UHP. When the current I enters at x_1 and leaves at x_2 the complex potential is

$$F(z) = U(z) + iV(z) = -\rho\int dz J(z)$$

$$= -\frac{\rho I}{\pi}\ln\left(\frac{z-x_1}{z-x_2}\right) \tag{6.166}$$

The mapping $w(z)$ is inverted to give $z(w)$ so that $x_1 \to w_1 = w(x_1)$. The above expression gives

$$F(w) = -\frac{\rho I}{\pi} \ln\left(\frac{z(w) - x_1}{z(w) - x_2}\right) \tag{6.167}$$

For example, the mapping $w = \sqrt{z}$ converts the UHP of z space to the first quadrant $(u \geqslant 0, v \geqslant 0)$ of w space as shown in Fig. 6.12(c).

$$F(w) = -\frac{\rho I}{\pi} \ln\left(\frac{w^2 - x_1}{w^2 - x_2}\right) = -\frac{\rho I}{\pi} \ln\left(\frac{w^2 - w_1^2}{w^2 - w_2^2}\right) \tag{6.168}$$

$$U(w_4) - U(w_3) = -\frac{\rho I}{\pi} \ln\left[\frac{(w_4^2 - w_1^2)(w_3^2 - w_2^2)}{(w_4^2 - w_2^2)(w_3^2 - w_1^2)}\right]$$

$$= -\frac{\rho I}{\pi} \ln\left[\frac{(x_4 - x_1)(x_3 - x_2)}{(x_4 - x_2)(x_3 - x_1)}\right] \tag{6.169}$$

$$R_{12:34} = \frac{U(w_4) - U(w_3)}{dI} = -\frac{\rho}{d\pi} \ln\left[\frac{(x_4 - x_1)(x_3 - x_2)}{(x_4 - x_2)(x_3 - x_1)}\right] \tag{6.170}$$

The result for the resistance is unchanged by the mapping. The van der Pauw theorem which was proved for all leads along a single straight edge is valid for an arbitrarily shaped figure that can be attained by a Schwartz–Christoffel transformation. The theorem is not valid for platelets that have holes in them.

A key feature of the theorem is that the voltage difference, such as $U_4 - U_3$, is unchanged by the mapping. The voltage $U(w) = U(z(w))$. However, the formula for the current is altered by the mapping. In the z and w spaces we have

$$J(z) = -\sigma \frac{d}{dz} U(z) \tag{6.171}$$

$$J(w) = -\sigma \frac{d}{dw} U(w) \tag{6.172}$$

$$J(w) = -\sigma \frac{d}{dw} U[z(w)] = -\sigma \left[\frac{d}{dz} U(z)\right] \frac{dz}{dw} \tag{6.173}$$

$$J(w) = J(z) \frac{dz}{dw} \tag{6.174}$$

The last identity is required to preserve the potential difference between two points as evaluated by the line integral of the current,

$$U_4 - U_3 = -\rho \int_{w_3}^{w_4} dw J(w) = -\rho \int_{w_3}^{w_4} dw J(z) \frac{dz}{dw}$$

$$= -\rho \int_{z_3}^{z_4} dz J(z) \tag{6.175}$$

In a mapping from z space to w space the potential $U(z)$ becomes $U(w) = U[z(w)]$. However, the current changes according to Eq. (6.174). For the mapping $w = \sqrt{z}$ the current is

$$J(w) = 2w \frac{I}{\pi} \left(\frac{1}{w^2 - w_1^2} - \frac{1}{w^2 - w_2^2} \right) \tag{6.176}$$

Mapping can be used in two dimension to find the current distribution for two cases, (a) the boundary conditions are fixed voltages and (b) the boundary conditions are the injection of a current I through external wire leads. These two methods of injecting current are useful for a variety of problems.

References

1. R. V. Churchill, *Introduction to Complex Variables and Applications.* McGraw-Hill, 1948.
2. P. M. Morse and H. Feshbach, *Methods of Theoretical Physics,* vol. 1 and 2. McGraw-Hill, 1953.
3. L. J. van der Pauw (1958), Philips Research Report **13**, 1–9.

Problems

1. Give the region in w space obtained from the following mappings. In each case the space in z is the upper-half plane ($v \geqslant 0$).

   (a) $w = \ln(z^2)$

   (b) $w = \dfrac{i - z}{i + z}$

   (c) $w = \pi i + z - \ln(z)$

2. For the problem of a fin of potential U_0 above a grounded plate, in fig. 6.6, find the formula for the surface charge on the grounded plate.

3. For the mapping that gave two fins, find $V(u, v)$.

4. For the problem with two fins, set $U_0 = \pi$:

   (a) Find the value of V, $\partial U/\partial v$ at the point $(u = 2, v = 0)$.
   (b) Find the value of v on the v-axis (where $u = 0$) which has the same value of V as the point in (a).
   (c) Find $\partial U/\partial u$ at the point in (b).
   (d) Does ∇U have the same value at all points along the stream function?

5. Along the real axis, let $U(x,0) = + U_0$, $a \leqslant x \leqslant b$, and $U(x, 0) = - U_0$, $-b \leqslant x \leqslant -a$, and $U(x, 0) = 0$ elsewhere.

   (a) Find $U(x, y)$ in the UHP.
   (b) Find $J_x = - \sigma \partial U/\partial x$ along the y-axis.
   (c) Find the total current I between the two potential sources by integrating J_x along the y-axis over the interval $(0, \infty)$.
   (d) Find the total resistance of the system: $R = 2U_0/I$ for current flow between the negative and positive electrode.

6. Find the potential in the first quadrant ($u \geqslant 0$, $v \geqslant 0$, see fig. 6.13a) subject to the boundary condition that $U = 0$ on all boundaries (u- and v-axes) except: $U = U_0$ for $v = 0$, $0 \leqslant u \leqslant a$; $u = 0$, $0 \leqslant v \leqslant a$. Then find the potential along the line $u = v$.

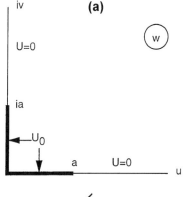

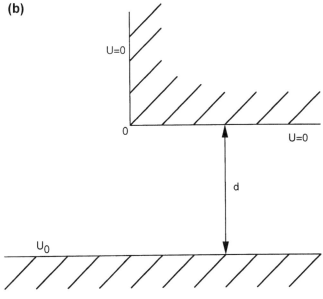

Figure 6.13 (a) and (b).

7. Assume that the w-space is a strip over all values of u but confined $0 \leqslant v \leqslant b$.

(a) Do a Schwartz-Christoffel transformation with one singular point $x_1 = 0$, $u_1 \to \infty$, which should give $w = w_0 + A \ln(z)$.

(b) Do a Schwartz-Christoffel transformation with two singular points $x_j = \pm 1$, $u_j = \pm \infty$. What is $w(z)$ in this case?

(c) Assume in w-space that the upper boundary is at a potential U_0 and the lower boundary is at a potential $U = 0$. How to these lines map onto z-space in the above two mappings?

8. Give the Schwartz-Christoffel transformation for fig. 6.13(b).

(a) Give $w = f(z)$ including all constants.
(b) Find $F(z) = U + iV$ and the inverse $z(F)$.
(c) Use the results of (b) in (a) to get $w = g(F)$.
(d) Plot out the line of constant $U = U_0/2$

9. In the van der Pauw method, prove that $R_{ijkl} = R_{kl,ij}$

10. Let the w-space be the first quadrant ($u \geqslant 0$, $v \geqslant 0$). A current I enters at $w_1 = ia$ and departs at $w_2 = a$.

    (a) Derive an expression for the current $J(w)$.
    (b) Evaluate J_u, J_v along the line $u = v$.

11. Let a current I enter the UHP of z-space at $x_1 = -1$ and depart at $x_2 = +1$. Then do the mapping $w = \sin^{-1}(z)$.
    (a) Draw the map of w-space and indicate the position of the current leads.
    (b) Derive $J(w)$.
    (c) How do J_u, J_v depend upon v at $u = 0$?

12. Repeat the above problem for $w = \sin^{-1}(4z)$.

13. Let the w-space be the strip of all values of u and $0 \leqslant v \leqslant b$. A current I enters a lead at ($u = 0$, $v = 0$) and exits at ($u = 0$, $v = b$). Find the current (J_u, J_v) everywhere in the strip.

Markov Averaging

Markov refers to a type of statistical averaging that is widely used for many different kinds of problems. The easiest problem is to have a particle moving in one dimension with a random walk. Then one can calculate the probability $p_N(x)$ that the particle has moved a distance x after N steps. Alternately, the other particles may exert a force F on the central particle, depending upon the positions of these other particles. Then one might try to calculate the probability $p_N(F)$ of having the force F if there are N other particles. The central theme of these problems, defines the average as being Markovian, is that the other particles act independently of each other. Similarly, in the random walk, each step proceeds independently of any earlier or later step. This chapter will give three examples of Markov averaging for several different problems. Three are chosen to give different final distributions: Gaussian, exponential, and Lorentzian. The main ideas are contained in a review article by Chandrasekhyar (1).

7.1. Random Walk

Consider a particle undergoing random walk along a line in one dimension. The problem is to determine how far the particle has moved after N steps. The steps can either be forward or backwards, with equal probability. The distance of the i-th step is denoted as x_i, that can have either sign. The steps have a probability distribution $P(x_i)$ that is the probability of having a step of length x_i. It is assumed that this distribution is normalized to unity

$$\int_{-\infty}^{\infty} dx_i P(x_i) = 1 \tag{7.1}$$

Several different examples of $P(x_i)$ will be considered in the text and homework problems.

Define $p_N(x)$ as the probability of the particle going a distance x after N-steps. It depends on the choice of $P(x_i)$. The distance x is the summation of all of the steps

$$x = \sum_{i=1}^{N} x_i \tag{7.2}$$

This expression is correct, but is rather useless since each term on the right-hand side is a statistical quantity that needs to be averaged. The way of doing this averaging is described below. The proper expression is given for one, two, and then N steps:

$$p_1(x) = \int dx_1 P(x_1)\delta(x - x_1) = P(x) \tag{7.3}$$

$$p_2(x) = \int dx_1 P(x_1) \int dx_2 P(x_2)\delta(x - x_1 - x_2) = \int dx_1 P(x_1)P(x - x_1) \tag{7.4}$$

$$p_N(x) = \left[\Pi_{i=1}^{N} \int dx_i P(x_i) \right] \delta\left(x - \sum_{n=1}^{N} x_n\right) \tag{7.5}$$

The result for $p_1(x)$ is rather obvious, it is $P(x)$. The result for $p_2(x)$ is also intuitive, since it is the convolution of two step probabilities. The result for N-steps is the $N - 1$ convolution of N steps. The results are derived in an easy fashion by using the concept of the delta function to force the summation of the steps to be x. The multiple integrals over $P(x_i)dx_i$ allow the proper statistical averaging. In each expression for $p_N(x)$ there is only one delta function, but N-averages. Also note that the probability integrates to unity

$$\int dx p_N(x) = \left[\Pi_{i=1}^{N} \int dx_i P(x_i) \right] \int dx \delta\left(x - \sum_n x_n\right)$$

$$= \left[\Pi_{i=1}^{N} \int dx_i P(x_i) \right] = 1 \tag{7.6}$$

The solution to the general problem can be evaluated in a simple fashion by using the integral definition of the delta function

$$\delta\left(x - \sum_n x_n\right) = \int_{-\infty}^{\infty} \frac{dt}{2\pi} e^{it(x - \sum_n x_n)} \tag{7.7}$$

$$p_N(x) = \int \frac{dt}{2\pi} e^{itx} \Pi_{i=1}^{N} \left[\int dx_i P(x_i) e^{-itx_i} \right]$$

$$= \int \frac{dt}{2\pi} e^{itx} M(t)^N \tag{7.8}$$

$$M(t) = \int dx_i P(x_i) e^{-itx_i} \tag{7.9}$$

By using the integral form of the delta function, the factor of $\exp(-itx_i)$ enters the integrals over the statistical distribution $P(x_i)$. The Markov assumption is that all these factors $M(t)$ are exactly alike. The general solution for any value of N is obtained by a two step process: (i) evaluate the integral $M(t)$, and (ii) evaluate the t-integral to obtain $p_N(x)$.

A more general description would start by defining the probability distribution $P_N(x_1, x_2, \ldots, x_n)$ having N-steps, of which the first is x_1, the second is x_2, and so on. The Markov average is to assume that all these probabilities are statistically independent: $P_N = P(x_1)P(x_2)\ldots P(x_N)$. The answer has the simple form in Eqs. (7.8–7.9).

The first example uses a Gaussian distribution for $P(x_i)$

$$P(x_i) = \frac{1}{l\sqrt{2\pi}} e^{-x_i^2/2l^2} \tag{7.10}$$

$$\int_{-\infty}^{\infty} dx_i P(x_i) = 1 \tag{7.11}$$

$$\int_{-\infty}^{\infty} dx_i x_i P(x_i) = 0 \tag{7.12}$$

$$\int_{-\infty}^{\infty} dx_i x_i^2 P(x_i) = l^2 \tag{7.13}$$

A distribution is "Gaussian" if it is an exponential of $-x^2$. The first integral demonstrates that the distribution $P(x_i)$ integrates to unity. The second integral shows that the average step is zero, since there are equal numbers of steps in either direction. The third integral shows that the square of the average step is l^2. The "k-th moment" of a distribution is its integral over x^k. In order to be consistent, all distributions $P(x_i)$ will be normalized to have these same three moments.

In this case the integral for $M(t)$ is simple

$$M(t) = \int \frac{dx}{l\sqrt{2\pi}} \exp\left[-\frac{x^2}{2l^2} - itx\right]$$

$$= \int \frac{dx}{l\sqrt{2\pi}} \exp\left[-\frac{1}{2l^2}(x + itl^2)^2 - t^2l^2/2\right] \tag{7.14}$$

Change variables of integration to $x' = x + itl^2$ and the integral over dx' is another simple Gaussian. The term obtained by completing the square is the only part that remains

$$M(t) = e^{-t^2l^2/2} \tag{7.15}$$

$$p_N(x) = \int \frac{dt}{2\pi} e^{itx - Nt^2l^2/2} \tag{7.16}$$

The last integral is another Gaussian that is evaluated by completing the square in the exponent:

$$itx - \frac{Nt^2l^2}{2} = -\frac{Nl^2}{2}\left[t - \frac{ix}{Nl^2}\right]^2 - \frac{x^2}{2Nl^2} \tag{7.17}$$

Changing variables $s = t - ix/(Nl^2)$ gives a simple integral for ds. The final result is

$$p_N(x) = \frac{e^{-x^2/(2Nl^2)}}{l\sqrt{2\pi N}} \tag{7.18}$$

The above formula is the exact probability $p_N(x)$ when the individual x_i have the Gaussian distribution $P(x_i)$. Examine a few moments of $p_N(x)$:

$$\int dx\, p_N(x) = 1 \tag{7.19}$$

$$\int dx\, x p_N(x) = 0 \tag{7.20}$$

$$\int dx\, x^2 p_N(x) = Nl^2 \tag{7.21}$$

The zeroth moment is the normalization that agrees with the general rule Eq. (7.6). The first moment vanishes since the particle has equal probability of going either direction. The second moment gives Nl^2.

These results can be derived in a simple fashion. Let brackets denote the average

$$\langle x_i^n \rangle = \int dx_i x_i^n P(x_i) \tag{7.22}$$

Write

$$x = \sum_i x_i, \quad \langle x \rangle = \sum_i \langle x_i \rangle = 0 \tag{7.23}$$

$$x^2 = \sum_{ij} x_i x_j = \sum_i x_i^2 + \sum_i \sum_{j \neq i} x_i x_j \tag{7.24}$$

$$\langle x^2 \rangle = N \langle x_i^2 \rangle = Nl^2 \tag{7.25}$$

One assumes that $\langle x_i x_j \rangle = 0$ whenever $i \neq j$ and $\langle x_i^2 \rangle = l^2$. Then the average $\langle x^2 \rangle = Nl^2$.

It is interesting to select another choice of $P(x_i)$ in order to establish how the final formula changes for $p_N(x)$. Another choice is to have a fixed step length, that is either $\pm l$.

$$P(x_i) = \tfrac{1}{2}[\delta(x_i - l) + \delta(x_i + l)] \tag{7.28}$$

This choice maintains the moments listed in Eqs. (7.11–7.13). The function $M(t)$ is also simple

$$M(t) = \int \frac{dx_i}{2} [\delta(x_i - l) + \delta(x_i + l)] e^{-itx_i} = \cos(tl) \tag{7.27}$$

$$M(t)^N = \cos^N(tl) \tag{7.28}$$

$$p_N(x) = \int \frac{dt}{2\pi} e^{itx} \cos^N(tl) \tag{7.29}$$

The expression for $p_N(x)$ is a summation of delta functions. Some examples are

$$p_1(x) = \tfrac{1}{2}[\delta(x - l) + \delta(x + l)] \tag{7.30}$$

$$p_2(x) = \tfrac{1}{4}[2\delta(x) + \delta(x - 2l) + \delta(x + 2l)] \tag{7.31}$$

$$p_N(x) = \frac{1}{2^N} \sum_{j=0}^{N} \binom{N}{j} \delta[x - (N-2j)l] \tag{7.32}$$

The result for a general value of N is expressed as a summation of delta functions weighted by the binomial coefficient. At first sight, this expression seems to be very different from the earlier result for a Gaussian. However, the two results are very similar. They can be shown to be similar by smearing out the delta functions, that makes $p_N(x)$ a continuous distribution. The continuous distribution is obtained by drawing a smooth curve through the height of every delta function. To derive the result of this smoothing treat Eq. (7.29) as a saddle point integral. Let $x \to \bar{x}$ so that $t \to z = x + iy$, and x is not confused with $\bar{x}$

$$p_N(\bar{x}) = \int \frac{dz}{2\pi} e^{F(z)} \tag{7.33}$$

$$F(z) = iz\bar{x} + N \ln[\cos(zl)] \tag{7.34}$$

The position z_0 of the saddle point is given by

$$F'(z_0) = 0 = i\bar{x} - Nl \tan(z_0 l) \tag{7.35}$$

The points that can satisfy this equation are along the imaginary axis. The solution is found by setting $z_0 = iy_0$ giving

$$\frac{x}{Nl} = \tanh(y_0 l) = y_0 l - \frac{(y_0 l)^3}{3} + \cdots \tag{7.36}$$

On the right are the first two terms in the Taylor series for the hyperbolic tangent. At large value of N, the first term gives $y_0 \approx x/(Nl^2)$ and higher order terms in the Taylor series are neglected since they are $O(1/N^3)$. A good approximation to the function $F(z)$ is

$$F(z) \approx iz\bar{x} + N \ln[1 - (zl)^2/2] \approx iz\bar{x} - N(zl)^2/2 \tag{7.37}$$

The saddle point integral is done along the x axis. So set $z = x + iy_0$ and

$$F(z) \approx -\frac{\bar{x}^2}{2Nl^2} - \frac{Nx^2l^2}{2} \tag{7.38}$$

$$p_N(\bar{x}) \approx e^{-\bar{x}^2/(2Nl^2)} \int \frac{dx}{2\pi} e^{-Nl^2x^2/2} \tag{7.39}$$

$$= \frac{e^{-\bar{x}^2/(2Nl^2)}}{l\sqrt{2\pi N}} \tag{7.40}$$

The result is approximately correct, subject to corrections in the form of an asymptotic series in $O(1/N)$.

The final formula is identical to Eq. (7.18). This agreement is not a coincidence. In the limit of large N, all forms of $P(x_i)$ give the same formula for $p_N(x)$. To prove this assertion evaluate $M(t)$ as a power series in t. Start from Eq. (7.9) and expand the exponent in a Taylor series

$$M(t) = \int dx_i P(x_i) \left[1 - itx_i + \frac{(-itx_i)^2}{2!} \cdots \right]$$

$$= 1 - \frac{t^2l^2}{2} + O(t^4) \tag{7.41}$$

$$M(t)^N = \exp[N \ln(1 - t^2l^2/2 + \cdots)] \approx e^{-Nt^2l^2/2} \tag{7.42}$$

$$p_N(x) \approx \frac{e^{-\bar{x}^2/(2Nl^2)}}{l\sqrt{2\pi N}} \tag{7.43}$$

All forms of $P(x_i)$ give the same result for $p_N(x)$ in the limit of large N. The only constraint is that $P(x_i)$ be chosen to obey the moments in Eqs. (7.11–7.13).

The diffusion coefficient D has the units of meters^2 per second. Chandrasekhyar showed that in one dimension a diffusing particle has an average value at long time (t) of

$$\lim_{t \to \infty} \langle x^2 \rangle = 2Dt \tag{7.44}$$

Random walk is a diffusion process, so the two expressions for $\langle x^2 \rangle$ should be equal

$$\langle x^2 \rangle = 2Dt = Nl^2 \tag{7.45}$$

Denote by τ the average time it takes to do one step in the process of random walking. Then $t = N\tau$ and

$$D = \frac{Nl^2}{2t} = \frac{Nl^2}{2N\tau} = \frac{l^2}{2\tau} \tag{7.46}$$

The diffusion coefficient is the value of l^2 divided by twice the average time interval for each step. One can redefine $p_N(x)$ in Eq. (7.18) as the average

$$p(x, t) = \frac{e^{-x^2/(4Dt)}}{2\sqrt{\pi Dt}} \tag{7.47}$$

This formula will be derived again in a later chapter as a solution to the diffusion equation in one dimension.

Most distribution functions $P(x_i)$ give the same Gaussian distribution for $p_N(x)$. The only requirement is that the first three moments are Eqs. (7.11)–(7.13). Any distribution with these three moments give a Gaussian probability. In that sense the topic is dull, since one always gets the same result for $p_N(x)$. There are a few cases in which different distribution functions are obtained because the moments Eqs. (7.11)–(7.13) do not exist. Some of these other cases are discussed in the remaining sections of the chapter.

7.2. Speckle

Speckle is the interference pattern from the scattering of coherent light by a random collection of objects. Most students have had some optics, and are familiar with the Fraunhofer pattern of light from a single slit, or else Young's diffraction from a pair of parallel slits.

Imagine the diffraction pattern from a collection of N slits that are located randomly on the scattering plane. Although that experiment is rarely done, a similar idea is to scatter from a collection of N random objects. The objects could be water droplets in a cloud, or else colloidal particles in a solution. Huygen's principle is that when scattering from random objects, the light waves coherently form a wave front in the forward direction. The light maintains its coherence, and the only effect of the scattering is to slow down the velocity, that is expressed as a refractive index. However, Huygen's principle only applies when the average distance d between scattering centers is much smaller than the wavelength of light λ. Alternately, $kd \ll 1$, where $k = 2\pi/\lambda$ is the wave vector of light, so $d \ll \lambda$. Huygen's principle does not apply in the other limit $kd \gg 1$. The limit $kd \gg 1$ produces a speckle pattern in the forward direction, rather than a continuous wave front. Speckle appears as alternate regions of light and dark, in random arrangements, when projected on a screen behind the film or fluid doing the scattering.

Markov methods are used to calculate the intensity distribution $P(I)$ of the light. This distribution is found experimentally by measuring the light and dark regions, and then creating a distribution of intensities. Instead of working directly with intensities, start by considering the pattern of electric fields.

$$\mathbf{E} = e^{-i\omega t} \sum_{j=1}^{N} \mathbf{E}_j e^{i\phi_j} \tag{7.48}$$

$$I = |\mathbf{E}|^2 \tag{7.49}$$

$$I_0 = \sum_j |\mathbf{E}_j|^2 = N\langle E_j^2 \rangle \tag{7.50}$$

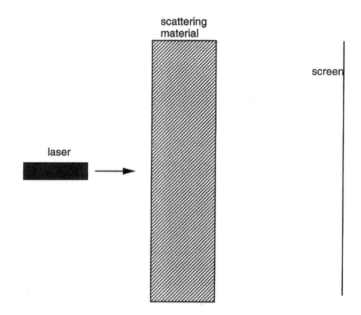

Figure 7.1 Experimental setup to observe speckle.

The total electric field at a point on a screen $\mathbf{E}$ is the sum of the individual fields coming from different scattering centers. The number of scattering centers N is limited by the aperature of the measuring device, and the finite extent of the scattering chamber. A typical experimental arrangement is shown in Fig. 7.1. The individual field components have a real amplitude $\mathbf{E}_j$ and a phase factor. The phase factor $\phi_j = kL_j + \Sigma_i \delta_i$ consists of two terms. One is the wave vector k times the pathlength L_j of that wave between the coherent light source and the detector. Since each light ray random walks through the medium, the pathlengths L_j are typically much larger than the dimensions of the system. Secondly, there are also phase shifts δ_i that occur each time the light scatters from a particle in the random medium. The outcome of all of these contributions is that the phases ϕ_j have a wide distribution of values. Since only values modulus 2π are important, it is assumed the phases ϕ_j have a continuous and random distribution of values over the range of $0 \leqslant \phi_j \leqslant 2\pi$.

$$P(\phi_j) = \frac{1}{2\pi}, \quad \int_0^{2\pi} d\phi_j P(\phi_j) = 1 \tag{7.51}$$

It is assumed the scattering is elastic. The amplitudes of the individual field components $\mathbf{E}_j$ will also have a distribution in value. Now it is assumed that all electric field amplitudes $\mathbf{E}_j$ have the same value E_0. Later a second calculation is performed that includes variations in these amplitudes. Since they are vectors, they can also point in random directions. This feature could be included by doing another Markov average over the possible vector orientations. Here we consider only the random phases, and the variations in directions are ignored. The electric field is treated as a scalar quantity. This viewpoint is rigorous if the scalars are the field components in the different phases.

First consider the Markov average of the function Λ

$$E = E_0 \Lambda \tag{7.52}$$

$$\Lambda = \sum_j e^{i\phi_j} \tag{7.53}$$

$$p(\Lambda) = \left\langle \delta \left(\Lambda - \sum_j e^{i\phi_j} \right) \right\rangle$$

$$= \int_{-\infty}^{\infty} \frac{dt}{2\pi} e^{it\Lambda} \Pi_j \int_0^{2\pi} \frac{d\phi_j}{2\pi} e^{-ite^{i\phi_j}}$$

$$= \int_{-\infty}^{\infty} \frac{dt}{2\pi} e^{it\Lambda} M^N(t) \tag{7.54}$$

$$M(t) = \int_0^{2\pi} \frac{d\phi_j}{2\pi} e^{-ite^{i\phi_j}} \tag{7.55}$$

The average over angles ϕ_j assume they have equal probability of having any value. The last integral gives one, since expanding the exponential gives

$$\int_0^{2\pi} \frac{d\phi_j}{2\pi} e^{-ite^{i\phi_j}} = \sum_l \frac{(-it)^l}{l!} \int_0^{2\pi} \frac{d\phi_j}{2\pi} e^{il\phi_j} = 1 \tag{7.56}$$

All the integrals give zero except for $l = 0$. This result gives $p(\Lambda) = \delta(\Lambda)$. This answer is correct. The average value of Λ is zero, and the distribution has no width.

The actual quantity of interest is the intensity $I = E_0^2 |\Lambda|^2$.

$$\Lambda^* \Lambda = \sum_j [\cos(\phi_j) + i\sin(\phi_j)] \sum_l [\cos(\phi_l) - i\sin(\phi_l)]$$

$$= \left[\sum_j \cos(\phi_j) \right]^2 + \left[\sum_j \sin(\phi_j) \right]^2 + i \sum_{jl} \sin(\phi_l - \phi_j)$$

$$= \left[\sum_j \cos(\phi_j) \right]^2 + \left[\sum_j \sin(\phi_j) \right]^2 \tag{7.57}$$

The complex term averages to zero, which it must since $\Lambda^* \Lambda$ is real. The term $|\Lambda|^2$ is always positive so it cannot average to zero. The fact that Λ averages to zero is irrelevant. Instead, average $|\Lambda|^2$, which is accomplished by separately averaging $\cos(\phi_j)$ and $\sin(\phi_j)$.

$$\Lambda_c = \sum_j \cos(\phi_j) \tag{7.58}$$

$$p(\Lambda_c) = \left\langle \delta\left[\Lambda_c - \sum_j \cos(\phi_j)\right]\right\rangle$$

$$= \int_{-\infty}^{\infty} \frac{dt}{2\pi} e^{it\Lambda_c} \Pi_j \int_0^{2\pi} \frac{d\phi_j}{2\pi} e^{-it\cos(\phi_j)}$$

$$= \int_{-\infty}^{\infty} \frac{dt}{2\pi} e^{it\Lambda_c} J_0(t)^N \tag{7.59}$$

$$J_0(t) = \int_0^{2\pi} \frac{d\phi_j}{2\pi} e^{-it\cos(\phi_j)} \tag{7.60}$$

$$\Lambda_s = \sum_j \sin(\phi_j) \tag{7.61}$$

$$p(\Lambda_s) = \int_{-\infty}^{\infty} \frac{dt}{2\pi} e^{it\Lambda_s} J_0(t)^N \tag{7.62}$$

The distributions for cosine and sine functions are identical, and are given in terms of Bessel's function $J_0(t)$. Since $\Lambda \sim N \gg 1$, saddle point methods are used to evaluate the integrals. For small values of t then

$$J_0(t) \approx 1 - \frac{t^2}{4} + O(t^4) \tag{7.63}$$

$$J_0(t)^N \approx \exp[N\ln(1 - t^2/4 + \cdots +)] \approx \exp\left[-\frac{Nt^2}{4}\right] \tag{7.64}$$

$$p(\Lambda_{c,s}) = \frac{e^{-\Lambda_{c,s}^2/N}}{\sqrt{\pi N}} \tag{7.65}$$

The next term in the expansion is of $O(t^4)$. In the saddle point integral, one has values of t such that $Nt^2 \sim 1$ which means that $Nt^4 \sim (Nt^2)^2/N \sim O(1/N)$ that can be neglected. These results are used to obtain the distributions of intensity in the speckle pattern. Define $I_0 = NE_0^2$ as the average intensity

$$I_0 = E_0^2\langle|\Lambda|^2\rangle = E_0^2 \left\langle \sum_{i=1}^N 1 + \sum_{i \neq j} e^{i(\phi_i - \phi_j)} \right\rangle = NE_0^2 \tag{7.66}$$

The second term in brackets averages to zero.

$$P(I) = \langle\delta[I - E_0^2(\Lambda_c^2 + \Lambda_s^2)]\rangle$$

$$= \int_{-\infty}^{\infty} d\Lambda_c p(\Lambda_c) \int_{-\infty}^{\infty} d\Lambda_s p(\Lambda_s)\delta[I - E_0^2(\Lambda_c^2 + \Lambda_s^2)]$$

$$= \frac{2}{\pi I_0} e^{-I/NE_0^2} \int_0^{\sqrt{I/E_0}} \frac{d\Lambda_c}{\sqrt{I/E_0^2 - \Lambda_c^2}} \tag{7.67}$$

Evaluate the last integral by changing variables $\Lambda_c = (\sqrt{I}/E_0) \sin(\theta)$ turns it into $\int d\theta = \pi/2$.

$$P(I) = \frac{e^{-I/I_0}}{I_0} \tag{7.68}$$

The last equation is the final result. The intensity I must be positive. Its statistical distribution follows a simple exponential law, rather than a Gaussian. The integral over the distribution gives unity:

$$\int_0^\infty dIP(I) = 1 \tag{7.69}$$

$$\int_0^\infty IdIP(I) = I_0 \tag{7.70}$$

The mostly probable distribution is $I = 0$ and the average is I_0. The probability of having large values of I are vanishingly small. See Goodman (2) for more information. He shows how to calculate the average dimension of the light and dark regions of the speckle pattern.

So far all of the averages have been over a single variable. It is possible to average several random variables. An example is done for the speckle problem. The average is taken over the phase ϕ_j and the electric field amplitude E_j. The later average is

$$P(E) = \frac{e^{-E/E_0}}{E_0} \tag{7.71}$$

This choice is somewhat arbitrary, and other choices are possible. Note that this choice assumes $E \geqslant 0$. Negative values of electric field are provided by phase values $\phi_j \sim \pi$.

The cosine average $M_c(t)$ is found by averaging over both electric field and phase angle

$$M_c(t) = \int_0^{2\pi} \frac{d\phi}{2\pi} \int_0^\infty \frac{dE}{E_0} \exp\left[-\frac{E}{E_0} - itE\cos(\phi) \right]$$

$$= \int_0^{2\pi} \frac{d\phi}{2\pi} \frac{1}{1 + itE_0\cos(\phi)} = \frac{1}{\sqrt{1 + (tE_0)^2}} \tag{7.72}$$

Both integrals are easy, and give a simple result. The latter integral is evaluated by letting $z = \exp(i\phi)$ that gives a circular contour of unit radius and the integrand has a simple pole inside this contour. Then the distribution function is

$$p(\Lambda_c) = \int_{-\infty}^{\infty} \frac{dt}{2\pi} \, e^{it\Lambda_c - (N/2)\ln[1 + (tE_0)^2]}$$

$$\approx \int_{-\infty}^{\infty} \frac{dt}{2\pi} \, e^{it\Lambda_c - (N/2)(tE_0)^2}$$

$$\approx \frac{e^{-\Lambda_c^2/I_0}}{\sqrt{\pi I_0}}, \quad I_0 = 2NE_0^2 \tag{7.73}$$

$$P(I) = \langle \delta(I - \Lambda_c^2 - \Lambda_s^2) \rangle = \frac{e^{-I/I_0}}{I_0} \tag{7.74}$$

The distribution $p(\Lambda_s) = p(\Lambda_c)$. The same result is obtained for $P(I)$. The result for $p(\Lambda_c)$ is also still a Gaussian. Averaging the field intensity E_j had little effect on the nature of the answer.

7.3. Inhomogeneous Broadening

Many different kinds of experiments in physics measure a function of energy $F(E)$ of the type shown in Fig. 7.2 There is a resonance curve, of Lorentzian shape, whose center is at E_0 and whose width is γ

$$F(E) = \frac{A}{(E - E_0)^2 + \gamma^2} \tag{7.75}$$

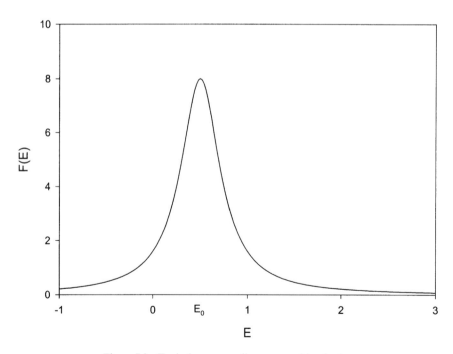

Figure 7.2 Typical resonance line measured in physics.

The usual interpretation is that the system has a resonance at the energy E_0. The width γ is due to lifetime broadening. This interpretation is correct in most cases.

There is, however, another process that will produce a Lorentzian resonance line, where the width has nothing at all to do with the lifetime of the resonance state. Instead, the width is due to *inhomogeneous broadening*. In this case one is measuring the response of a collection of states. Each state i has a resonance centered at E_i. The probability distribution of resonance energies E_i is Lorenzian. The curve in Fig. 7.2 measures the distribution of resonance positions E_i, and each state has an infinite lifetime. Inhomogeneous broadening occurs in systems that have a degree of randomness in the various states that are measured.

The basic premise is that a particle in the system experiences a potential energy $V(\mathbf{R}_0)$ due to the interaction $v(\mathbf{R}_i - \mathbf{R}_0)$ with the other similar particles at distances $\mathbf{R}_i$

$$V(\mathbf{R}_0) = \sum_i v(\mathbf{R}_i - \mathbf{R}_0) \tag{7.76}$$

The other particles are distributed randomly in space. There are N of them in a volume Ω with an average density $n = N/\Omega$. The probability $P(\mathbf{R}_i)$ is taken to be a constant $P = 1/\Omega$. This assumption is the key to the Markov approximation. Correlations between the positions of the particles are neglected. Since all points are alike, set $\mathbf{R}_0 = 0$. Using the previous formalism, the distribution of potential values is

$$p_N(V) = \int \frac{dt}{2\pi} e^{itV} M(t)^N \tag{7.77}$$

$$M(t) = \int d^3 R P(R) e^{-itv(R)} = \frac{1}{\Omega} \int d^3 R e^{-itv(R)} \tag{7.78}$$

The formalism for random walks is being carried over to random potentials.

Next perform some simple manipulations on the above equation. The function $M(t)$ is written as

$$M(t) = \frac{1}{\Omega} \int d^3 R [1 + (e^{-itv(R)} - 1)]$$

$$= 1 - \frac{1}{\Omega} \phi(t) \tag{7.79}$$

$$\phi(t) = \int d^3 R [1 - e^{-itv(R)}] \tag{7.80}$$

This last result is now raised to the power of N

$$M^N = \exp\left\{ N \ln\left(1 - \frac{\phi}{\Omega}\right)\right\} \approx \exp\left\{ -\frac{N}{\Omega} \phi + O(1/N)\right\} \tag{7.81}$$

$$M(t)^N = e^{-n\phi(t)} \tag{7.82}$$

The log function was expanded in a Taylor series. Since Ω is of $O(N)$ and $\phi(t)$ is of $O(1)$ then terms of order $N\phi^2/\Omega^2$ are of order $O(1/N)$ and are neglected.

This series of steps only makes sense if the function $\phi(t)$ behaves mathematically. That is true for the cases considered here.

The distribution is a Gaussian if the function $\phi(t)$ can be expanded as a power series in t, such as

$$\phi(t) = \int d^3R\left[1 - \left(1 - itv - \frac{v^2t^2}{2} - \cdots\right) - \right]$$

$$= it\bar{v} + \gamma t^2 + \cdots + \tag{7.83}$$

$$\bar{v} = \int d^3Rv(R) \tag{7.84}$$

$$\gamma = \frac{1}{2}\int d^3Rv(R)^2 \tag{7.85}$$

$$p_N(V) = \frac{1}{2\sqrt{\pi n\gamma}}\exp[-(V - n\bar{v})^2/4n\gamma] \tag{7.86}$$

Many distributions are Gaussian because the moments $\bar{v}$ and γ exist. However, some of the most interesting cases are where this expansion is invalid because the moments $\int d^3Rv^n$ diverge.

As a simple example, consider that the potential $v(R)$ is a square well of depth v_0 for a distance a. Then the function is

$$\phi(t) = \int_{r<a} d^3r[1 - e^{-itv_0}] = \frac{4\pi a^3}{3}[1 - e^{-itv_0}] \tag{7.87}$$

$$n\phi(t) = g(1 - e^{-itv_0}), \quad g = \frac{4\pi na^3}{3} \tag{7.88}$$

$$p_N(V) = \int \frac{dt}{2\pi}e^{itV-g(1-e^{-itv_0})} = e^{-g}\sum_m \frac{g^m}{m!}\int \frac{dt}{2\pi}e^{it(V-mv_0)} \tag{7.89}$$

$$p_N(V) = e^{-g}\sum_{m=0}^{\infty} \frac{g^m}{m!}\delta(V - mv_0) \tag{7.90}$$

The result is a series of delta functions for energy conservation, distributed according to a Poisson distribution. The parameter g is the average number of particles in the volume determined by the radius a. The Poisson distribution $\exp(-g)g^m/m!$ is the probability of having m particles in the spherical volume. This distribution becomes a Gaussian at large values of g and m. In this example, the probabilty $P(R)$ was assumed to be unaffected by the attractive potential. A better model would have $P(R) \sim \exp[-v_0/k_BT]$ for $R < a$.

As an example of a non-Gaussian distribution, consider a gas of CO molecules. Carbon monoxide has a fixed dipole moment $\mathbf{p}$. Choose one of these molecules at random, and try to calculate the electric field acting upon it, from the other CO molecules. The chosen molecule is at the origin. The others are assumed randomly distributed in space, according to Markov statistics. The

$$E_\mu = \sum_j \left[\frac{p_{j\mu}}{R_j^3} - 3 \frac{R_{j\mu} \mathbf{p} \cdot \mathbf{R}_j}{R_j^5} \right] \tag{7.91}$$

In order to simplify the calculation, assume that all the dipole moments point in the same direction, taken as "z". The effective interaction is then $v(\mathbf{R}) = p^2(1 - 3v^2)/R^3$ where $v = \cos(\theta)$. The angle θ is between the direction of $\mathbf{R}$ and the z-axis. Markov methods will be used to calculate the distribution of electric fields. The important quantity is

$$\phi(t) = \int d^3 R(1 - \exp[-itp^2(1 - 3v^2)/R^3])$$

$$= 2\pi \int_{-1}^{1} dv \int_0^\infty R^2 dR(1 - \exp[-itp^2(1 - 3v^2)/R^3]) \tag{7.92}$$

Denote $\lambda = p^2 t(1 - 3v^2)$ and $x = R^3$ so that $dx = 3R^2 dR$. The integral can be rewritten as

$$\phi(t) = \frac{2\pi}{3} \int_{-1}^{1} dv \int_0^\infty dx(1 - e^{-i\lambda/x}) \tag{7.93}$$

It is important to note that parameter λ can be either positive or negative: the symbol t has both plus and minus values, and the factor of $(1 - 3v^2)$ can have either sign since $-1 < v < 1$. The above integral cannot be usefully expanded in moments, since none of them converge. For example, the second moment is

$$I_2 = \frac{2\pi}{3} \int_{-1}^{1} dv \lambda^2 \int_0^\infty \frac{dx}{x^2} \tag{7.94}$$

The integral over dx diverges as $x \to 0$. As a consequence, the final distribution is not a Gaussian.

Change variables $\int_{-1}^{1} dv = 2 \int_0^1 dv$, and consider the integral

$$I = n\phi = \frac{4\pi n}{3} \int_0^1 dv L(\lambda) \tag{7.95}$$

$$L = \lim_{a \to 0} \int_a^{1/a} dx[1 - e^{-i\lambda/x}] \tag{7.96}$$

The limits of $(a, 1/a)$ are included to handle the two limits of $x \to 0, \infty$ when $a \to 0$ at the end. First, integrate by parts

$$\int u\,dv = uv - \int v\,du \tag{7.97}$$

$$u = [1 - e^{-i\lambda/x}] \tag{7.98}$$

$$v = x \tag{7.99}$$

$$L = i\lambda \left[1 + \lim_{a \to 0} \int_a^{1/a} \frac{dx}{x} e^{-i\lambda/x} \right] \tag{7.100}$$

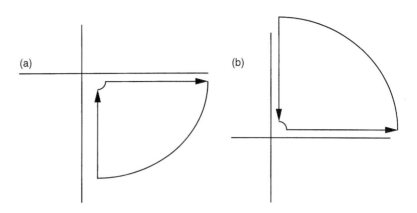

Figure 7.3 Integration contours for: (a) $\lambda > 0$ and (b) $\lambda < 0$.

The factor '1' in the bracket comes from the constant term uv evaluated as $1/a \rightarrow \infty$ and $u \rightarrow i\lambda/x$. In the remaining integral change variables $x = 1/z$ giving

$$L = i\lambda \left\{ 1 + \lim_{a \rightarrow 0} \int_a^{1/a} \frac{dz}{z} e^{-iz\lambda} \right\} \tag{7.101}$$

The evaluation of this integral depends on the sign of λ as shown in Fig. 7.3:

- When $\lambda > 0$ then rotate the contour of integration to the negative imaginary axis using $z = -iy$. There is also a segment along the quarter circle using $z = ae^{i\theta}$, $dz/z = id\theta$ for θ from zero to $-\pi/2$. the result of this rotation is

$$(\lambda > 0): \quad L = i\lambda \left[1 - i\frac{\pi}{2} + \int_a^{1/a} \frac{dy}{y} e^{-\lambda y} \right] \tag{7.102}$$

  where the term in brackets $-i\pi/2$ comes from the quarter circle at small z.
- When $\lambda < 0$ then rotate the contour of integration towards the positive imaginary axis using $z = iy$. There is also a segment along the quarter circle $z = a\exp(i\theta)$ for the angles θ from zero to $+\pi/2$:

$$(\lambda < 0): \quad L = i\lambda \left[1 + i\frac{\pi}{2} + \int_a^{1/a} \frac{dy}{y} e^{\lambda y} \right] \tag{7.103}$$

These two results are summarized by

$$L = \frac{\pi}{2} |\lambda| + i\lambda \left[1 + \lim_{a \rightarrow 0} \int_a^{1/a} \frac{dy}{y} e^{-|\lambda| y} \right] \tag{7.104}$$

In the integral, change variables to $s = |\lambda| y$ and the factor of λ disappears from the integral except in the limits. The upper limit is no problem and can be set equal to infinity. To handle the lower limit, write this term as ($s_0 = a|\lambda|$)

$$\int_{s_0}^\infty \frac{ds}{s} e^{-s} = \left[\int_{s_0}^1 ds + \int_1^\infty ds \right] \frac{e^{-s}}{s} \approx -\ln(s_0) + \int_1^\infty \frac{ds}{s} e^{-s} + O(s_0) \tag{7.105}$$

Also note that

$$\int_0^1 dv\, \lambda = p^2 t \int_0^1 dv(1 - 3v^2) = 0 \tag{7.106}$$

The integral of $\int dv\, \lambda = 0$. Any factor in the integral vanishes if it depends on a single power of λ. The integral Eq. (7.104) has two terms that do not integrate to zero

$$\int_0^1 dv\, L = \int_0^1 dv \left[\frac{\pi}{2} |\lambda| - i\lambda \ln|1 - 3v^2| \right] \tag{7.107}$$

The integrals gives

$$I = \gamma|t| - i\Delta t \tag{7.108}$$

$$\gamma = \frac{2\pi^2 n}{3} p^2 \int_0^1 dv|1 - 3v^2| = \frac{\pi^2}{3^3\sqrt{3}} p^2 n \tag{7.109}$$

$$\Delta = \frac{4\pi}{3} np^2 \int_0^1 dv(1 - 3v^2) \ln|1 - 3v^2| \tag{7.110}$$

The final integral determines the distribution of potential energies

$$p_N(V) = \int \frac{dt}{2\pi} e^{it(V + \Delta) - \gamma|t|}$$

$$= \frac{\gamma/\pi}{(V + \Delta)^2 + \gamma^2} \tag{7.111}$$

The above formula is the final result. The distribution of potentials is Lorentzian with a width γ and a center at $-\Delta$. This width has nothing to do with the lifetime of the particle, and is due to inhomogeneous broadening. The occurance of a Lorentzian is unusual, and only occurs when the Gaussian is not allowed because the moments diverge. This calculation was first done in 1919 by von Holtsmark (3).

References

1. S. Chandrasekhyar, *Rev. Mod. Phys.* **15**, 1 (1942).
2. J. W. Goodman, *Statistical Optics* (John Wiley, 1985).
3. J. von Holtsmark, *Annalen der Physik*, **4**, 38 (1919).

Problems

1. Suppose that $P(x_i)$ was given by a Gaussian not centered at zero:

$$P(x_i) = \frac{1}{l\sqrt{2\pi}} e^{-(x_i - \Delta)^2/2l^2} \tag{7.112}$$

Determine $p_N(x)$ in this case.

2. Suppose that $P(x_i)$ is given by the Lorentzian

$$P(x_i) = \frac{l/\pi}{x_i^2 + l^2} \tag{7.113}$$

Note that for this function the second moment diverges. Nevertheless, find the distribution $p_N(x)$, that is well-behaved mathematically. Show that $p_N(x)$ is not Gaussian, which is related to the fact that the second moment diverges.

3. Calculate the distribution of potential energies $p_N(V)$ for a gas of CO molecules that have a fixed dipole moment p. Assume that the direction of the moment can be either up or down with equal probability. How is this case different from the one worked out in the notes, where the moments are always up?

4. Consider in one dimension a gas of $N/2$ charges $+q$ and $N/2$ charges $-q$ spread uniformly over a length L where $n = N/L$. Calculate the random potential on one charge from the potential of the others. Show that

$$n\phi(t) = n \int dx[1 - \cos(q^2 t/x)] = \gamma|t|, \ \gamma = \pi n q^2 \tag{7.114}$$

$$p_N(V) = \frac{\gamma/\pi}{V^2 + \gamma^2} \tag{7.115}$$

Fourier Transforms

Transforms are useful for a wide variety of mathematical problems. This chapter will discuss three kinds of transforms: Fourier transforms, Laplace transforms, and wavelet transforms. The Laplace transform has only one type and is rather simple. The Fourier transform has many different varieties that depend on the space under investigation. The method depends on the dimensions of the space and whether it is bounded. Wavelets are a new type of transform.

8.1. Fourier Transforms

The phrase "Fourier transform" means to use transforms with functions such as $\sin(kx)$, $\cos(kx)$, $\exp(ikx)$ or $\exp(-ikx)$. Which of these trigometric functions is relevant depends on the problem to be solved. Various choices are discussed below.

8.1.1. Unbounded Space

The most common Fourier transform is used when the physical space is unbounded. The first example is in one dimension. Assume the function $f(x)$ has values over the interval $-\infty \leqslant x \leqslant \infty$. The Fourier transform $F(k)$ depends on a variable k that has the dimensions of wave vector

$$F(k) = \int_{-\infty}^{\infty} dx f(x) e^{ikx} \tag{8.1}$$

$$f(x) = \frac{1}{2\pi} \int_{-\infty}^{\infty} dk F(k) e^{-ikx} \tag{8.2}$$

The first integral is called the *Fourier transform*, while the second integral is called the *inverse transform*. The factor of $1/(2\pi)$ can go in front of either integral. In fact, some authors define this transform with the factor of $1/\sqrt{2\pi}$ in front of both integrals. All of these alternates are correct. The definition needs a net overall factor of $1/(2\pi)$. The second remark is that the first exponent is defined as $+ikx$ while the second exponent is defined as $-ikx$. These signs can be reversed. Since the k-integral includes both signs of k, it matters not which sign convention is made for the exponents. It is important to have different signs in the two exponents.

That this transform is *complete* can be demonstrated by nesting the two definitions, and then inverting the order of the integrals

$$f(x) = \frac{1}{2\pi} \int_{-\infty}^{\infty} dk e^{-ikx} \int_{-\infty}^{\infty} dx' e^{ikx'} f(x')$$

$$= \int_{-\infty}^{\infty} dx' f(x') \Lambda(x - x') \tag{8.3}$$

$$\Lambda(x - x') = \frac{1}{2\pi} \int_{-\infty}^{\infty} dk e^{ik(x'-x)} = \delta(x - x') \tag{8.4}$$

The integral for $\Lambda(x - x')$ gives a delta function. The transform is complete if it works for any function $f(x)$, that is ensured by the delta function.

Let us do some examples. The first one is $f(x) = 1/x$. Then

$$F(k) = \int_{-\infty}^{\infty} \frac{dx}{x} e^{ikx} = \int_{-\infty}^{\infty} \frac{dx}{x} [\cos(kx) + i \sin(kx)] \tag{8.5}$$

The first integral, of $\cos(kx)/x$, equals zero. The integrand is an odd function of x, and the result is zero due to the cancellation of parts from negative and positive values of x. The integral over the other function gives the answer. The result from a previous chapter is

$$\int_{0}^{\infty} \frac{dx}{x} \sin(kx) = \frac{\pi}{2} \operatorname{sgn}(k) \tag{8.6}$$

$$F(k) = i\pi \operatorname{sgn}(k) \tag{8.7}$$

where $\operatorname{sgn}(k)$ is the sign of k: it is plus one if $k > 0$ and minus one if $k < 0$. The inverse transform is

$$f(x) = \frac{i\pi}{2\pi} \left[\int_{0}^{\infty} dk e^{-ikx} - \int_{-\infty}^{0} dk e^{-ikx} \right] \tag{8.8}$$

Change variables of integration ($k \to -k$) in the second integral

$$f(x) = \int_{0}^{\infty} dk \sin(kx) = -\frac{\cos(kx)}{x} \bigg|_{0}^{\infty} = \frac{1}{x} \tag{8.9}$$

The function $\cos(kx)$ is assumed to vanish in the limit of large (kx). The inverse transform recovers the original ansatz for $f(x)$.

Another example is the discontinuous function

$$f(x) = \text{sgn}(x)e^{-p|x|} \tag{8.10}$$

$$F(k) = -\int_{-\infty}^{0} dx e^{x(p+ik)} + \int_{0}^{\infty} dx e^{-x(p-ik)}$$

$$= -\frac{1}{p+ik} + \frac{1}{p-ik} = \frac{2ki}{p^2 + k^2} \tag{8.11}$$

The inverse transform is written as

$$f(x) = \frac{1}{2\pi} \int_{-\infty}^{\infty} dk e^{-ikx} \frac{2ki}{k^2 + p^2}$$

$$= \frac{i}{2\pi} \int_{-\infty}^{\infty} dk e^{-ikx} \left[\frac{1}{k+ip} + \frac{1}{k-ip} \right] \tag{8.12}$$

The factor of $\exp(-ikx)$ determines the form of the contour integral. For $x < 0$ the above integral must be closed in the UHP that gives $f = -\exp(px)$. For $x > 0$ it must be closed in the LHP that gives a sign change $f = \exp(-px)$. These two cases give the original function in Eq. (8.10).

8.1.2. Half Space

Another Fourier transform is used when the problem is confined to a half space, such as the region $x \geq 0$ in one dimension. Then the values of x and k are confined to positive values. There are two possible Fourier transforms in this case and are called the cosine and the sine transforms,

$$F_s(k) = \int_{0}^{\infty} dx f(x) \sin(kx) \tag{8.13}$$

$$f(x) = \frac{2}{\pi} \int_{0}^{\infty} dk F_s(k) \sin(kx) \tag{8.14}$$

$$F_c(k) = \int_{0}^{\infty} dx f(x) \cos(kx) \tag{8.15}$$

$$f(x) = \frac{2}{\pi} \int_{0}^{\infty} dk F_s(k) \cos(kx) \tag{8.16}$$

Both transforms work for most values of $f(x)$. For example, we may think that the factor of $\sin(kx)$ means that a function defined by a sine transform must vanish at $x = 0$. This conclusion is false, as will be shown below as an example. Also, the factor of $2/\pi$ can be put in front of either integral.

First it is shown that the transform is complete. It is proved for one case, say the sine transform. Again the transform and its inverse are nested and give

$$f(x) = \frac{2}{\pi} \int_0^\infty dk \sin(kx) \int_0^\infty dx' \sin(kx') f(x')$$

$$= \int_0^\infty dx' f(x') \Lambda(x, x') \tag{8.17}$$

$$\Lambda(x, x') = \frac{2}{\pi} \int_0^\infty dx \sin(kx) \sin(kx') \tag{8.18}$$

The integral for Λ is evaluated by using the trigometric identity that the product of two sine functions is the difference of two cosines,

$$\Lambda(x, x') = \frac{1}{\pi} \int_0^\infty dk \{\cos[k(x - x')] - \cos[k(x + x')]\}$$

$$= \delta(x - x') - \delta(x + x') \tag{8.19}$$

The cosine integral is equivalent to the integral for the delta function. There are two delta functions, however, the second has no effect. Since both x and x' are positive, then $x + x'$ can never be zero. Then $\Lambda(x, x') = \delta(x - x')$ and the transformation is complete. It should be valid for any function $f(x)$. The same steps can be used to prove that the cosine transformation is also complete. In fact, both transforms can be used for many problems. In a particular problem, the choice of which transform to use depends on the differential equation and the boundary conditions. Examples are given in later chapters. The two transforms cannot be mixed, and if we use a sine transform to get $F_s(k)$ the sine transform must be used to get back to $f(x)$.

As an example, consider the function $f(x) = \exp(-px)$. The sine transform gives

$$F_s(k) = \int_0^\infty dx e^{-px} \sin(kx) = \frac{1}{2i} \int_0^\infty dx e^{-px}[e^{ikx} - e^{-ikx}]$$

$$= \frac{1}{2i}\left[\frac{1}{p - ik} - \frac{1}{p + ik}\right] = \frac{k}{p^2 + k^2} \tag{8.20}$$

and the inverse sine transform is

$$f(x) = \frac{2}{\pi} \int_0^\infty dk \sin(kx) \frac{k}{k^2 + p^2} = \frac{1}{\pi} \int_{-\infty}^\infty dk \sin(kx) \frac{k}{k^2 + p^2}$$

$$= \frac{1}{2\pi i} \int_{-\infty}^\infty \frac{k \, dk}{k^2 + p^2}(e^{ikx} - e^{-ikx}) = e^{-px} \tag{8.21}$$

The contour integral is closed in the UHP for $\exp(ikx)$ and in the LHP for $\exp(-ikx)$. Both integrals give $\exp(-px)/2$.

The cosine transform of the same function gives

$$F_c(k) = \int_0^\infty dx e^{-px} \cos(kx) = \frac{1}{2} \int_0^\infty dx e^{-px} [e^{ikx} + e^{-ikx}]$$

$$= \frac{1}{2} \left[\frac{1}{p - ik} + \frac{1}{p + ik} \right] = \frac{p}{p^2 + k^2} \tag{8.22}$$

and the inverse cosine transform is

$$f(x) = \frac{2}{\pi} \int_0^\infty dk \cos(kx) \frac{p}{k^2 + p^2} = \frac{1}{\pi} \int_{-\infty}^\infty dk \cos(kx) \frac{p}{k^2 + p^2}$$

$$= \frac{p}{2\pi} \int_{-\infty}^\infty \frac{dk}{k^2 + p^2} (e^{ikx} + e^{-ikx}) = e^{-px} \tag{8.23}$$

Both transforms work for this simple exponential function.

When solving problems using Fourier transforms, the immediate decision is whether to use cosine or sine transforms for the half space problem. The choice depends on whether at the end of the system $x = 0$, the value of $f(x)$ is known, or its derivative is known. This difference arises from the transform of the second derivative that is written out for both sine and cosine transform. The transform of the derivatives are evaluated in successive integration by parts.

$$F_c(k) = \int_0^\infty dx \cos(kx) f(x) \tag{8.24}$$

$$\int_0^\infty dx \cos(kx) \frac{\partial^2}{\partial x^2} f(x) = -k^2 F_c(k) - \left(\frac{\partial f}{\partial x} \right)_{x=0} \tag{8.25}$$

$$F_s(k) = \int_0^\infty dx \sin(kx) f(x) \tag{8.26}$$

$$\int_0^\infty dx \sin(kx) \frac{\partial^2}{\partial x^2} f(x) = -k^2 F_s(k) + kf(0) \tag{8.27}$$

For the cosine transform, the transform of the second derivative requires knowledge of df/dx at the origin. If it is not known beforehand, the problem is hard to solve in this way. The sine transform requires a knowledge of $f(0)$ and the choice of the sine or cosine transform depends on the choice of boundary conditions. The cosine transform is easier to use in problems where the derivative is known at the boundary, while the sine transform is easier if the function is specified at the boundary.

8.1.3. Finite Systems in 1D

Consider a system that is a bar of length L. How does one set up a Fourier transform for a function $f(x)$ that exists over the length of the bar? The answer is that there are several ways of doing it, that appear to be different but give the same answer. This situation is confusing, however, it really means there is more than one way of doing it correctly.

Our preference is to set the origin of $x = 0$ at the center of the bar, so that values span $-L/2 \leqslant x \leqslant L/2$. This convention is used here. The Fourier transform has the generic form

$$f(x) = \sum_n [a_n \cos(k_n x) + b_n \sin(k_n x)] \tag{8.28}$$

The important questions are to determine the values of k_n, a_n, and b_n. First, note that the function $\cos(k_n x)$ is a symmetric function of x. If the function $f(x)$ is symmetric, then set $b_n = 0$ and only use the symmetric functions $\cos(k_n x)$. Similarly, the terms with $\sin(k_n x)$ are antisymmetric in x. If the function $f(x)$ is perfectly antisymmetric, then set $a_n = 0$ and only use the sine functions. Both terms, $\sin(k_n x)$ and $\cos(k_n x)$, are retained for a function of low symmetry.

The choice of k_n is dictated by some constraints. Since each term in the Fourier series is supposed to be independent of the others, it means that all functions should be orthogonal. This condition is

$$\int_{-L/2}^{L/2} dx \cos(k_n x) \cos(k_m x) = \frac{L}{2} \delta_{nm} \tag{8.29}$$

$$\int_{-L/2}^{L/2} dx \sin(k_n x) \sin(k_m x) = \frac{L}{2} \delta_{nm} \tag{8.30}$$

$$\int_{-L/2}^{L/2} dx \cos(k_n x) \sin(k_m x) = 0 \tag{8.31}$$

Two choices of k_n satisfy these conditions. The properties of the trigometric functions with these choices are:

1. $k_n = 2\pi n/L$ for all positive n and $n = 0$.
   - $\cos(k_n L/2) = (-1)^n$. The cosine functions with this k_n are good for symmetric functions $f(x)$ that do not vanish at the ends of the bar.
   - $\sin(k_n L/2) = 0$. The sine functions with this k_n are good for antisymmetric functions $f(x)$ that vanish at the ends of the bar.
   - For $n = 0$ the normalization integrals are different.
2. $k_n = \pi(2n + 1)/L$ for all positive values of n including $n = 0$.
   - $\cos(k_n L/2) = 0$. The cosine functions with this k_n are good for symmetric functions $f(x)$ that vanish at the ends of the bar.
   - $\sin(k_n L/2) = (-1)^n$. The sine functions with this k_n are good for antisymmetric functions $f(x)$ that do not vanish at the ends of the bar.
3. However, for most problems it does not matter which choice is made for k_n.

To prove the above assertion, examine the integral in Eq. (8.29). When $n = m$ the integral is

$$\int_{-L/2}^{L/2} dx \frac{1}{2}[1 + \cos(2k_n x)] = \frac{L}{2} + \frac{\sin(2k_n x)}{4k_n}\Big|_{-L/2}^{L/2} = \frac{L}{2} + \frac{\sin(k_n L)}{2k_n} \tag{8.32}$$

where $\sin(k_n L) = 0$ whenever $k_n L$ is π times any integer, odd or even. Equation (8.38) is correct when $n = m$. For the case that $m \neq n$ the integral equals

$$\int_{-L/2}^{L/2} dx \frac{1}{2} \{\cos[x(k_n - k_m)] + \cos[x(k_m + k_n)]\}$$

$$= \frac{\sin[(k_n - k_m)L/2]}{k_n - k_m} + \frac{\sin[(k_n + k_m)L/2]}{k_n + k_m} \tag{8.33}$$

The above terms vanish as long as $L(k_n \pm k_m) = 2\pi l$ where l is an integer. It vanishes for each choice of k_n. It does not vanish if k_n is an odd and k_m is an even integer. It only works if both are odd or both are even. These steps prove Eq. (8.29) for $n \neq m$. A similar proof applies to the integral with two sine functions. The integral with one sine and one cosine function is always zero since the integrand is an odd function of x.

It is useful to give an example. Consider the symmetric function

$$f(x) = \frac{L}{2} - |x| \tag{8.34}$$

where the bracket on x means absolute value. The Fourier transform of this function is found using each choice for k_n.

1. $k_n = 2\pi n/L$. Since the function is symmetric, write the Fourier transform as

$$f(x) = a_0 + \sum_{n=1}^{\infty} a_n \cos(k_n x) \tag{8.35}$$

The first term on the right is for $n = 0$. The Fourier coefficients are

$$a_0 = \frac{1}{L} \int_{-L/2}^{L/2} dx f(x) = \frac{L}{2} - \frac{2}{L} \int_0^{L/2} x \, dx = \frac{L}{4} \tag{8.36}$$

$$a_n = \frac{2}{L} \int_{-L/2}^{L/2} dx f(x) \cos(k_n x) = \frac{4}{L} \int_0^{L/2} dx \left[\frac{L}{2} - x\right] \cos(k_n x)$$

$$= \frac{4}{L} \left[\left(\frac{L}{2} - x\right) \frac{\sin(k_n x)}{k_n} - \frac{\cos(k_n x)}{k_n^2}\right]_0^{L/2}$$

$$= \frac{4}{Lk_n^2} [1 - \cos(k_n L/2)] = \left(\frac{8}{Lk_n^2}\right)_{n=\text{odd}} \tag{8.38}$$

where $[1 - \cos(k_n L/2)] = [1 - (-1)^n] = 0$ for even n and equals two for odd n. The final expression for the Fourier series is

$$f(x) = \frac{L}{4} + \frac{2L}{\pi^2} \sum_{n=0}^{\infty} \frac{1}{(2n+1)^2} \cos\left[\frac{2\pi x}{L}(2n+1)\right] \tag{8.39}$$

Note that the constant term $a_0 = L/4$ is the average value of the function. The other terms provide fluctuations around the average. A useful series is

$$\sum_{n=0}^{\infty} \frac{1}{(2n+1)^2} = 1 + \frac{1}{3^2} + \frac{1}{5^2} + \cdots + = \frac{\pi^2}{8} \tag{8.40}$$

Using this series one can prove that $f(0) = L/2$ and $f(L/2) = 0$. In fact, only a few terms in the series are needed to give a good approximation to the linear behavior for $f(x)$.

2. $k_n = \pi(2n + 1)/L$. For this case k_n does not vanish for any value of n, positive, negative, or zero, so the Fourier series does not have a constant term such as a_0. The series has the form

$$f(x) = \sum_{n=0}^{\infty} a_n \cos(k_n x) \tag{8.41}$$

$$a_n = \frac{2}{L} \int_{-L/2}^{L/2} dx f(x) \cos(k_n x) = \frac{4}{L} \int_0^{L/2} dx \left[\frac{L}{2} - x \right] \cos(k_n x)$$

$$= \frac{4}{L} \left[\left(\frac{L}{2} - x \right) \frac{\sin(k_n x)}{k_n} - \frac{\cos(k_n x)}{k_n^2} \right]_0^{L/2}$$

$$= \frac{4}{L k_n^2} [1 - \cos(k_n L/2)] = \frac{4}{L k_n^2} \tag{8.42}$$

$$f(x) = \frac{4L}{\pi^2} \sum_{n=0}^{\infty} \frac{1}{(2n+1)^2} \cos\left[\frac{\pi x}{L} (2n + 1) \right] \tag{8.43}$$

This derivation is similar to the previous one. In this case, $\cos(k_n L/2) = 0$ for all values of n, that gives a slightly different series than in the prior case. Again, the series has the correct result that $f(L/2) = 0$ and also $f(0) = L/2$. In this case, an accurate approximation to $f(x)$ is also achieved with only a few terms in the series.

The above two series are different. Both are rigorous Fourier series for the function $f(x)$. There are two ways to obtain a Fourier expansion for this function.

8.2. Laplace Transforms

Laplace transforms are usually applied to functions of time $f(t)$ that are zero before some reference time t_0. Since the results depend only on $(t - t_0)$ it is customary to set $t_0 = 0$. The function $f(t) = 0$ for $t < 0$ and is generally nonzero for $t > 0$. Define the Laplace transform and inverse transform as

$$F(p) = \int_0^\infty dt\, e^{-pt} f(t) \tag{8.44}$$

$$f(t) = \int_{\varepsilon - i\infty}^{\varepsilon + i\infty} \frac{dp}{2\pi i} F(p) e^{pt} \tag{8.45}$$

The expression for $F(p)$ seems straightforward. The integral should converge since p appears to be real. However, in the inverse transform, the integral over p runs up the imaginary axis. The symbol p should always be regarded as a complex number and the inverse transform as a contour integral.

The acronyms RHP and LHP represent the right- and left-hand planes. The integral in the inverse transform runs up the imaginary axis. A closed contour is made by closing the integral in the RHP or in the LHP. It depends on the sign of t. If $t < 0$ then the factor of $\exp(pt)$ only converges if the real part of p is positive, that is the RHP. Therefore, for $t < 0$ close the contour in the RHP.

THEOREM. *$F(p)$ is analytic in the RHP.*

The constraint on $f(t)$ is that it is zero for $t < 0$. Therefore, when closing the contour in the RHP for $t < 0$, there can be no poles or branch cuts in this region. If there were, the contour integral would be nonzero, which disagrees with the ansatz that $f = 0$ for $t < 0$. Therefore, $F(p)$ must be analytic in the RHP. It is not analytic in the LHP. If it were, then F would be a constant, and $f(t) \propto \delta(t)$.

If $F(p)$ has a simple pole at $p = 0$ then the inverse transform gives a term that has $f = $ constant. This solution is permitted. The inverse transform has the factor of $\varepsilon > 0$ in the path of integration to make it go just to the right of the imaginary axis. Then if there is a pole at $p = 0$, the contour of integration can be completed in the LHP that includes this pole.

The Laplace transform is complete, in that it can be used to express any function $f(t)$ that vanishes at negative times, and has reasonable convergence properties at large times. Again, nest the definitions of the transform and inverse transform, and reverse the order of the two integrals

$$f(t) = \int_0^\infty dt' f(t') \Lambda(t - t') \tag{8.46}$$

$$\Lambda(t - t') = \int_{\varepsilon - i\infty}^{\varepsilon + i\infty} \frac{dp}{2\pi i} e^{p(t-t')} = \delta(t - t') \tag{8.47}$$

The integral becomes a delta function with the change of variable $p = i\omega$ and proves completeness. Some examples follow.

1. $f(t) = Ate^{-\alpha t}$. The transform and inverse transform are

$$F(p) = A \int_0^\infty dt\, t e^{-t(p+\alpha)} = \frac{A}{(p + \alpha)^2} \tag{8.48}$$

$$f(t) = \frac{A}{2\pi i} \int_{-i\infty}^{i\infty} \frac{dp}{(p + \alpha)^2} e^{pt} \tag{8.49}$$

There is a double pole at $p = -\alpha$. Close the integration contour in the LHP and the residue is evaluated by a single derivative

$$f(t) = A\left(\frac{d}{dp}\,e^{pt}\right)_{p=-\alpha} = Ate^{-\alpha t} \tag{8.50}$$

that agrees with the original function.

2. $f(t) = A/\sqrt{t}$. The transform and inverse transform are

$$F(p) = A\int_0^\infty \frac{dt}{\sqrt{t}}\,e^{-pt} = \frac{A\sqrt{\pi}}{\sqrt{p}} \tag{8.51}$$

$$f(t) = \frac{A\sqrt{\pi}}{2\pi i}\int_{-i\infty}^{i\infty} \frac{dp}{\sqrt{p}}\,e^{pt} \tag{8.52}$$

The factor of $\sqrt{p}$ gives a branch point at $p = 0$. It requires a branch cut to infinity in some direction. The positive real axis is not permitted by causality, since otherwise $f(t)$ would have values for $t < 0$. So put the branch cut along the negative real axis. The contour of integration can be deformed to wrap around this branch cut as shown in Fig. 8.1. Let $p = -\rho$, where $\sqrt{p} = i\sqrt{\rho}$ above the axis and $\sqrt{p} = -i\sqrt{\rho}$ below the axis. This result makes the two integrals, above and below the negative axis, identical. Therefore the inverse transform is

$$f(t) = \frac{A}{\sqrt{\pi}}\int_0^\infty \frac{d\rho}{\sqrt{\rho}}\,e^{-\rho t} = \frac{A}{\sqrt{t}} \tag{8.53}$$

The Laplace transform is particularly useful in transient diffusion problems. In those cases, the function $F(p)$ usually depends in some way on $\sqrt{p}$ so that a similar branch cut is required. The above example is a prototype for many of the cases to follow.

Laplace transform theory works quite well, although it is somewhat of a swindle. In writing the transform as $\int dt\,e^{-pt}f(t)$ it is assumed that the parameter p has a real part that is positive. Then the integral converges for most forms of $f(t)$. In evaluating the inverse transform, the contour integral over p space is evaluated in the region where the real part of p is negative. The

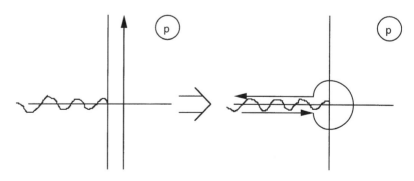

Figure 8.1 Integration contour for inverse Laplace transform.

integral $\int dt e^{-pt} f(t)$ does not converge if the real part of p is negative. The reason why the theory works is that $F(p)$ is an analytic function of p and integrals over it can be continued into different regions of complex space.

8.3. Wavelets

The previous two sections discuss two methods of transforming a function $f(t)$ of time. The Laplace transform is useful if the function is known for all values of time greater than some reference times such as zero. The Fourier transform is useful if the function is known for all values of time $-\infty < t < \infty$. Wavelets are a different method of transforming $f(t)$. It is useful when the function $f(t)$ develops without end.

Suppose you wish to transform, as a function of time, a stock market index such as the Dow Jones Average. In most weeks fresh data accumulates for five days. How is this fresh data treated? Using a Fourier series, one has to retransform the data again for all time. Wavelets are a method of transforming each new data set that arrives daily. Many branches of science and engineering have data sets that accumulate without end. Geologists have seismology data, astronomers have infrared and X-ray data, power companies have noise on their transmission lines, and meterologists have temperature data. All need a method of transforming the new data that arrives daily. Wavelets provide such a transform and are also useful for recording transients. Most of the wavelet transform have the general form of

$$W_{k,b}(f) = \int dt \psi^*_{k,b}(t) f(t) \tag{8.54}$$

$$\psi_{k,b}(t) = \sqrt{k}\, \phi[k(t-b)] \tag{8.55}$$

All integrals over time extend over $(-\infty < t < \infty)$ unless noted otherwise. The function $\phi(x)$ is specified below. It provides the basic shape of the wavelet. The constant b is the center of the transform in time. The function $\phi(x)$ vanishes for large values of $|x|$ so that the transform is centered in time about the point b. As time develops, this point moves forward, say for each day. The second parameter k gives the width of the transform. It has the units of frequency. The wavelet transform requires two parameters, that are denoted as (k, b). The Laplace and Fourier transforms each had one parameter, such as p or ω. Two parameters are required for wavelets, since one localizes in time, while the other controls the width of the wavelet. The advantage of the wavelet method is that it permits a wide variety of functions $\phi(x)$.

The wavelet transform is different from traditional windowing used in signal processing. There the typical window function is a Gaussian. An oscillatory term is added to provide a Fourier transform, that gives the typical form

$$g_{k,b,\omega}(t) = \sqrt{k}\, e^{-i\omega t} e^{-k^2(t-b)^2} \tag{8.56}$$

$$G_{k,b,\omega} = \int dt f(t) g^*_{k,b,\omega}(t) \tag{8.57}$$

There are now three parameters: k, b, ω. Furthermore, the function $g_{k,b,\omega}(t)$ is complex. The wavelet function $\psi_{k,b}(t)$ is real, and it has only two parameters. The wavelet transform is actually simpler than the traditional window transform.

Define N^2 as the normalization integral on $|\phi(x)|^2$. Define the average position $\langle t \rangle$ and width Δ_ϕ of the function as $\{x = k(t - b)\}$

$$N^2 = \int dt\, |\psi_{k,b}(t)|^2 = \int dx\, |\phi(x)|^2 \tag{8.58}$$

$$\langle t \rangle = \frac{1}{N^2} \int dt\, t\, |\psi_{k,b}(t)|^2 = b + t^* \tag{8.59}$$

$$t^* = \frac{1}{kN^2} \int dx\, x\, |\phi(x)|^2 \tag{8.60}$$

$$\Delta_\phi^2 = \frac{k^2}{N^2} \int dt\, (t - b)^2 |\psi_{k,b}(t)|^2 = \frac{1}{N^2} \int dx\, x^2 |\phi(x)|^2 \tag{8.61}$$

Most choices of wavelets show $|\phi(-x)|^2 = |\phi(x)|^2$ which makes $t^* = 0$. That symmetry is assumed here. The above transform then samples $f(t)$ during the interval $b - \Delta_\phi/k < t < b + \Delta_\phi/k$ that provides the desired localization in time.

Denote by a carat the Fourier transform of a function, so that

$$\hat{\phi}(\xi) = \int dx\, e^{-ix\xi} \phi(x) \tag{8.62}$$

$$\hat{f}(\omega) = \int dt\, e^{-it\omega} f(t) \tag{8.63}$$

$$f(t) = \int \frac{d\omega}{2\pi} e^{it\omega} \hat{f}(\omega) \tag{8.64}$$

$$
\begin{aligned}
W_{k,b}(f) &= \int dt\, \psi_{k,b}^*(t)\, f(t) = \int \frac{d\omega}{2\pi} \hat{f}(\omega) \int dt\, e^{i\omega t} \sqrt{k}\, \phi^*[k(t - b)] \\
&= \frac{1}{2\pi\sqrt{k}} \int d\omega\, \hat{f}(\omega) e^{i\omega b} \int dx\, \phi^*(x) e^{ix(\omega/k)} \\
&= \frac{1}{2\pi\sqrt{k}} \int d\omega\, \hat{f}(\omega) e^{i\omega b}\, \hat{\phi}^*(\omega/k) \tag{8.65}
\end{aligned}
$$

The Fourier transform of the wavelet has a center ξ^* and a width $\Delta_{\hat{\phi}}$

$$\xi^* = \frac{1}{N^2} \int d\xi\, \xi\, |\hat{\phi}(\xi)|^2 \tag{8.66}$$

$$\Delta_{\hat{\phi}}^2 = \frac{1}{N^2} \int d\xi\, (\xi - \xi^*)^2 |\hat{\phi}(\xi)|^2 \tag{8.67}$$

The Fourier transform $\hat{f}(\omega)$ is sampled in the frequency region of

$k(\xi^* - \Delta_{\hat\phi}) < \omega < k(\xi^* + \Delta_{\hat\phi})$. The parameter k controls the width of the sampling in both the time space and the frequency space.

8.3.1. Continuous Wavelet Transform

The basic form of the wavelet transform is defined in Eq. (8.54). The transform can be inverted to recover the original function $f(t)$,

$$f(t) = \frac{1}{C_\phi} \int_0^\infty dk \int_{-\infty}^\infty db \, W_{k,b}(f) \psi_{k,b}(t) \tag{8.68}$$

$$C_\phi = \int_0^\infty \frac{d\xi}{\xi} |\hat\phi(\xi)|^2 \tag{8.69}$$

In order to prove completeness we again nest the two integrals [Eqs. (8.54), (8.68)],

$$f(t) = \int dt' f(t') K_\phi(t, t') \tag{8.70}$$

$$K_\phi(t, t') = \frac{1}{C_\phi} \int_0^\infty dk \int_{-\infty}^\infty db \psi_{k,b}^*(t') \psi_{k,b}(t) \tag{8.71}$$

The parameter k controls the width of the wavelet, and is assumed to be positive. The proof of the above formula requires that $K_\phi(t, t') = K_\phi(t - t') = \delta(t - t')$. Change variables of integration to $x = k(t - b)$, $db = -k\,dx$ and find

$$K_\phi(t) = \frac{1}{C_\phi} \int_0^\infty dk \int_{-\infty}^\infty dx \phi^*(x) \phi(x + tk) \tag{8.72}$$

Fourier transform $\phi(x)$ to $\hat\phi(\xi)$

$$K_\phi(t) = \frac{1}{C_\phi} \int_0^\infty dk \int_{-\infty}^\infty \frac{d\xi}{2\pi} |\hat\phi(\xi)|^2 e^{i\xi kt} \tag{8.73}$$

The integration variable ξ is divided into the two intervals $(-\infty < \xi < 0)$ and $(0 < \xi < \infty)$. In the first interval change $\xi \to -\xi$ and assume that $|\hat\phi(-\xi)|^2 = |\hat\phi(\xi)|^2$.

$$K_\phi(t) = \frac{1}{C_\phi} \int_0^\infty dk \int_0^\infty \frac{d\xi}{2\pi} |\hat\phi(\xi)|^2 [e^{i\xi kt} + e^{-i\xi kt}]$$

$$\int_0^\infty dk [e^{i\xi kt} + e^{-i\xi kt}] = 2\pi\delta(\xi t) = \frac{2\pi}{\xi} \delta(t) \tag{8.74}$$

$$K_\phi(t) = \delta(t) \tag{8.75}$$

that completes the proof.

During the derivation the function $\phi(x)$ was assumed to have some mathematical properties. These are now collected in order to ascertain the type of function needed for the wavelet transform.

1. The constant C_ϕ exists. Its integral definition in Eq. (8.69) has two possible divergences, one is at small values of ξ where convergence requires that

$$\lim_{\xi \to 0} \hat{\phi}(\xi) = \int dx \phi(x) = 0 \tag{8.76}$$

The right-hand equality is the most important constraint regarding the function $\phi(x)$. It requires that it be a function with positive and negative regions in equal measure. The further requirement that it be restricted to small values of x means that $\phi(x)$ has the shape of a small wave. Two forms are shown in Fig. 8.2. They resemble small wave forms, the origin of the word *wavelet*. The second constraint on the integral for C_ϕ is that it converges at large values of ξ that gives (with $s = x\xi$)

$$\lim_{\xi \to \infty} \hat{\phi}(\xi) = \frac{1}{\xi} \int ds e^{is} \phi(s/\xi) \approx \frac{\phi(0)}{\xi} \int ds e^{is} = 0 \tag{8.77}$$

The integral vanishes as long as $\phi(0)$ is not divergence. Most reasonable functions satisfy this latter constraint.

2. $|\hat{\phi}(-\xi)|^2 = |\hat{\phi}(\xi)|^2$. This condition is obeyed for all real functions $\phi(x)$, in which case $\hat{\phi}(-\xi) = \hat{\phi}^*(\xi)$.

The wavelet transform works for all real functions $\phi(x)$ that obey the constraint

$$\int_{-\infty}^{\infty} dx \phi(x) = 0, \tag{8.78}$$

$$\int_{-\infty}^{\infty} dt \psi_{k,b}(t) = 0. \tag{8.79}$$

There is an infinite set of functions that satisfy these conditions. Usually the best choice of $\phi(x)$ depends on the anticipated form of the signal $f(t)$. Since $\phi(x)$ is assumed to be a real function, the complex conjugate notation is unneeded. Since $\phi(x)$ gives zero when integrated over all values of x, then $\psi_{k,b}(t)$ gives zero when integrated over all t. We might think, from Eq. (8.68), that we can only use functions $f(t)$ whose integral $\int dt f(t)$ integrates to zero. This contraint is not required, although it appears that when integrating Eq. (8.68), $\int dt$ on both sides of the equals sign gives $\int dt f(t) = 0$. This derivation is invalid since the integration $\int dt'$ on the right-hand side does not interchange with the integral $\int db$. The above procedure can be applied to functions with a nonzero value of $\int dt f(t)$.

As a simple example, consider the choice

$$\phi(x) = xe^{-x^2/2} \tag{8.80}$$

$$\hat{\phi}(\xi) = \int dx x e^{-i\xi x - x^2/2} = -i\sqrt{2\pi}\, \xi e^{-\xi^2/2} \tag{8.81}$$

$$C_\phi = \int_0^\infty \frac{d\xi}{\xi} |\hat{\phi}(\xi)|^2 = 2\pi \int_0^\infty \xi\, d\xi e^{-\xi^2} = \pi \tag{8.82}$$

$$N^2 = \frac{\sqrt{\pi}}{2}, \qquad \Delta_\phi^2 = \frac{3}{2} = \Delta_{\hat{\phi}}^2 \tag{8.83}$$

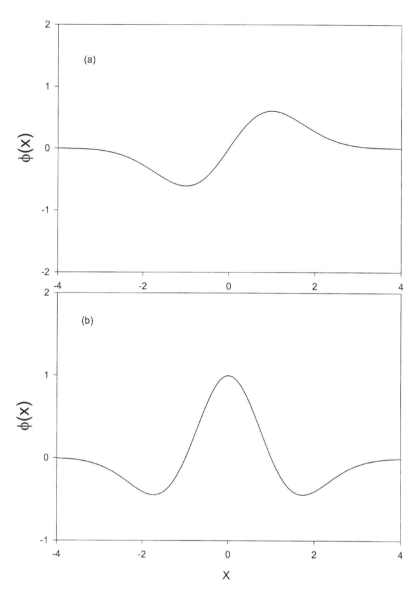

Figure 8.2 Several forms of wavelets.

This choice for $\phi(x)$ has the form shown in Fig. 8.2(a). It can be used to find
the wavelet transform of many functions $f(t)$ such as $f(t) = \cos(\omega t)$

$$W_{k,b}(f) = \sqrt{k} \int dt \, \cos(\omega t)[k(t - b)] \exp\left[-\frac{k^2}{2}(t - b)^2\right]$$

$$= \frac{1}{\sqrt{k}} \text{Re}[e^{i\omega b}\hat{\phi}(-\omega/k)]$$

$$= -\omega \sqrt{\frac{2\pi}{k^3}} e^{-(\omega/k)^2/2} \sin(b\omega) \tag{8.84}$$

The transform oscillates when b is varied, but is constrained in k to values near $k \sim \omega$. Other examples are given in the problems. Figure 8.2(b) shows another choice of wavelet

$$\phi(x) = (1 - x^2)e^{-x^2/2} \tag{8.85}$$

which is called "Mexican Hat."

The transform is called *continuous* since (k, b) are assumed to have continuous values. This transform is most useful for analog signals. Note that the transform is not an orthogonal set, since

$$M(k, k'; b, b') = \int dx \psi_{k,b}(x) \psi_{k',b'}(x) \tag{8.86}$$

is not zero when $k \neq k'$ or $b \neq b'$. It is a complete set, since any function $f(t)$ can be cast exactly as an expansion of the form Eq. (8.68).

8.3.2. Discrete Transforms

Discrete transforms are useful for digital signals. Wavelet transforms can be defined when the variables (k, b) are not continuous. The usual method is to write the transform as

$$\psi_{m,n}(t) = a^{m/2}\phi(a^m t - t_0 n) \tag{8.87}$$

$$W_{m,n}(f) = \int dt f(t)\psi_{m,n}(t) \tag{8.88}$$

$$f(t) = \sum_{m,n} W_{m,n}\psi_{m,n}(t) \tag{8.89}$$

The parameters (a, t_0) are both assumed to be positive, while the integers (m, n) are either positive or negative. The integer n controls the center of the wavelet. If t_0 is a typical timescale for the digital signal, then $b = nt_0$ is the center of the wavelet. The integer m controls the width of the timescale. It spans different decades in time as m is varied. The Fourier transform of this function is

$$\hat{\psi}_{m,n}(\omega) = \int dt e^{-it\omega} a^{m/2}\phi(a^m t - t_0 n)$$

$$= a^{-m/2}e^{-int_0(\omega/a^m)}\hat{\phi}(\omega/a^m) \tag{8.90}$$

The factor of a^m also controls the frequency scale. The geometrical form allows one to sample many octaves.

Daubechies[2] noted that this form of the wavelet is an orthonormal set when $a = 2$ and $\phi(x)$ is a function first proposed by Haar

$$\phi(x) = \begin{cases} 1 & 0 \leqslant x < \tfrac{1}{2} \\ -1 & \tfrac{1}{2} < x \leqslant 1 \\ 0 & \text{elsewhere} \end{cases} \tag{8.91}$$

$$\psi_{m,n}(t) = 2^{m/2}\phi(2^m t - n) \tag{8.92}$$

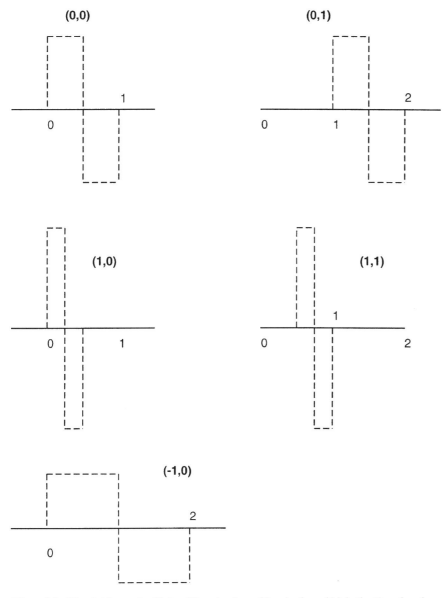

Figure 8.3 Wavelet forms $\psi_{m,n}(t)$ for different values of (m, n) where $\phi(x)$ is the Haar function.

A dimensionless scale is used for time, with $t_0 = 1$. The interesting feature of this function is that it is completely orthogonal in both variables

$$\int dt \psi_{m,n}(t) \psi_{m',n'}(t) = \delta_{mm'} \delta_{nn'} \tag{8.93}$$

This orthogonality can be proved by inspection. Figure 8.3 shows $\psi_{m,n}(t)$ for five values of (m, n). Each one has a region of a positive constant and a region of a negative constant. If they overlap with another function, their orthogonality is due to the fact that one function is constant over the region where the other is varying between plus and minus one. Often their nonzero regions do not overlap. Having an orthogonal set for the wavelet basis has many advantages.

The Haar function obeys the condition that $\int dx\phi(x) = 0$ which also means that

$$\hat{\phi}(\xi = 0) = \int dx\phi(x) = 0 \tag{8.94}$$

$$\left(\frac{d\hat{\phi}}{d\xi}\right)_{\xi=0} = -i\int dxx\phi(x) \neq 0 \tag{8.95}$$

Sometimes it is useful to have additional smoothness such that $(d\hat{\phi}/d\xi)_0 = 0$ that can be obtained by insisting that

$$\int dxx^r\phi(x) = 0, \quad \text{for } r = 0, 1 \tag{8.96}$$

A function with these properties is

$$\phi(x) = \begin{cases} 1 & 0 \leqslant x < \frac{1}{4} \\ -1 & \frac{1}{4} < x \leqslant \frac{3}{4} \\ 1 & \frac{3}{4} \leqslant x < 1 \\ 0 & \text{elsewhere} \end{cases} \tag{8.97}$$

that can be verified by direct calculation. We can also show that using this function to construct $\psi_{m,n}(t)$ gives an orthogonal set. Daubechies showed how to construct orthonormal functions that obey Eq. (8.96) for values of $0 < r < M$, which make $\hat{\phi}(\xi)$ and its first M derivatives vanish at $\xi = 0$.

An example is provided for representing a function $f(t)$ as a summation over Haar wavelets

$$f(t) = \bar{f} + \sum_{m,n} c_{m,n}\psi_{m,n}(t) \tag{8.98}$$

$$c_{m,n} = \int dtf(t)\psi_{m,n}(t) \tag{8.99}$$

$$\bar{f} = \int dtf(t) \tag{8.100}$$

Figure 8.4(a) shows a function $f(t)$ over a finite interval. Divide the interval into N steps of width t_0 and normalized by setting $t_0 = 1$. It is convenient to set $N = 2^J$ where J is an integer. The time span is $0 \leqslant t \leqslant 2^J$. The final formula is

$$f(t) = \bar{f} + \sum_{l=0}^{J} \sum_{n=0}^{2^l-1} c_{-J+l,n}\psi_{J-l,n}(t) \tag{8.101}$$

$$\bar{f} = \int_0^{2^J} dtf(t) \tag{8.102}$$

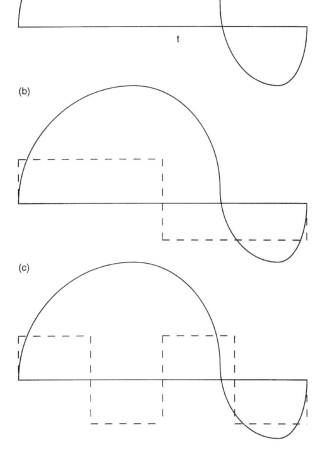

Figure 8.4 Decomposing $f(t)$ into Haar wavelets. (a) The solid line is the original function and the dashed line is the average value of $\bar{f}$. (b) The dashed line is the Haar wavelet $\psi_{-J,0} = 2^{-J/2}\phi(t/2^J)$. (c) The dashed lines are the two Haar wavelet $\psi_{1-J,0}$ and $\psi_{1-J,1}$, side-by-side, which span the time interval.

The first step is to calculate the average value $\bar{f}$ of $f(t)$ over the time interval. The function $f(t) - \bar{f}$ has an average value of zero over the time interval.

The second step is to construct the Haar function $\psi_{-J,0}(t) = 2^{-J/2}\phi(t/2^J)$. As shown in Fig. 8.4(b), it is positive over the interval $0 \leqslant t \leqslant 2^{J-1}$ and negative over the interval $2^{J-1} \leqslant t \leqslant 2^J$. Its coefficient is

$$c_{-J,0} = \int_0^{2^J} dt[f(t) - \bar{f}]\psi_{-J,0}(t) = \int_0^{2^J} dt f(t)\psi_{-J,0}(t) \qquad (8.103)$$

The integral over the constant term $\bar{f}$ is zero.

The third step is to include the two Haar functions $\psi_{1-J,0}(t)$ and $\psi_{1-J,1}(t)$. The first one spans the interval $0 \leqslant t \leqslant 2^{J-1}$ while the latter spans the interval $2^{J-1} \leqslant t \leqslant 2^{J}$. They are shown side-by-side in Fig.8.4(c). Their coefficients are evaluated using the integral in Eq. (8.99). These steps have so far created the series

$$f(t) = \bar{f} + c_{-J,0}\psi_{-J,0}(t) + c_{1-J,0}\psi_{1-J,0}(t) + c_{1-J,1}\psi_{1-J,1}(t) + \cdots + \quad (8.104)$$

The fourth step is to divide the interval into four segments that are spanned by $\psi_{2-J,n}(t)$, $n = 0, 1, 2, 3$. Their coefficients $c_{2-J,n}$ are evaluated using Eq. (8.99). Continue on with eight Haar functions, then sixteen, and so on. The process derives the terms in the series Eq. (8.100).

References

1. C. K. Chui, *An Introduction to Wavelets*. Academic Press, 1992.
2. I. Daubechies, *Ten Lectures on Wavelets*. Society Industrial Applied Mathematics, 1992.
3. I. Daubechies, ed., *Different Perspectives on Wavelets*, Proc. Symp. Appl. Math. Vol 47. American Mathematical Society, 1993.
4. G. Erlebacher, M. Y. Hussaini, and L. M. Jameson, eds., *Wavelets: Theory and Applications*. Oxford University Press, 1996.
5. Y. Meyer, translated by R. D. Ryan, *Wavelets, Algorithms and Applications*, Soc. Industrial and Appl. Math., 1993.

Problems

1. In 1D find the unbounded Fourier transform of

$$f(x) = -V_0, \quad |x| < a \quad (8.105)$$

$$= 0, \quad |x| > a \quad (8.106)$$

   Show that the inverse transform restores one to $f(x)$.

2. In 1D use the Fourier cosine transform for the half space $x \geqslant 0$ to evaluate $f(x) = x \exp(-px)$. Evaluate the inverse transform to find the original $f(x)$.

3. In 1D find the Fourier transform of the function $f(x) = x$ over the interval $-L/2 < x < L/2$.

4. In 3D find the Fourier transform of the function $f(\mathbf{r}) = \exp(-r^2 k_s^2)$ where k_s is a constant.

5. Find the Laplace transforms of the following functions

   (a) $f(t) = \delta(t - t_0)$, $[t_0 > 0]$

   (b) $f(t) = e^{-\alpha t}\sin(\omega t)$

   Evaluate the inverse transform for each case, to retrieve the original function.

6. Given the Laplace transform $F(p)$ listed below, evaluate the inverse transform and determine the original function $f(t)$.

   (a) $F(p) = \dfrac{e^{-bp}}{p}$

(b) $F(p) = e^{-b\sqrt{p}}$

(c) $F(p) = \dfrac{1 - e^{-t_0(a+p)}}{p + a}$

7. Show that the traditional windowing transform also gives completeness for a suitable choice of D:

$$K_g(t, t') = \frac{1}{D} \int d\omega \int db\, g_{k,b,\omega}(t)^* g_{k,b,\omega}(t') = \delta(t - t') \qquad (8.107)$$

8. Find $W_{k,b}$ when $\phi(x) = x \exp(-x^2/2)$ and $f(t) = \cos(\omega t) \exp(-\alpha^2 t^2/2)$. Also integrate $\int dt\, f(t)$ and find the value of ω for which this integral vanishes.

9. Let $f(t) = t(1 - t)$ for $0 < t < 1$ and is zero elsewhere. Let $\phi(t)$ be the Haar function that is 1 for $0 < t < 1/2$ and is -1 for $1/2 < t < 1$. Evaluate

$$W_{m,n} = \int_0^1 dt f(t) 2^{m/2} \phi(2^m t - n) \qquad (8.108)$$

for the three cases of $(m, n) = (0, 0)$, $(m, n) = (1, 0)$, and $(m, n) = (1, 1)$.

10. For the Mexican hat function $\phi(x) = (1 - x^2) \exp(-x^2/2)$:
    (a) Show $\int dx\, \phi(x) = 0$.
    (b) Derive the Fourier transform $\hat{\phi}(\xi)$.
    (c) Derive C.
    (d) Find the transform of $f(t) = \cos(\omega t)$.

11. For the Haar function for $\phi(x)$ find N^2, Δ_ϕ^2, $\hat{\phi}(\xi)$, C, then find $W_{n,m}$ when $f(t) = \sin(\omega t)$.

Equations of Physics

The remainder of this book will discuss the method of solving some differential equations. These equations are those encountered in a variety of physics problems:

$$\text{Laplaces Equation} \quad \nabla^2\phi = 0 \tag{9.1}$$

$$\text{Helmholtz Equation} \quad (\nabla^2 + k^2)\phi = 0 \tag{9.2}$$

$$\text{Poisson Equation} \quad \nabla^2\phi = -\frac{e}{\varepsilon_0}n(\mathbf{r}) \tag{9.3}$$

$$\text{Diffusion Equation} \quad \left(D\nabla^2 - \frac{\partial}{\partial t}\right)\phi = 0 \tag{9.4}$$

$$\text{Wave Equation} \quad \left(v^2\nabla^2 - \frac{\partial^2}{\partial t^2}\right)\phi = 0 \tag{9.5}$$

The symbol ϕ could represent any kind of variable. D is the diffusion coefficient that has units of m^2/s. k denotes a wave vector, and v is the wave velocity. An equation with two different kinds of derivatives is called a *partial differential equation.*

There are many other equations of physics besides those listed here. Some important ones are: Maxwell's equations, Newton's equations, Schrödinger's equation, Euler's equation among others. Many of these other equations have entire courses devoted to solving them for a variety of problems. The present course is devoted to solving Eqs. (9.1–9.5).

Laplace's equation is encountered in electrostatics, where ϕ is the potential function. Laplace's equation is also found in static diffusion problems. Helmholtz equation is usually encountered in solving other equations. If the time dependence of the wave equation is $\exp(-i\omega t)$ then the wave equation becomes Helmholtz equation, where $k = \omega/v$. Similarly, if the time dependence of the diffusion equation is $\exp(-t/\tau)$, then the diffusion equation becomes a Helmholtz equation with $k^{-2} = \tau D$.

The diffusion equation is usually encountered as the diffusion of particles ($\phi = n$), where $n(\mathbf{r}, t)$ is the particle density, or else as the diffusion of heat ($\phi = T$), where $T(\mathbf{r}, t)$ is the temperature.

These equations often have a source term $S(\mathbf{r}, t)$ on the right-hand side of the equal sign, instead of a zero. However, usually they are solved as written, and the source terms are inserted as boundary conditions.

9.1. Boundary and Initial Conditions

These equations are solved in one, two, and three dimensions for both static and transient problems. In each case, additional information is required:

1. **Boundary Conditions:** If the system has a boundary, the solution depends upon the behavior at the boundary. Quite often this boundary condition is specified as part of the problem. The system is often clamped at the boundary $\phi = 0$ for example. In general, boundary conditions are usually of the following types:

   - Dirichlet: Here the function ϕ is specified at the boundary.
   - Neumann: here the gradient of the function $(\nabla\phi)$ is specified at the boundary.
   - Cauchy: Here the boundary conditions are mixed between Dirichlet and Neumann. A typical example is

$$\phi + A\hat{n} \cdot \nabla\phi = 0 \tag{9.6}$$

   where $\hat{n}$ is the unit vector that is perpendicular to the boundary surface.

2. **Initial Conditions:** In transient problems, it is required to state the values of the function ϕ at the start, that is usually taken to be $t = 0$. The number of initial conditions depends on the number of time derivatives in the differential equation.

   - For the diffusion equation, it is required to define $\phi(\mathbf{r}, t = 0)$ at the start of the transient behavior.
   - For the wave equation, it is required to state both $\phi(\mathbf{r}, t)$ and $\partial\phi(\mathbf{r}, t)/\partial t$ at $t = 0$.

For the problems in this book, the initial and boundary conditions are always stated.

9.2. Boltzmann Equation

The diffusion equation describes nonequilibrium behavior. The fundamental equation for nonequilibrium behavior is the Boltzmann equation. It comes in two varieties, known as the Quantum Boltzmann Equation (QBE), and the Boltzmann Equation (BE). The QBE is more fundamental, but its solutions are hard and it is seldom solved. Most work is based upon the BE, which is discussed here.

The Boltzmann equation is an equation of motion for the distribution function $f(\mathbf{r}, \mathbf{v}, t)$. The symbol $\mathbf{r}$ denotes position, $\mathbf{v}$ denotes velocity, while t denotes time. The distribution function is the probability, per unit phase space, that the system has particles at position $\mathbf{r}$, with velocity $\mathbf{v}$ at time t. Phase space is defined as $d^3r d^3v$. The integral over velocity, with various arguments, produces various macroscopic quantities.

$$\text{Particle Density} \quad n(\mathbf{r}, t) = \int d^3v f(\mathbf{r}, \mathbf{v}, t) \qquad (9.7)$$

$$\text{Current density} \quad \mathbf{J}(\mathbf{r}, t) = e \int d^3v \mathbf{v} f(\mathbf{r}, \mathbf{v}, t) \qquad (9.8)$$

$$\text{Energy Density} \quad U = \int d^3v \varepsilon_v f(\mathbf{r}, \mathbf{v}, t) \qquad (9.9)$$

$$\text{Energy Current} \quad \mathbf{J}_E(\mathbf{r}, t) = \int d^3v \mathbf{v} \varepsilon_v f(\mathbf{r}, \mathbf{v}, t) \qquad (9.10)$$

The dimensions of these variables are: $n(\#/m^3)$, $\mathbf{J}$ (Amp/m^2), U (Joule/m^3), and $\mathbf{J}_E$ (Watts/m^2), where m is meter. The symbol $\varepsilon_v = mv^2/2$ is the kinetic energy, and e is the charge of the particle.

If the distribution f is the solution for nonequilibrium processes, then it is stable under small variations of its coordinates. The distribution should be unchanged by taking a complete derivative with respect to time. This derivative is on the explicit time variable, but also on all other variables

$$\frac{\delta f}{\delta t} = 0 = \left(\frac{\partial}{\partial t} + \sum_j \dot{q}_j \frac{\partial}{\partial q_j} \right) f(t, \{q_j\}) \qquad (9.11)$$

For the BE, the variables q_j are the position variables (x, y, z) and the velocity variables (v_x, v_y, v_z). Of course, $\dot{q}_j$ for the space variables are just the components of the velocity. Similarly, $\ddot{q}_j$ for the velocity variables gives the components of acceleration $\mathbf{a}$. Newtons Second Law ($\mathbf{F} = m\mathbf{a}$) relates acceleration to the force on the particles. The BE is

$$0 = \frac{\partial}{\partial t} f + \mathbf{v} \cdot \nabla_r f + \frac{1}{m} \mathbf{F} \cdot \nabla_v f + \left(\frac{df}{dt} \right)_s \qquad (9.12)$$

The last term is from the scattering processes. If the particles scattering from the walls of the container, or from other particles, these processes could alter the distribution function.

9.2.1. Moment Equations

Usually, for a system out of equilibrium, it is too laborious to solve for $f(\mathbf{r}, \mathbf{v}, t)$ as a function of all of its variables. There are seven variables in three dimensions and five variables in two dimensions. Instead, it is practical to have a set of equations for the macroscopic variable such as n, $\mathbf{J}$, and so on. These macroscopic equations are derived from Boltzmann's equation. The process is called "taking moments." Denoting Eq. (9.12) as "BE", a set of macroscopic equations is generated by taking

$$0 = \int d^3v G(\mathbf{v})\{\text{BE}\} \tag{9.13}$$

Each different function $G(\mathbf{v})$ gives a different macroscopic equation;

1. $G = 1$ generates the *equation of continuity*

$$0 = \int d^3v \left\{ \frac{\partial}{\partial t} f + \mathbf{v} \cdot \nabla_r f + \frac{1}{m} \mathbf{F} \cdot \nabla_v f + \left(\frac{df}{dt} \right)_s \right\}$$

$$= \frac{\partial n}{\partial t} + \frac{1}{e} \nabla \cdot \mathbf{J} \tag{9.14}$$

In macroscopic equations, the gradient symbol always denotes derivative with respect to position. The equation of continuity comes from the first two terms in Boltzmann's equation. The third term gives zero since it is assumed that the force depends only upon position, and not on velocity. The remaining integral can be changed to a surface integral using Gauss's theorem

$$\mathbf{F} \cdot \int d^3v \nabla_v f = -\mathbf{F} \cdot \int_{\text{boundary}} d^2v \hat{n} f = 0 \tag{9.15}$$

The boundary of f is at infinite velocities, and it is assumed that f is zero at these large values. The scattering term vanishes, since it is assumed that the scattering does not change the number of particles.

The equation of continuity is a statement of particle conservation. In any element of volume d^3r, the number of particles will change $\dot{n}$ only by flow through the walls of that element. Such flow is given by the term $\nabla \cdot \mathbf{J}$.

2. The second moment has $G(\mathbf{v}) = e\mathbf{v}$. The moment equation is

$$0 = \frac{\partial \mathbf{J}}{\partial t} + \frac{2e}{3m} \nabla U - \frac{en}{m} \mathbf{F} + \frac{\mathbf{J}}{\tau_j} \tag{9.16}$$

Each term in BE gives a term in this moment equation. In the second term, there are two factors of velocity $\langle v_\mu v_v \rangle$ so the evaluation give a tensor. It is assumed that f is isotropic, so that this tensor has nonzero components along the diagonal: $\mu = v$, Then the integrand has $v_\mu^2 = v^2/3$. This factor, when multiplied by $3m/2$, gives the kinetic energy, and its integral is the internal energy per unit volume U. In the third term, the integral is evaluated by parts

$$\mathbf{F} \cdot \int d^3v \mathbf{v} \nabla_v f = -\mathbf{F} \cdot \int d^3v f \nabla_v \mathbf{v} = -\mathbf{F}n \tag{9.18}$$

Finally, in the last term, the scattering term leads to a relaxation time τ_j of the current. If a current down a wire is driven by a battery, and suddenly the battery is switched-off, it takes an average time τ_j for the current to stop. A typical time in solids is nanoseconds. The time is very short.

The time dependence of this equation is assumed to be governed by $\exp(-i\omega t)$, in which case the equation is written as

$$[1 - i\omega\tau_j]\mathbf{J} = -\frac{2e\tau_j}{3m}\nabla U + \frac{en\tau_j}{m}\mathbf{F} \tag{9.18}$$

In transport then $\omega\tau_j \ll 1$ and this term can be ignored. Laboratory experiments use frequencies such as kHz or MHz, and neither is large enough to surmount the fact that τ_j is very small. The current equation is usually treated as instantaneous. However, for frequencies in the optical regime, $\omega\tau_j > 1$ and the frequency dependence is important.

The internal energy, per unit volume, of a classical gas is $U = (3/2)k_B Tn$. For this choice, the above expression is approximately $(\mathbf{F} = -e\nabla V)$

$$\mathbf{J} = -\sigma[\nabla V + S\nabla T] - eD\nabla n \tag{9.19}$$

$$\sigma = \frac{e^2 n\tau_j}{m} \tag{9.20}$$

$$D = \frac{\tau_j}{m} k_B T \tag{9.21}$$

$$S = \frac{k_B}{e} \tag{9.22}$$

the symbol σ is the electrical conductivity in units of Siemans per meter. The symbol D is the diffusion coefficient in units of m^2/s, and S is the Seebeck coefficient in units of volts/degree. Equation (9.19) is the usual way of writing the equation for the current, in terms of the gradients with respect to potential (V), temperature (T) or density (n). The Seebeck coefficient is generally $S = (k_B/e)$ times a dimensioness number.

Equation (9.19) has a number of familiar features. If the potential is a constant $(\nabla V = 0)$ and the temperature is a constant $(\nabla T = 0)$ the resulting equation is *Fick's Law*

$$\mathbf{J} = -eD\nabla n \tag{9.23}$$

Another famous equation occurs when $(\nabla T = 0, \nabla n = 0)$. The electric field is $\mathbf{E} = -\nabla V$ and the result is *Ohm's Law*.

$$\mathbf{J} = \sigma\mathbf{E} \tag{9.24}$$

3. $G(\mathbf{v}) = \varepsilon_v$. When the scattering conserves energy, the energy relaxation time is infinite, since the energy is unchanged. That will be assumed here. There are cases where the total energy of a set of particles will change due to scattering. These cases are where there are more than one kind of particle. If two or more sets of particles are at different temperatures, they come to equilibrium by exchanging energy. That case is often important, but is not considered here.

The other terms in the BE for energy relaxation are:

$$0 = \frac{\partial U}{\partial t} + \mathbf{V} \cdot \mathbf{J}_E - \frac{1}{e} \mathbf{J} \cdot \mathbf{F} \qquad (9.25)$$

$$\frac{\partial U}{\partial t} = \frac{dU}{dT} \frac{\partial T}{\partial t} = C \frac{\partial T}{\partial t} \qquad (9.26)$$

$$C \frac{\partial T}{\partial t} = -\mathbf{V} \cdot \mathbf{J}_E + \frac{1}{e} \mathbf{J} \cdot \mathbf{F} \qquad (9.27)$$

where C is the heat capacity per unit volume [J/(m³K)]. The last equation is the conservation of energy. The temperature changes at a point ($\dot{T}$) are due to local variations in heat flow ($\mathbf{V} \cdot \mathbf{J}_E$) or else to local Joule heating ($\mathbf{J} \cdot \mathbf{F}/e$) from current flow.

 4. $G(\mathbf{v}) = \mathbf{v}\varepsilon_v$ This choice produces an equation for the time derivative of the energy current $\dot{\mathbf{J}}_E$. For diffusive properties in solids, this time derivative is negligible. Then the moment equation gives an equation for the energy current that is

$$\mathbf{J}_E = -K'\mathbf{V}T - ST[\sigma\mathbf{V}V + eD\mathbf{V}n] \qquad (9.28)$$

The term in brackets is given by the current Eq. (9.20), so that

$$\mathbf{J} + \sigma S\mathbf{V}T = -[\sigma\mathbf{V}V + eD\mathbf{V}n] \qquad (9.29)$$

$$\mathbf{J}_E = ST(\mathbf{J} + \sigma S\mathbf{V}T) - K'\mathbf{V}T \qquad (9.30)$$

$$\mathbf{J}_E = ST\mathbf{J} - K\mathbf{V}T \qquad (9.31)$$

$$K = K' - \sigma TS^2 \qquad (9.32)$$

Note that K is the thermal conductivity at constant current, while K' is the thermal conductivity at constant density and voltage. It is assumed that the system is locally isotropic. Equation (9.31) shows that the flow of heat can be induced by an electrical current. It can force heat to flow from a cold to a hot reservoir, that is the basis of solid state thermoelectric refrigerators.

 The macroscopic equations have been derived here starting from Boltzmann's equation. A more fundamental derivation is provided using nonequilibrium statistical mechanics. This topic is covered in standard references. A problem with generating moment equations is that each new moment introduces an additional unknown variable:

- The equation for $\dot{n}$ depends on $\mathbf{J}$.
- The equation for $\dot{\mathbf{J}}$ depends on U.
- The equation for $\dot{U}$ depends on $\mathbf{J}_E$.

There must be a method of truncating the series. Equation (9.29) made an assumption regarding the stress tensor that was suitable for electrons. In the next section, similar equations are derived for the flow of fluids, and the stress tensor has a different form. Moment equations always require that the set of equations be truncated at some order, to produce a closed set of equations. Complete books are written on different methods of performing this truncation.

9.2.2. Diffusion Equations

The diffusion equation for particles is derived by combining Eqs. (9.14) and (9.19). In the latter, it is assumed that $\nabla V = 0 = \nabla T$ so that $\mathbf{J} = -eD\nabla n$ that gives in the equation of continuity

$$0 = \frac{\partial n}{\partial t} + \frac{1}{e}\nabla\cdot[-eD\nabla n] = \frac{\partial n}{\partial t} - D\nabla^2 n \qquad (9.33)$$

that is the desired equation. The equation is valid only when it is justified to neglect the term $\omega\tau_j$ compared to one. Most diffusion is slow, and this approximation is very accurate. In some cases the diffusion coefficient depends on the density of the particles $D(n)$, and the equation becomes nonlinear.

The diffusion equation for heat comes by combining Eqs. (9.27) and (9.31) while assuming that $\mathbf{F} = 0 = \mathbf{J}$

$$C\frac{\partial T}{\partial t} + \nabla\cdot[-K\nabla T] = 0 \qquad (9.34)$$

$$C(T)\frac{\partial T}{\partial t} - \nabla\cdot[K(T)\nabla T] = 0 \qquad (9.35)$$

The diffusion coefficent for heat is $D = K/C$ and has the usual value of about 1 cm^2/s for many materials. However, since both $C(T)$ and $K(T)$ depend upon temperature, the equation is inherently nonlinear. In our applications, C, K, D are taken to be constants, so that the diffusion equation is linear.

9.2.3. Fluid Equations

Another set of equations is derived for the case of fluid flow. The important macroscopic variables are: fluid density ρ (units are kilograms/m^3), fluid velocity $\mathbf{u}$, internal energy ε and heat flow $\mathbf{q}$ that are all functions of $(\mathbf{r}, t)$. The distribution function still depends upon $f(\mathbf{r}, \mathbf{v}, t)$. However, assume the fluid is moving at $(\mathbf{r}, t)$ with an average velocity $\mathbf{u}(\mathbf{r}, t)$. Then the first two fluid moments are

$$\int d^3v f(\mathbf{r}, \mathbf{v}, t) = \rho(\mathbf{r}, t) \qquad (9.36)$$

$$\int d^3v \mathbf{v} f(\mathbf{r}, \mathbf{v}, t) = \rho\mathbf{u} \qquad (9.37)$$

The velocity $\mathbf{u}$ represents the center of mass motion of the fluid. Define the motion relative to the center of mass as $\mathbf{c} = \mathbf{v} - \mathbf{u}$. The average value of

$\langle \mathbf{c} \rangle = 0$, or

$$\int d^3vc f(\mathbf{r}, \mathbf{v}, t) = \int d^3v(\mathbf{v} - \mathbf{u}) f(\mathbf{r}, \mathbf{v}, t) = 0 \tag{9.38}$$

Next consider the average value of v^2

$$\frac{1}{2} \int d^3vv^2 f(\mathbf{r}, \mathbf{v}, t) = \frac{1}{2} \int d^3v(\mathbf{u} + \mathbf{c})^2 \, f(\mathbf{r}, \mathbf{v}, t) = \frac{1}{2} u^2 \rho + \rho\varepsilon$$

$$\rho\varepsilon = \frac{1}{2} \int d^3vc^2 f(\mathbf{r}, \mathbf{v}, t) \tag{9.39}$$

The internal energy $\varepsilon = (3/2)RT$ is from the motion relative to the center of mass. The gas constant is R. The average kinetic energy has a term $\rho u^2/2$ from the motion of the center-of-mass. Two additional moments of the relative velocity are defined below, where c_i, u_i are the (x,y,z) components of the vectors.

$$p_{ij} = \int d^3vc_i c_j f(\mathbf{r}, \mathbf{v}, t) \tag{9.40}$$

$$q_i = \frac{1}{2} \int d^3vc_i c^2 f(\mathbf{r}, \mathbf{v}, t) \tag{9.41}$$

$$\int d^3vv_i v_j f(\mathbf{r}, \mathbf{v}, t) = \int d^3v(u_i + c_i)(u_j + c_j) f(\mathbf{r}, \mathbf{v}, t) = \rho u_i u_j + p_{ij} \tag{9.42}$$

The average p_{ij} was set equal to zero when discussing electrons. Here it is nonzero, and represents the stress tensor of the fluid. A simple approximation for the diagonal components is to set $p_{ii} = p$, where the fluid pressure is p. By summing over all components, it is easy to prove $p = (2/3)\rho\varepsilon = \rho RT$ that is the ideal gas law. These various definitions and relationships are now used to evaluate the moments of the Boltzmann equation.

1. The zeroth moment of the BE is

$$0 = \frac{\partial}{\partial t} \rho + \mathbf{\nabla} \cdot (\rho\mathbf{u}) \tag{9.43}$$

   that is still the equation of continuity. An incompressible fluid has the density ρ as a constant, in which case the above equation is simply $\mathbf{\nabla} \cdot \mathbf{u} = 0$.

2. The velocity moment of the BE is

$$0 = \frac{\partial}{\partial t} (\mathbf{u}\rho) + \mathbf{\nabla} \cdot (\rho\mathbf{u}\mathbf{u}) + \sum_i \frac{\partial}{\partial r_i} p_{ij} + \rho g\hat{z} \tag{9.44}$$

$$\mathbf{\nabla} \cdot (\rho\mathbf{u}\mathbf{u}) \equiv \frac{\partial}{\partial x} (\rho u_x\mathbf{u}) + \frac{\partial}{\partial y} (\rho u_y\mathbf{u}) + \frac{\partial}{\partial z} (\rho u_z\mathbf{u}) \tag{9.45}$$

The term $\rho g \hat{z}$ is from the force on the fluid, where the force is assumed to be gravity ($g = 9.8$ m/s^2) that is the usual case. Note that in the case of an ideal fluid that $p_{ij} = \delta_{ij} p$. For a static fluid ($\mathbf{u} = 0$), there is just the last two terms $0 = \mathbf{\nabla} p + \rho g \hat{z}$ and is the usual equation for the variation of hydrostatic pressure with the depth in a fluid. Unlike the electron case, there is no relaxation time from the scattering of the particles.

3. The moment of $v^2/2$ in the BE gives

$$0 = \frac{\partial}{\partial t}\left[\rho\left(\frac{1}{2}u^2 + \varepsilon\right)\right] + \sum_i \frac{\partial}{\partial r_i}\left[\rho u_i\left(\frac{1}{2}u^2 + \varepsilon\right) + \sum_j p_{ij}u_j + q_i\right] + \rho g u_z$$

(9.46)

The equation represents the conservation of energy. The change of energy at a point is due to variations in the flow of energy. In the Navier–Stokes approximation

$$p_{ij} = \delta_{ij}(p - \lambda \mathbf{\nabla} \cdot \mathbf{u}) - \mu\left(\frac{\partial u_i}{\partial r_j} + \frac{\partial u_j}{\partial r_i}\right)$$

(9.47)

$$q_i = -K\frac{\partial T}{\partial r_i}$$

(9.48)

The last expression is the heat flow as given by the thermal conductivity K. The first equation shows that there are additional terms in p_{ij} due to the viscosity coefficients (μ, λ). The *Navier–Stokes* equation for a fluid is found by setting $q_i = 0$ and assuming the fluid is incompressible.

The main problem with deriving equations using moments of the Boltzmann equation is that each successive equation generates new moments such as $\mathbf{\nabla}(\rho \mathbf{uu})$. Evaluating these expressions requires additional moment equations. Truncating the series is difficult, and requires physical intuition. The Navier–Stokes equation is one such truncation, where the unknown terms such as p_{ij} are expressed as other quantities. The equations are nonlinear, which makes them hard to solve but causes a rich variety of behavior such as vortices and turbulence.

9.3. Solving Differential Equations

Here we provide an overview of how to solve differential equations. Most of this material should be familiar to the student from prior courses, and is presented as a review. The types of equations are broken down into several categories:

9.3.1. Homogeneous Linear Equations

The variable x will be used in all examples. It could represent any variable, such as position or time. Some examples follow:

- First order-equations having a single derivative. The function $\phi(x)$ is unknown, while $f(x)$ is assumed to be a known function.

$$\left[\frac{d}{dx} + f(x)\right]\phi(x) = 0 \qquad (9.49)$$

This equation can be solved in closed form. Let $F(x)$ be the first integral of $f(x)$,

$$F(x) = \int_{x_i}^{x} dx' f(x') \qquad (9.50)$$

$$\phi(x) = Ce^{-F(x)} \qquad (9.51)$$

where C is a constant determined by the boundary or initial conditions. The constant x_i is also unimportant, since changing it just alters C. There is only one unknown constant in first-order differential equations.

- Second-order equations are of the form

$$\left[\frac{d^2}{dx^2} + f(x)\right]\phi(x) = 0 \qquad (9.52)$$

Each equation of this kind is a special case. Different choices for $f(x)$ are solved in terms of different functions. They will be discussed systematically in later chapters. The solutions are easy if f is a positive or negative constant:

$$\left[\frac{d^2}{dx^2} + a^2\right]\phi(x) = 0, \qquad \phi = A\sin(ax) + B\cos(ax) \qquad (9.53)$$

$$\left[\frac{d^2}{dx^2} - a^2\right]\phi(x) = 0, \qquad \phi = A\sinh(ax) + B\cosh(ax) \qquad (9.54)$$

For f a constant, the solution is given either in terms of sines and cosines, or else their hyperbolic equivalent. There are two unknown constants A and B for any second-order differential equation. They are determined by the initial and boundary conditions.

- If $f(x)$ is a linear function of x such as $f = x + b$ then the solution is in terms of Airy functions

$$\left[\frac{d^2}{dx^2} - x - b\right]\phi(x) = 0 \qquad (9.55)$$

$$\phi(x) = CAi(x + b) + DBi(x + b) \qquad (9.56)$$

$$Ai(x) = \frac{1}{\pi}\int_0^{\infty} dt\,\cos\left(\frac{1}{3}t^3 + xt\right) \qquad (9.57)$$

$$Bi(x) = \frac{1}{\pi}\int_0^{\infty} dt\left[\sin\left(\frac{1}{3}t^3 + xt\right) + \exp\left(-\frac{1}{3}t^3 + xt\right)\right] \qquad (9.58)$$

Again note that the second-order differential equation has two independent solutions. The final result depends on the boundary conditions that
determine the constants (C, D). The asymptotic expansion of $Ai(x)$ was
discussed in Chapter 4.

- Another simple case is when $f(x) = k^2/x^2$

$$\left[\frac{d^2}{dx^2} - \frac{k^2}{x^2} \right] \phi(x) = 0 \tag{9.59}$$

The important feature of this equation is that both terms in the bracket
have the same power of x. This feature requires that the solution be a
power law such as x^s. Inserting this ansatz into the equation gives

$$s(s-1)x^{s-2} - k^2 x^{s-2} = [s(s-1) - k^2]x^{s-2} = 0 \tag{9.60}$$

$$s = \frac{1}{2} \pm \sqrt{k^2 + \frac{1}{4}} \equiv s_{1,2} \tag{9.61}$$

$$\phi = Ax^{s_1} + Bx^{s_2} \tag{9.62}$$

Again there are two solutions, and the constants (A, B) are determined by
the boundary conditions.

9.3.2. Inhomogeneous Linear Equations

Inhomogeneous equations have a source term $C(x)$ on the righthand
side of the equals sign. It is asssumed that $C(x)$ is a known function.
Inhomogeneous equations always have two different kinds of solutions. They
are expressed as $\phi = \phi_1 + \phi_2$. If the operator $\mathcal{L} = [d^n/dx^n + f(x)]$, then

$$\mathcal{L}\phi_1 = C(x) \tag{9.63}$$

$$\mathcal{L}\phi_2 = 0 \tag{9.64}$$

1. ϕ_1 is a solution that involves the source term $C(x)$. Usually it has no
   adjustable constants. For example, if $f(x)$ and $C(x)$ are both constants,
   then $\phi_1 = C/f$.
2. ϕ_2 is a solution of the homogeneous equation. They were discussed in the
   prior subsection. Their amplitude has adjustable constants that are set
   by the initial and boundary conditions. Note that ϕ_2 is not a solution of
   the differential equation by itself. However, since ϕ_1 is a solution of the
   equation, then we can add arbitrary amounts of ϕ_2 to the final solution.
   The amounts are not actually arbitrary, since the constants are set by the
   initial and boundary conditions.

Some examples are given next of inhomogeneous equations and their two
solutions

- Equations with first-order derivatives can be solved exactly. Again $F(x) = \int dx' f(x')$

$$\left[\frac{d}{dx} + f(x) \right] \phi(x) = C(x) \tag{9.65}$$

$$\frac{d}{dx} (e^{F(x)} \phi(x)) = e^{F(x)} C(x) \tag{9.66}$$

$$\phi(x) = A e^{-F(x)} + e^{-F(x)} \int^x dx' C(x') \, e^{F(x')} \tag{9.67}$$

The first term on the right is ϕ_2 and the second is ϕ_1. It is not necessary to specify the lower limit on the integral, since that choice changes the constant A.

- Second-order equations with constant coefficients f and C. Some examples are

$$C = \left[\frac{d^2}{dx^2} + a^2 \right] \phi \tag{9.68}$$

$$\phi(x) = \frac{C}{a^2} + A \sin(xa) + B \cos(xa) \tag{9.69}$$

$$C = \left[\frac{d^2}{dx^2} - a^2 \right] \phi \tag{9.70}$$

$$\phi(x) = -\frac{C}{a^2} + A \sinh(xa) + B \cosh(xa) \tag{9.71}$$

- Second-order equations with constant f but variable $C(x)$. In this case the best approach is through a Fourier or Laplace transform. Overbars denote the transformed variables. An example for an unbounded system is

$$\bar{\phi}(k) = \int_{-\infty}^{\infty} dx e^{-ikx} \phi(x) \tag{9.72}$$

$$\bar{C}(k) = \int_{-\infty}^{\infty} dx e^{-ikx} C(x) \tag{9.73}$$

$$\bar{\phi}(k) = \frac{\bar{C}(k)}{f - k^2} \tag{9.74}$$

$$\phi(x) = \int_{-\infty}^{\infty} \frac{dk}{2\pi} e^{ikx} \frac{\bar{C}(k)}{f - k^2} \tag{9.75}$$

One could use another type of transform, depending on the boundary conditions.

- Green's Functions. The most general case has a second-order equation with both $f(x)$ and $C(x)$ depending on the variable x. These cases are harder, and each choice of $f(x)$ is a special case, that will be dealt with later. One strategy is to introduce a Green's function $G(x; x_0)$ that has the form

$$\left[\frac{d^2}{dx^2} + f(x)\right] G(x; x_0) = \delta(x - x_0) \tag{9.76}$$

$$\phi(x) = \int dx_0 G(x; x_0) C(x_0) \tag{9.77}$$

$$\left[\frac{d^2}{dx^2} + f(x)\right] \phi(x) = \int dx_0 \left[\frac{\partial^2}{\partial x^2} + f(x)\right] G(x; x_0) C(x_0)$$

$$= \int dx_0 \delta(x - x_0) C(x_0) = C(x) \tag{9.78}$$

The solution is found to the Green's function Eq. (9.76), which has a delta function as the source term. It is a homogeneous equation except at the point $x = x_0$. The second step is to perform the integral in Eq. (9.77) which gives the final answer for $\phi(x)$. This technique is discussed in later chapters. Its advantage is that the same $G(x; x_0)$ applies to all functions $C(x)$.

9.3.3. Nonlinear Equations

Most nonlinear equations cannot be solved analytically. Computers are quite useful for these problems. However, there are a few cases that can be solved by quadrature. Usually the goal is to obtain a solution $\phi(x)$. A solution by quadrature obtains $x(\phi)$ that is the inverse function. That is not quite as good as $\phi(x)$, but is often very useful.

1. First-order equations have the general structure

$$\frac{d}{dx} \phi = M(\phi) \tag{9.79}$$

where $M(\phi)$ is nonlinear: if expanded in a Taylor series, it has powers of ϕ^n with $n \geqslant 2$. The general method is to rearrange the derivatives into

$$\frac{d\phi}{M(\phi)} = dx \tag{9.80}$$

$$x = x_0 + \int^{\phi} \frac{d\phi'}{M(\phi')} \tag{9.81}$$

that is a typical solution by quadrature. As with all first-order equations, there is a constant term x_0.

As an example, consider the equation

$$\frac{d}{dx}\phi = -\frac{A}{\phi^2} \tag{9.82}$$

which is solved using the above steps

$$\phi^2 d\phi = -Adx \tag{9.83}$$

$$\phi(x)^3 = -3A(x - x_0) \tag{9.84}$$

If the boundary condition is $\phi(0) = 0$ then $x_0 = 0$ and the solution is $\phi(x) = -(3Ax)^{1/3}$.

2. Second-order nonlinear equations can also be solved by quadrature.

$$\frac{d^2}{dx^2}\phi(x) = M(\phi) \tag{9.85}$$

Here the trick is to multiply both sides of this equation by $d\phi/dx$. Let $N(\phi)$ be the integral of $M(\phi)$. The result is

$$\frac{d\phi}{dx}\frac{d^2\phi}{dx^2} = \frac{1}{2}\frac{d}{dx}\left(\frac{d\phi}{dx}\right)^2 = \frac{d\phi}{dx}M(\phi) = \frac{d}{dx}N(\phi) \tag{9.86}$$

$$N(\phi) = \int^\phi d\phi' M(\phi') \tag{9.87}$$

Cancel the common denominator of dx from both sides of the above identity, and the result is a perfect differential on both sides that can be integrated

$$2dN(\phi) = d\left(\frac{d\phi}{dx}\right)^2 \tag{9.88}$$

$$2N(\phi) + A = \left(\frac{d\phi}{dx}\right)^2 \tag{9.89}$$

$$\frac{d\phi}{dx} = \pm\sqrt{2N + A} \tag{9.90}$$

$$x = x_0 \pm \int^\phi \frac{d\phi'}{\sqrt{2N(\phi') + A}} \tag{9.91}$$

The second-order equation has generated two unknown constants, that are denoted as x_0 and A. The last step has a first-order equation, that is solved as shown above.

The above technique is actually familiar from Newtonian mechanics. The basic equation in one dimension is

$$m\ddot{x} = F(x) = -\frac{d}{dx} U(x) \tag{9.92}$$

The standard trick is to multiply Newton's Second Law by the velocity $\dot{x}$ and integrate as described above.

$$\frac{m}{2} \frac{d}{dt} \left(\frac{dx}{dt}\right)^2 = -\frac{d}{dt} U(x) \tag{9.93}$$

The resulting equation is the conservation of energy. The energy E is the constant resulting from the integral

$$\frac{m}{2} \dot{x}^2 + U(x) = E \tag{9.94}$$

$$\int^x \frac{dx'}{\sqrt{E - U(x')}} = \sqrt{\frac{2}{m}} (t - t_0) \tag{9.95}$$

The above integral can be solved to find the motion $t(x)$. This standard trick in mechanics works for every equation of the type $\phi'' = M(\phi)$. Two examples of this method follow.

Example 1: *Pendulum Equation.* The energy equation for the simple pendulum of moment of inertia I and torque τ is

$$I\ddot{\theta} + \tau \sin(\theta) = 0 \tag{9.96}$$

$$\frac{I}{2} (\dot{\theta})^2 + \tau(1 - \cos(\theta)) = E \tag{9.97}$$

$$\int_0^\theta \frac{d\theta'}{\sqrt{1 - \frac{\tau}{E} (1 - \cos(\theta'))}} = t \sqrt{\frac{2E}{I}} \tag{9.98}$$

The first equation is Newton's Law for the pendulum. The second equation is the conservation of energy obtained by integrating the first equation. The third equation is the solution by quadrature. It may be cast in conventional form by defining $\theta' = 2\alpha$, $(1 - \cos \theta') = 2 \sin^2 \alpha$ to write the answer as

$$2 \int_0^{\theta/2} \frac{d\alpha}{\sqrt{1 - m \sin^2 \alpha}} = t \sqrt{\frac{2E}{I}} \tag{9.99}$$

$$m = \frac{2\tau}{E} \tag{9.100}$$

The integral on the left is a standard form that is an *elliptic integral*. Some properties of elliptic integrals are given in section 9.4.

Example 2: *Space charge between two metal plates.* An interesting example of a nonlinear equation is to solve for the self-consistent potential energy $U(x)$ between two metal plates separated by a distance $2d$. Assume the metal plates are identical, and each has a charge density n_s at the surface due to thermal excitation from the interior over a work function $e\phi$. The nature of the solution is shown in Fig. 9.1. The density $n(x)$ falls towards the center of the opening, while the potential $U(x)$ increases. The solution is symmetric, $n(-x) = n(x)$, $U(-x) = U(x)$.

The potential energy obeys a Poisson equation in terms of the density $n(x)$, that in turn depends upon the potential

$$\frac{d^2}{dx^2} U(x) = -\frac{e^2 n(x)}{\varepsilon_0} = -\frac{e^2 n_0}{\varepsilon_0} \exp\left[-\frac{U(x)}{k_B T}\right] \tag{9.101}$$

$$n(d) = n_s = n_0 \exp\left[-\frac{U(d)}{k_B T}\right] = \frac{2}{\lambda^3} \exp\left[-\frac{e\phi}{k_B T}\right] \tag{9.102}$$

$$\frac{1}{\lambda^2} = \frac{m k_B T}{2\pi\hbar^2} \tag{9.103}$$

The last equality in Eq. (9.102) comes from statistical mechanics. It is convenient to normalize the potential $U(d) = e\phi$ so that the zero of potential is actually the chemical potential of the metal plates. Then the constant $n_0 = 2/\lambda^3$ and k_D is the Debye screening wave vector. It is convenient to use a

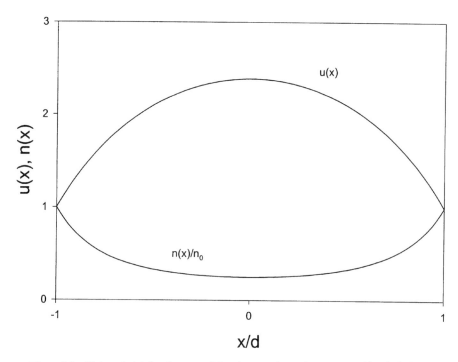

Figure 9.1 $U(x)$ and $n(x)$ for the potential and space charge between two identical plates.

dimensionless set of units, where

$$u(y) = \frac{U(x)}{k_B T} \tag{9.104}$$

$$y = k_D x = \left(\frac{e^2 n_0}{\varepsilon_0 k_B T}\right)^{1/2} x \tag{9.105}$$

$$\frac{d^2}{dy^2} u(y) = - e^{-u} \tag{9.106}$$

The first step in the solution is to multiply both sides of this equation by du/dy and then integrate

$$\frac{du}{dy}\frac{d^2u}{dy^2} = \frac{1}{2}\frac{d}{dy}\left(\frac{du}{dy}\right)^2 = -\frac{du}{dy}e^{-u} = \frac{d}{dy}e^{-u} \tag{9.107}$$

$$\left(\frac{du}{dy}\right)^2 = 2e^{-u} + A \tag{9.108}$$

Since the solution is symmetric, then $du/dy = 0$ at the point $y = 0$. This condition gives the value of the constant $A = -2\exp(-u_0)$, where $u_0 \equiv u(y = 0)$. Take the square root of the above equation, and arrange it into

$$\frac{du}{dy} = \pm\sqrt{2(e^{-u} - e^{-u_0})} = \pm\sqrt{2}e^{-u_0/2}\sqrt{e^{-\delta u} - 1} \tag{9.109}$$

$$\delta u = u - u_0 \tag{9.110}$$

$$\int^{\delta u} \frac{d\delta u'}{\sqrt{e^{-\delta u'} - 1}} = \pm 2q(y + y_0) \tag{9.111}$$

$$q = \frac{e^{-u_0/2}}{\sqrt{2}} \tag{9.112}$$

The integral on the left can be evaluated by changing variables to $s = \exp(\delta u/2)$. It is then the integral for the $\arccos(s)$

$$\int^{\delta u} \frac{d\delta u'}{\sqrt{e^{-\delta u'} - 1}} = 2\int \frac{ds}{\sqrt{1 - s^2}} = 2\cos^{-1}(s) \tag{9.113}$$

$$\cos^{-1}(s) = \pm q(y + y_0) \tag{9.114}$$

$$s = \cos[q(y + y_0)] = \exp(\delta u/2) \tag{9.115}$$

$$\delta u = 2\ln\{\cos[q(y + y_0)]\} \tag{9.116}$$

The last line is the solution. The constant y_0 must be determined. It is fixed by the condition that $\delta u(y = 0) = 0$ which sets $y_0 = 0$.

$$U(x) = U_0 + 2k_B T \ln\{\cos(qk_D x)\} \tag{9.117}$$

$$n(x) = n_0 e^{-u(y)} = \frac{n_0 e^{-u_0}}{\cos^2(qk_D x)} \tag{9.118}$$

The last constant to be determined is the maximum potential $u_0 = U_0/k_B T$. It is determined by a nonlinear equation, since the factor of q depends on U_0. At the point $x = d$ recall that $U(d) = e\phi$ so there is a self-consistent equation

$$e\phi = U_0 + 2k_B T \ln\left\{\cos\left(e^{-u_0/2}\frac{d}{d_0}\right)\right\} \tag{9.119}$$

$$d_0^{-1} = \frac{k_D}{\sqrt{2}} = \sqrt{\frac{e^2 n_0}{\varepsilon_0 k_B T}} \tag{9.120}$$

This equation is solved rather easily using a calculator or a computer. The constant d_0 is of atomic dimension.

In most experiments, the value of the plate separation $2d$ is macroscopic, which means that the ratio of d/d_0 can be very large: of the order of 10^6. The argument of the cosine must be less than $\pi/2$; otherwise, the cosine vanishes and the logarithm diverges to negative infinity. When d/d_0 is very large, the factor of $\exp(-u_0/2)$ becomes small. When U_0 becomes much larger than $e\phi$, the value of the logarithm must be large and negative. Large negative values of $\ln[\cos(qk_D d)]$ are obtained when the $\cos\theta \to 0$ that happens when $\theta = qk_D d = \pi/2$. In this case, an accurate estimate of U_0 for $d \gg d_0$ is

$$\frac{\pi}{2} = \left(\frac{d}{d_0}\right)e^{-U_0/2k_B T} \tag{9.121}$$

$$U_0 = 2k_B T \ln\left(\frac{2d}{\pi d_0}\right) \tag{9.122}$$

At room temperature ($T = 300\,\text{K}$), for a plate separation of $d = 1\,\text{mm}$, the potential U_0 reaches about $U_0 = 0.72$ eV which is a fairly large energy. The electrons between the two metal plates form a plasma, which creates a repulsive potential to the electrons that is much larger than the thermal energy $k_B T$. This solution was first published by Langmuir (5).

The above example shows how to solve a second-order nonlinear equation using quadrature. The basic procedure for other problems is similar. Always start by multiplying by dU/dx and then one gets an equation that can be integrated. The second integration is usually tougher. The above example is interesting since the second integration can be done analytically, that yields an exact solution.

There are several kinds of elliptic integrals.

1. *Complete elliptic integrals* come in two forms

$$K(m) = \int_0^{\pi/2} \frac{d\theta}{\sqrt{1 - m\sin^2\theta}} \tag{9.123}$$

$$E(m) = \int_0^{\pi/2} d\theta \sqrt{1 - m\sin^2\theta} \tag{9.124}$$

Some special values are: $K(0) = \pi/2$, $K(1) = \infty$, $E(0) = \pi/2$, $E(1) = 1$. The function $E(m)$ goes between $\pi/2$ and 1 as m is varied between zero and one. The function $K(m)$ diverges as $m \to 1$.

2. *Incomplete elliptic integrals* have the upper limit of integration ϕ less than $\pi/2$.

$$F(\phi|m) = \int_0^{\phi} \frac{d\theta}{\sqrt{1 - m\sin^2\theta}} \tag{9.125}$$

$$E(\phi|m) = \int_0^{\phi} d\theta \sqrt{1 - m\sin^2\theta} \tag{9.126}$$

Note that $F(\pi/2|m) = K(m)$, $E(\pi/2|m) = E(m)$.

References

1. C. Cercignani, *The Boltzmann Equation and Its Applications*, (Springer-Verlag, 1988).
2. H. B. Callen, *Thermodynamics* (John Wiley, 1960) Ch. 17.
3. R. Haase, *Thermodynamics of Irreversible Processes* (Dover, 1990), Ch. 4.
4. S. R. deGroot and P. Mazur, *Non-equilibrium Thermodynamics* (Dover, 1984).
5. I. Langmuir, *Phys. Rev.* **21**, 419 (1923).

Problems

1. Assume that the distribution function has the form of a *Drifted Maxwellian*, where $\mathbf{v}_0$ is the drift velocity:

$$f(\mathbf{r}, \mathbf{v}, t) = n \left(\frac{m}{2\pi k_B T}\right)^{3/2} \exp\left[-\frac{m}{2k_B T}(\mathbf{v} - \mathbf{v}_0)^2\right] \tag{9.127}$$

where n, T, and $\mathbf{v}_0$ depend upon $(\mathbf{r}, t)$.
(a) Evaluate the integrals that define n, $\mathbf{J}$, U, $\mathbf{J}_E$ and express these quantities in terms of n, T, $\mathbf{v}_0$.
(b) Evaluate the moment equations for the Boltzmann Equation using $G(\mathbf{v}) = 1$, $\mathbf{v}$, ε_v and express the equation in terms of n, T, $\mathbf{v}_0$.

2. Derive Euler's equation for a fluid

$$0 = \frac{\partial}{\partial t}\mathbf{u} + \mathbf{u} \cdot \nabla\mathbf{u} + \frac{\nabla P}{\rho} + g\hat{z} \qquad (9.128)$$

3. Derive Bernoulli's equation for an incompressible, irrotational fluid. An irrotational fluid obeys $\nabla \times \mathbf{u} = 0$.

4. Solve the following differential equation, with the boundary conditions that $\phi(0) = 0$, $\phi'(0) = 0$.

$$\left[\frac{d^2}{dx^2} - a^2\right]\phi(x) = x \qquad (9.129)$$

5. Solve the following nonlinear equation, where $\phi(0) = 0$

$$\frac{d\phi(x)}{dx} = Ae^{-\phi} \qquad (9.130)$$

6. Solve the following nonlinear equation

$$\frac{d\phi(x)}{dx} = \phi(x)[1 - \phi(x)] \qquad (9.131)$$

with the boundary condition that $\phi(0) = 1/2$.

7. Solve the differential equation

$$\frac{d^2}{dx^2}V(x) = -\frac{a^2}{V^3} \qquad (9.132)$$

subject to the conditions that: $V(0) = 0$, $dV/dx = 0$ when $V \to \infty$.

8. Assume there is a plane with a circular surface charge $\sigma > 0$ (Coulombs/m^2) of radius a at the center. A charge $e > 0$ of mass m starts at the center of the circle with initial velocity $v_i = 0$.
   (a) Find the force on the charge as a function of the distance z from the plane, for motion perpendicular to the plane.
   (b) Solve Newton's second law to find the final velocity of the particle at infinity.

One Dimension

10.1. Introduction

Some differential equations are solved in one dimension. The first is Laplace's equation

$$\frac{d^2}{dx^2}\phi(x) = 0 \tag{10.1}$$

Its only solution is $\phi(x) = a + bx$ where the constants a, b are set by the boundary conditions. A second order equation always has two constants. The solution is dull: it is a straight line.

The next equation is that of Helmholtz

$$\left[\frac{d^2}{dx^2} + k^2\right]\phi(x) = 0 \tag{10.2}$$

$$\phi(x) = a\cos(kx) + b\sin(kx) \tag{10.3}$$

which has a solution with trigometric functions with two constants a and b. This solution is also rather dull.

A more interesting problem is to solve Green's function of Laplace's equation

$$\frac{d^2}{dx^2}\phi(x, x_0) = -\delta(x - x_0) \tag{10.4}$$

where $0 \leqslant x$, $x_0 \leqslant L$ with the boundary conditions that $\phi = 0$ at the boundaries $x = 0, L$. The solution uses the technique described in the previous chapter. Space is divided into two intervals, $0 \leqslant x \leqslant x_0$ and $x_0 \leqslant x \leqslant L$. In each segment the solution is found to the homogeneous equation that obeys the boundary condition for that segment. The homogeneous solution is $ax + b$. Choosing a and b to satisfy the two boundary conditions $0 = \phi(0) = \phi(L)$ gives

$$\phi(x, x_0) = \begin{cases} Ax & 0 < x < x_0 \\ B(L - x) & x_0 < x < L \end{cases} \tag{10.5}$$

237

Matching at x_0 gives

$$Ax_0 = B(L - x_0) \tag{10.6}$$

$$\phi = \Theta(x_0 - x)Ax + \Theta(x - x_0)B(L - x) \tag{10.7}$$

$$\frac{d}{dx}\phi = -\delta(x - x_0)[Ax_0 - B(L - x_0)] + A\Theta(x_0 - x) - B\Theta(x - x_0)$$

$$= A\Theta(x_0 - x) - B\Theta(x - x_0) \tag{10.8}$$

$$\frac{d^2}{dx^2}\phi = -(A + B)\delta(x - x_0) \tag{10.9}$$

The derivative of $\Theta(x_0 - x)$ is $-\delta(x - x_0)$ while the derivative of $\Theta(x - x_0)$ is $+\delta(x - x_0)$. There is no delta function in the first derivative since the factor in brackets cancel. Comparing this with the original equation gives $A + B = 1$. Solving the two equations for (A, B) gives

$$\phi(x, x_0) = \frac{1}{L}\begin{cases} x(L - x_0) & 0 < x < x_0 \\ x_0(L - x) & x_0 < x < L \end{cases} \tag{10.10}$$

that is the easiest way to write the solution. This solution is plotted in Fig. 10.1. The function has a cusp at $x = x_0$. The cusp is a change in slope and the first derivative of a cusp is a step function. The second derivative of a cusp is a delta function. The delta function $\delta(x - x_0)$ is produced by introducing the theta function. This is an alternative, but an equivalent method to taking the integral $\int dx$ of Eq. (10.4) over the interval $(x_0 - \varepsilon, x_0 + \varepsilon)$ that gives

$$-1 = \left(\frac{d\phi}{dx}\right)_{x_0^+} - \left(\frac{d\phi}{dx}\right)_{x_0^-} = -(A + B) \tag{10.11}$$

Either method of treating the delta function works well.

Green's function can also be expressed as a Fourier series. Choose the function $\sin(k_n x)$ with $k_n = \pi n/L$, that vanishes at the end points $x = 0, L$.

$$\phi(x, x_0) = \sum_{n=1}^{\infty} a_n(x_0) \sin(k_n x) \tag{10.12}$$

$$a_n = \frac{2}{L}\int_0^L dx \sin(k_n x)\phi(x, x_0)$$

$$= \frac{2}{L^2}\left[(L - x_0)\int_0^{x_0} dx\, x \sin(k_n x) + x_0 \int_{x_0}^L dx(L - x)\sin(k_n x)\right]$$

$$= \frac{2}{Lk_n^2}\sin(k_n x_0) \tag{10.13}$$

$$\phi(x, x_0) = \frac{2}{L}\sum_{n=1}^{\infty}\frac{1}{k_n^2}\sin(k_n x)\sin(k_n x_0) \tag{10.14}$$

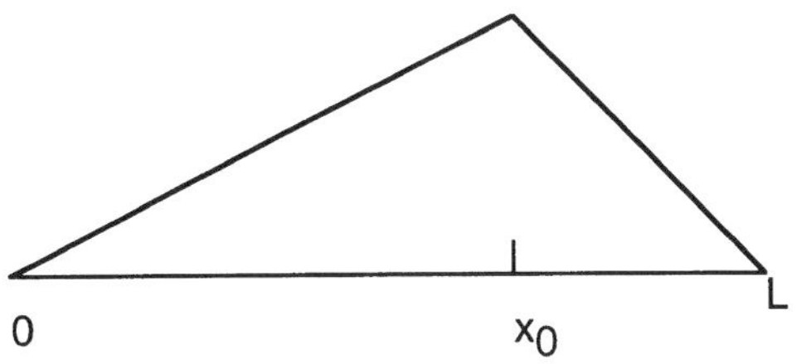

Figure 10.1 Graph of $\phi(x, x_0)$ in Eq. (10.10).

The series representation is useful for related problems. The above formula is used to prove an interesting result. Taking a double derivative of the series solution gives

$$\frac{d^2}{dx^2} \phi(x, x_0) = -\frac{2}{L} \sum_{n=1}^{\infty} \sin(k_n x) \sin(k_n x_0) = -\delta(x - x_0) \qquad (10.15)$$

The last equality expresses the delta function as a Fourier series. This identity will recur in numerous examples. To prove the series definition of the delta function, write it as

$$\delta(x - x_0) = \frac{2}{L} \sum_{n=1}^{\infty} \sin(k_n x) \sin(k_n x_0)$$

$$= \frac{1}{L} \sum_{n=1}^{\infty} \{\cos[k_n(x - x_0)] - \cos[k_n(x + x_0)]\} \qquad (10.16)$$

Consider the first term that is written as $\cos(n\theta)$, $\theta = \pi(x - x_0)/L$. A factor of $\exp(-n\varepsilon)$ is inserted to aid convergence, and later limit $\varepsilon \to 0$.

$$\sum_{n=1}^{\infty} \cos(n\theta)e^{-n\varepsilon} = \frac{1}{2} \sum_{n} (e^{in\theta} + e^{-in\theta})e^{-n\varepsilon}$$

$$= \frac{1}{2} \left(\frac{e^{i\theta - \varepsilon}}{1 - e^{i\theta - \varepsilon}} + \frac{e^{-i\theta - \varepsilon}}{1 - e^{-i\theta - \varepsilon}} \right)$$

$$= \frac{\cos(\theta) - e^{-\varepsilon}}{e^{\varepsilon} + e^{-\varepsilon} - 2\cos(\theta)} \qquad (10.17)$$

If $\theta \neq 0$, then taking the limit $\varepsilon \to 0$ gives

$$\frac{\cos(\theta) - 1}{2[1 - \cos(\theta)]} = -\frac{1}{2} \qquad (10.18)$$

This limit is independent of θ and is the same for both terms in Eq. (10.16). Both terms cancel and the summation is zero. This result is correct if $x \neq \pm x_0$. Next, consider the case that θ becomes small as ε becomes small. Consider both terms as small. Using $\cos(\theta) = 1 - \theta^2/2$, $e^{\pm\varepsilon} = 1 \pm \varepsilon + \varepsilon^2/2$ gives

$$\sum_n \cos(n\theta) \to \frac{\varepsilon}{\varepsilon^2 + \theta^2} - \frac{1}{2} \to \pi\delta(\theta) - \frac{1}{2}$$

$$= L\delta(x - x_0) - \frac{1}{2} \qquad (10.19)$$

The latter result is obtained in the limit that $\varepsilon \to 0$. These steps prove

$$\frac{2}{L} \sum_{n=1}^{\infty} \sin(k_n x) \sin(k_n x_0) = \delta(x - x_0) - \delta(x + x_0) \qquad (10.20)$$

The second delta function $\delta(x + x_0)$ has no effect since both (x, x_0) are positive and their sum is never zero. This term can be dropped.

10.2. Diffusion Equation

The diffusion equation in one dimension has the general form

$$\left[D\frac{\partial^2}{\partial x^2} - \frac{\partial}{\partial t} \right] T(x, t) = S(x, t) \qquad (10.21)$$

where the source term $S(x, t)$ on the right-hand side is assumed to be known.

Example 1. *Spread of heat pulse:* The first problem has the source term $S(x, t) = -C\delta(x)\delta(t)$

$$\left[D\frac{\partial^2}{\partial x^2} - \frac{\partial}{\partial t} \right] T(x, t) = -C\delta(x)\delta(t) \qquad (10.22)$$

The system is a bar of infinite length. The source term provides a heat impulse at the origin $(x = 0)$ at some initial time $(t = 0)$. The constant C has units of degrees/meter. The diffusion equation describes how the heat pulse spreads to the rest of the bar. The bar is assumed to have an initial temperature $T(x, 0) = T_i$. After the heat pulse, the solution will have the form $T(x, t) = T_i + \delta T(x, t)$. The second term is the transient, that is the function of interest.

Problems with two different kinds of derivatives are partial differential equations. The usual method of solution is to use a transform. The space variable (x) could be Fourier transformed, while the time variable (t) could be Laplace transformed, or both could be transformed. For most problems, one transform is usually enough.

First, use a Fourier transform on the space coordinate. Multiply the entire equation by $\exp(-ikx)$ and then integrate over all values of dx

$$\bar{T}(k, t) = \int_{-\infty}^{\infty} dx\, e^{-ikx} \delta T(x, t) \qquad (10.23)$$

$$\left[Dk^2 + \frac{\partial}{\partial t} \right] \bar{T}(k, t) = C\delta(t) \qquad (10.24)$$

The transform of the second derivative is evaluated by two integration by parts, which produces the factor of $-k^2\bar{T}$. The equation with a first derivative in time is simple to solve

$$\bar{T}(k, t) = C\Theta(t)e^{-tDk^2} \tag{10.25}$$

$$\frac{\partial}{\partial t}\bar{T}(k, t) = C[\delta(t) - \Theta(t)Dk^2e^{-tDk^2}] \tag{10.26}$$

$$\delta T(x, t) = C\Theta(t)\int_{-\infty}^{\infty}\frac{dk}{2\pi}\exp[-tDk^2 + ikx] \tag{10.27}$$

The integral has the Gaussian form. Completing the square gives an exponent $[-tD(k - ix/2tD)^2 - x^2/4tD]$ and the answer is

$$\delta T(x, t) = \frac{C\Theta(t)}{2\sqrt{\pi t D}}e^{-x^2/(4Dt)} \tag{10.28}$$

Note that the derivative of the step function is a delta function, $d\Theta(t)/dt = \delta(t)$. Since there is no signal for $t < 0$ then $\bar{T}(x, t < 0) = 0$. This dependence is achieved using a step function. The inclusion of the step function solves the problem of how to accommodate the delta function in the source term.

The last integral is a Gaussian. The final solution is a simple Gaussian function of x, whose width is determined by $4Dt$. Since the diffusion coefficient D has dimensions m^2/s, the quantity $x^2/(4Dt)$ is dimensionless. Dimensional analysis shows that the result must depend on the combination of x^2/Dt. Also note that the spatial integral is a constant

$$\int_{-\infty}^{\infty}dx\delta T(x, t) = C\Theta(t) \tag{10.29}$$

The total heat injected remains a constant in time.

The same problem is worked by using a Laplace transform. Multiply Eq. (10.22) by $\exp(-pt)$ and integrate dt. The Laplace transform of a time derivative is evaluated using an integration by parts

$$\bar{T}(x, p) = \int_0^{\infty}dt\, T(x, t)e^{-pt} \tag{10.30}$$

$$\int_0^{\infty}dt e^{-pt}\frac{\partial}{\partial t}T(x, t) = T(x, t)e^{-pt}\Big|_0^{\infty} + p\int_0^{\infty}dt\, T(x, t)e^{-pt}$$

$$= p\bar{T}(x, p) - T(x, 0) \tag{10.31}$$

$$\left[\frac{\partial^2}{\partial x^2} - \frac{p}{D}\right]\bar{T}(x, p) = -\frac{T(x, 0)}{D} - \frac{C}{D}\delta(x) \tag{10.32}$$

The integration by parts yields a constant term $T(x, 0) = T_i$ that is the initial temperature. The second comment is in regard to the integral over the delta function

$$? = \int_0^\infty dt\delta(t) \tag{10.33}$$

The ambiguity is that the delta function is at $t = 0$ where the integral begins. The proper way to evaluate the integral is $\int dt\delta(t - \varepsilon) = 1$ in the limit that $\varepsilon \to 0$. The entire delta function goes under the integral.

The solution to the homogeneous equation is something like $(a^2 = p/D)$

$$0 = \left(\frac{d^2}{dx^2} - a^2\right)f \tag{10.34}$$

$$f = A\sinh(ax) + B\cosh(ax) \tag{10.35}$$

Neither of these solutions is satisfactory since they both diverge as x goes to either $\pm\infty$. The function that converges at either infinity is

$$f(x) = Ae^{-|x|a} \equiv A[\Theta(x)e^{-ax} + \Theta(-x)e^{ax}] \tag{10.36}$$

Next, examine how this function behaves when taking the second derivative. The derivative of $\Theta(x)$ is $\delta(x)$ while the derivative of $\Theta(-x)$ is $-\delta(x)$

$$\frac{d}{dx}f(x) = \delta(x)[e^{-ax} - e^{ax}] - aA[\Theta(x)e^{-ax} - \Theta(-x)e^{ax}]$$

$$= -aA[\Theta(x)e^{-ax} - \Theta(-x)e^{ax}] \tag{10.37}$$

The first term vanishes, since $\delta(x)$ only has value at $x = 0$ and its coefficient is zero at that point. The second derivative gives

$$\frac{d^2}{dx^2}f = -aA\delta(x)[e^{-ax} + e^{ax}] + a^2A[\Theta(x)e^{-ax} + \Theta(-x)e^{ax}]$$

$$= -2aA\delta(x) + a^2f \tag{10.38}$$

The second derivative of $f(x)$ produces a delta function with a nonzero coefficient. This delta function provides the source term on the right. The solution to Eq. (10.32) is

$$\bar{T}(x, p) = \frac{T_i}{p} + A\exp\left[-|x|\sqrt{\frac{p}{D}}\right] \tag{10.39}$$

$$\frac{\partial^2}{\partial x^2}\bar{T}(x, p) = \frac{p}{D}A\exp\left[-|x|\sqrt{\frac{p}{D}}\right] - 2A\sqrt{\frac{p}{D}}\delta(x) \tag{10.40}$$

Equating the coefficients of $\delta(x)$ gives $A = C/(2\sqrt{pD})$ and the inverse Laplace transform is

$$T(x, t) = \int \frac{dp}{2\pi i} e^{pt} \left[\frac{T_i}{p} + \frac{C}{2\sqrt{pD}} e^{-|x|\sqrt{p/D}} \right] \tag{10.41}$$

The contour of integration is closed to the left when $t > 0$. The first term in brackets has a simple pole at $p = 0$ that gives the constant term T_i. The second term has a branch point at $p = 0$, and a branch cut is taken along the negative real axis. In this term, change variables to $p = -y$ where $\sqrt{p} = i\sqrt{y}$ above the branch cut and $\sqrt{p} = -i\sqrt{y}$ below.

$$T(x, t) = T_i + \frac{C}{2\pi\sqrt{D}} \int_0^\infty \frac{dy}{\sqrt{y}} e^{-yt} \cos\left(x\sqrt{\frac{y}{D}} \right) \tag{10.42}$$

Change variables of integration to $s^2 = yt$ and the resulting integral is simple

$$T(x, t) = T_i + \frac{C\Theta(t)}{\pi\sqrt{tD}} \int_0^\infty ds\, e^{-s^2} \cos(sx/\sqrt{Dt})$$

$$= T_i + \frac{C\Theta(t)}{2\sqrt{\pi tD}} e^{-x^2/(4Dt)} \tag{10.43}$$

that is the same answer obtained by using the Fourier transform. Note that this problem is also the solution to Green's function if $C = 1$ and $T_i = 0$.

$$\left[D\frac{\partial^2}{\partial x^2} - \frac{\partial}{\partial t} \right] G(x, x_0; t) = -\delta(t)\delta(x - x_0) \tag{10.44}$$

$$G(x, x_0; t) = \Theta(t) \frac{e^{-(x-x_0)^2/(4Dt)}}{2\sqrt{\pi Dt}} \tag{10.45}$$

that is useful for a number of problems.

Example 2. *Green's function for half space:* The above problem is solved for the case of a semi-infinite system, where $0 \leqslant x \leqslant \infty$. Green's function is a solution of the differential equation

$$\left[D\frac{\partial^2}{\partial x^2} - \frac{\partial}{\partial t} \right] G(x, x_0; t) = -\delta(t)\delta(x - x_0) \tag{10.46}$$

The initial condition is $G = 0$ for $t < 0$. A boundary condition must also be imposed on the end of the bar at $x = 0$. Two possible choices are

1. $G(0, x_0, t) = 0 = G(x, 0; t)$. This boundary condition applies when the temperature is fixed at the end.
2. $(\partial G/\partial x)_{x=0} = 0 = (\partial G/\partial x_0)_{x_0=0}$. This boundary condition applies when no heat can flow from the end of the bar.

For the present example, select the second boundary condition. One can also show that $G(x, x_0; t) = G(x_0, x; t)$ and the result is symmetric in the two space variables. Use the Laplace transform for this example. The Laplace transform of the above equation is

$$\left[D \frac{\partial^2}{\partial x^2} - p \right] \bar{G}(x, x_0; p) = -\delta(x - x_0) \tag{10.47}$$

The presence of the term $\delta(x - x_0)$ divides the space into the two regions $x < x_0$ and $x > x_0$. In each space the differential equation is homogeneous. The solution that obeys the feature that $\bar{G}' = 0$ at the origin and $\bar{G}$ converges as $(x, x_0) \to \infty$, is

$$\bar{G}(x, x_0; p) = A \cosh \left(x \sqrt{\frac{p}{D}} \right), \quad 0 \leqslant x < x_0$$

$$= B e^{-x\sqrt{p/D}}, \quad x > x_0 \tag{10.48}$$

Matching these two solutions at $x = x_0$ gives ($\alpha \equiv x_0 \sqrt{p/D}$)

$$A \cosh(\alpha) = B e^{-\alpha} \tag{10.49}$$

$$\bar{G}(x, x_0; p) = A \left[\Theta(x_0 - x) \cosh \left(x \sqrt{\frac{p}{D}} \right) + \Theta(x - x_0) \cosh(\alpha) e^{-(x - x_0)\sqrt{p/D}} \right] \tag{10.50}$$

Take two derivatives of this expression in order to obtain the coefficient of the term with $\delta(x - x_0)$

$$\frac{\partial}{\partial x} \bar{G}(x, x_0; p) = A \sqrt{\frac{p}{D}} \left[\Theta(x_0 - x) \sinh \left(x \sqrt{\frac{p}{D}} \right) \right.$$

$$\left. - \Theta(x - x_0) \cosh(\alpha) e^{-(x - x_0)\sqrt{p/D}} \right] \tag{10.51}$$

$$D \frac{\partial^2}{\partial x^2} \bar{G}(x, x_0; p) = p \bar{G}(x, x_0; p) - A \sqrt{pD} \, e^{\alpha} \delta(x - x_0) \tag{10.52}$$

The coefficient of the delta function must be -1, that gives the value of A. The remaining task is to evaluate the inverse Laplace transform

$$A = \frac{e^{-\alpha}}{\sqrt{pD}} \tag{10.53}$$

$$G(x, x_0; t) = \int \frac{dp}{4\pi i} \frac{e^{pt}}{\sqrt{pD}} \{ \Theta(x_0 - x)[e^{-(x_0 - x)\sqrt{p/D}} + e^{-(x + x_0)\sqrt{p/D}}]$$

$$+ \Theta(x - x_0)[e^{-(x - x_0)\sqrt{p/D}} + e^{-(x + x_0)\sqrt{p/D}}] \} \tag{10.54}$$

The answer will depend only on $(x \pm x_0)^2$ and the two square brackets give identical answers. They can be combined, using $\Theta(x) + \Theta(-x) = 1$, that simplifies the notation. The integral over complex variable p is evaluated by closing the contour in the LHP for $t > 0$. There is a branch cut along the negative real axis. Combining the contributions from above and below the branch cuts gives $(p = -y)$

$$
G(x, x_0; t) = \frac{\Theta(t)}{2\pi\sqrt{D}} \int_0^\infty \frac{dy}{\sqrt{y}} e^{-yt} \left\{ \cos\left[(x - x_0)\sqrt{\frac{p}{D}} \right] + \cos\left[(x + x_0)\sqrt{\frac{p}{D}} \right] \right\}
$$

(10.55)

change variables to $s^2 = yt$, $dy = 2s\,ds/t$, this gives

$$
G(x, x_0; t) = \frac{\Theta(t)}{\pi\sqrt{Dt}} \int_0^\infty ds\, e^{-s^2} \left\{ \cos\left[s\frac{(x - x_0)}{\sqrt{Dt}} \right] + \cos\left[s\frac{(x + x_0)}{\sqrt{Dt}} \right] \right\}
$$

$$
= \frac{\Theta(t)}{2\sqrt{\pi Dt}} [e^{-(x-x_0)^2/4Dt} + e^{-(x+x_0)^2/4Dt}]
$$

(10.56)

The first term in the square bracket is identical to the result for the infinite bar. The second term in the square bracket is present to keep the Green's function symmetric in the variables (x, x_0). A symmetric function has a zero derivative at $x = 0$. The second term in brackets appears to be a heat pulse that is spreading from the point $-x_0$. Students who have studied electrostatics are familiar with the concept of the *image charge*. A charge outside of a dielectric surface has a potential that is well represented by putting an image charge inside the surface. The same image concept works here, except there is an *image source*. The boundary condition of no heat flow through the point $x = 0$ is easily accomplished by putting the image source of x_0 at $-x_0$. Each source spreads equally, and their heat flows are equal and opposite at the origin. There is no net heat flow through the point $x = 0$, and that satisfies the boundary condition.

The other boundary condition listed at the beginning of this example showed Green's function vanishing at the point $x = 0$. This is accomplished using image theory by giving the source the opposite sign. Green's function for a semi-infinite bar has a solution of

$$
G(x, x_0; t) = \frac{\Theta(t)}{2\sqrt{\pi Dt}} [e^{-(x-x_0)^2/4Dt} - e^{-(x+x_0)^2/4Dt}]
$$

(10.57)

The heat diffusion from the two source points of $(x_0, -x_0)$ exactly cancel at $x = 0$, for all values of time. Image theory is a useful method of solving Green's function near boundaries.

Example 3. *Heating end of bar:* The next diffusion problem is for a semi-infinite bar. It stretches over $0 \leqslant x \leqslant \infty$. Initially the bar is at $T(x, 0) = T_i$. At $t = 0$ the end at $x = 0$ is put in contact with a reservoir at temperature T_f. The problem has no sources, only initial and boundary conditions.

$$\left[D\frac{\partial^2}{\partial x^2} - \frac{\partial}{\partial t}\right]T(x, t) = 0 \tag{10.58}$$

with $T(x, 0) = T_i$, $T(0, t) = T_f$. Again the problem will be solved two ways using either Fourier or Laplace transforms.

When solving by Fourier transforms on the x variable, the immediate decision is whether to use cosine or sine transforms for the half space problem. The correct answer is to use sine transform since the value of $T(0, t)$ is known. The cosine transform is used in problems where the derivative is known at the boundary.

Taking the sine transform, gives

$$\left[Dk^2 + \frac{\partial}{\partial t}\right]\bar{T}_s(k, t) = kDT_f \tag{10.59}$$

$$\bar{T}_s(k, t) = \frac{T_f}{k} + A(k)\exp[-tDk^2] \tag{10.60}$$

The top equation gives the differential equation in time, while the second equation gives its most general solution. Note that there is *no* factor of $\Theta(t)$ multiplying the transient term. This factor is omitted when the change of temperature is due to boundary conditions.

The next step is to Fourier transform the initial conditions $T(x, 0) = T_i$

$$\bar{T}_s(k, 0) = T_i \int_0^\infty dx \, \sin(kx) = \frac{T_i}{k} = \frac{T_f}{k} + A \tag{10.61}$$

that gives the solution $A = -\Delta T/k$, $\Delta T = T_f - T_i$

$$\bar{T}_s(k, t) = \frac{1}{k}\{T_f - \Delta T \exp[-tDk^2]\} \tag{10.62}$$

$$T(x, t) = \frac{2}{\pi}\int_0^\infty \frac{dk}{k}\sin(kx)[T_f - \Delta T e^{-tDk^2}] \tag{10.63}$$

The first integral gives T_f. In the second integral, change variables to $s = k\sqrt{tD}$ giving

$$T(x, t) = T_f - \Delta T f\left(\frac{x}{2\sqrt{tD}}\right) \tag{10.64}$$

$$f(\xi) = \frac{2}{\pi}\int_0^\infty \frac{ds}{s}\sin(2\xi s)e^{-s^2} \tag{10.65}$$

$$\xi = \frac{x}{2\sqrt{tD}} \tag{10.66}$$

The last function is evaluated by taking its first derivative. This step yields an integral that can be evaluated

$$\frac{df(\xi)}{d\xi} = \frac{4}{\pi}\int_0^\infty ds\,\cos(2\xi s)e^{-s^2} = \frac{2e^{-\xi^2}}{\sqrt{\pi}} \tag{10.67}$$

Using the fact that $f(0) = 0$ the derivative can be integrated to find

$$f(\xi) = \frac{2}{\sqrt{\pi}}\int_0^\xi dt\,e^{-t^2} \equiv \mathrm{erf}(\xi) \tag{10.68}$$

The resulting integral is defined as an *error function*. There is a tendency to give names to integrals that cannot be evaluated analytically. The error function is certainly an example. The name "error function" comes from probability theory. The final result is

$$T(x, t) = T_f - \Delta T\,\mathrm{erf}\left(\frac{x}{2\sqrt{Dt}}\right) \tag{10.69}$$

The error function has the properties

$$\mathrm{erf}(0) = 0 \tag{10.70}$$

$$\mathrm{erf}(\infty) = 1 \tag{10.71}$$

There is also a *complimentary error function* denoted by $\mathrm{erfc}(x)$ that is sometimes useful.

$$\mathrm{erfc}(x) = 1 - \mathrm{erf}(x) = \frac{2}{\sqrt{\pi}}\int_x^\infty dt\,e^{-t^2} \tag{10.72}$$

$$\mathrm{erfc}(0) = 1 \tag{10.73}$$

$$\mathrm{erfc}(\infty) = 0 \tag{10.74}$$

The solution may also be written as

$$T(x, t) = T_i + \Delta T\,\mathrm{erfc}\left(\frac{x}{2\sqrt{Dt}}\right) \tag{10.75}$$

that is the final result. The error function is tabulated in reference books or can be evaluated in seconds on a computer. Note that the solution again depends on the combination of factors $x^2/4Dt$, as required by dimensional analysis.

Now solve the same problem by Laplace transform. Many of the steps are similar to those for the infinite bar. The first step is to Laplace transform the diffusion equation to

$$\left[D\frac{\partial^2}{\partial x^2} - p\right]\bar{T}(x, p) = -T_i \tag{10.76}$$

$$\bar{T}(x, p) = \frac{T_i}{p} + Ae^{-x\sqrt{p/D}} + Be^{x\sqrt{p/D}} \tag{10.77}$$

The first line is the differential equation for the transformed function, while the second line is the solution of this differential equation in terms of two unknown constants (A, B). Set $B = 0$ since it multiplies a solution that behaves badly at $x \to \infty$. Since the physical solution is well-behaved in this limit, the last term must be absent.

There remains the task of determining the coefficient A. It is found from the boundary condition at $x = 0$, where $T(0, t) = T_f$. When solving the transform equation, one must also Laplace transform the boundary condition. The Laplace transform of a constant T_f is T_f/p. At $x = 0$ the boundary condition is

$$\frac{T_f}{p} = \frac{T_i}{p} + A \tag{10.78}$$

$$A = \frac{\Delta T}{p} \tag{10.79}$$

$$\bar{T}(x, p) = \frac{1}{p}[T_i + \Delta T e^{-x\sqrt{p/D}}] \tag{10.80}$$

$$T(x, t) = \int \frac{dp}{2\pi i} \frac{e^{pt}}{p}[T_i + \Delta T e^{-x\sqrt{p/D}}] \tag{10.81}$$

The inverse transform again involves a contour integral. There is a pole at $p = 0$ whose residue is $T_i + \Delta T = T_f$. There is a branch cut along the negative axis, and the integration along this cut gives the other contribution

$$T(x, t) = T_f - \Delta T \frac{2}{\pi} \int_0^\infty \frac{ds}{s} e^{-s^2} \sin\left(\frac{sx}{\sqrt{Dt}}\right) \tag{10.82}$$

$$= T_f - \Delta T \operatorname{erf}\left(\frac{x}{2\sqrt{Dt}}\right) \tag{10.83}$$

The same solution is obtained as was derived using a Fourier transform.

Example 4. *Heating finite bar:* The next diffusion problem is for a bar of length L where the distance x spans $0 \leqslant x \leqslant L$. Initially the bar is at temperature T_i. At some time $(t = 0)$ a hot source of temperature T_f is put on the end at $x = 0$. If $T_f > T_i$ the bar will gradually heat up until the entire bar is at temperature T_f. The other end $(x = L)$ is assumed not to be attached to a reservoir, so that no heat can escape from it. The possible heat loss due to radiation can be neglected unless the temperatures are very large $(T > 1000 \text{ K})$. Since the heat current, in the absence of electric fields or currents, is given by $J_E = -K\partial T/\partial x$, then the statement of no heat flow is merely the Neumann condition that the derivative vanishes at $x = L$. Again the problem is solved twice, once with a Fourier transform of space, and once with a Laplace transform of time. The problem is stated as

$$\left[\frac{\partial^2}{\partial x^2} - \frac{1}{D}\frac{\partial}{\partial t}\right] T(x, t) = 0 \tag{10.84}$$

$$T(x, 0) = T_i, \quad T(0, t) = T_f, \quad \left(\frac{\partial T}{\partial x}\right)_{x=L} = 0 \tag{10.85}$$

The Fourier transform is done first. Write the solution as

$$T(x, t) = T_f + \delta T(x, t) \tag{10.86}$$

$$\delta T(x, t) = \sum_n a_n(t) \sin(k_n x) \tag{10.87}$$

Since the final temperature of the whole bar is T_f, then $T(x, t)$ has this value plus a transient $\delta T(x, t)$, where the transient vanishes at large time. For a system of finite length, the Fourier transform is either $\cos(k_n x)$ or $\sin(k_n x)$. In this case the choice of $\sin(k_n x)$ is a natural since it automatically obeys the boundary condition $\delta T(0, t) = 0$. The other boundary condition is $T'(L, t) = 0$. The derivative of $\sin(k_n x)$ is $k_n \cos(k_n x)$ so the boundary condition is satisfied if $\cos(k_n L) = 0$. The cosine vanishes at angles that are odd multiples of $90°$, resulting in

$$k_n = \frac{\pi}{2L}(2n + 1) \tag{10.88}$$

$$\delta T(x, t) = \sum_{n=0}^{\infty} a_n(t) \sin(k_n x) \tag{10.89}$$

$$\left[\frac{\partial^2}{\partial x^2} - \frac{1}{D}\frac{\partial}{\partial t}\right] \delta T(x, t) = -\sum_{n=0}^{\infty}\left[k_n^2 + \frac{1}{D}\frac{\partial}{\partial t}\right] a_n(t) \sin(k_n x) \tag{10.90}$$

$$0 = \left[k_n^2 + \frac{1}{D}\frac{\partial}{\partial t}\right] a_n(t) \tag{10.91}$$

$$a_n(t) = a_n(0)e^{-tDk_n^2} \tag{10.92}$$

The time variables give a simple differential equation for $a_n(t)$ that has a simple solution. There remains only the task of finding $a_n(0)$. The initial condition is

$$T(x, 0) = T_i = T_f + \sum_n a_n(0) \sin(k_n x) \tag{10.93}$$

The coefficient is found by multiplying the entire equation by $\sin(k_m x)$ and integrating between zero and L. Since the sine functions are orthogonal, this process picks out a_m as the only nonzero term in the summation,

$$\int_0^L dx \sin(k_n x) \sin(k_m x) = \frac{L}{2}\delta_{nm} \tag{10.94}$$

$$\int_0^L dx \sin(k_m x) = \frac{1 - \cos(k_m L)}{k_m} = \frac{1}{k_m} \tag{10.95}$$

$$a_m(0) = -\frac{2\Delta T}{Lk_m} = -\frac{4\Delta T}{\pi(2n + 1)} \tag{10.96}$$

$$\Delta T = T_f - T_i \tag{10.97}$$

$$T(x, t) = T_f - \frac{4\Delta T}{\pi}\sum_{n=0}^{\infty}\frac{1}{(2n + 1)}e^{-tDk_n^2}\sin(k_n x) \tag{10.98}$$

Although the summation runs over all nonnegative integers n, a numerically accurate answer requires only a few terms, except at the very earliest times. At long times, the factor of $Dk_n^2 = D(\pi/2L)^2(2n+1)^2$ gives a large value for large n, so that the exponent $-Dtk_n^2$ has a large negative value, whose exponential is negligible. Also, notice that the timescale for diffusion is set by the constants $t \sim L^2/D$. A solution of the diffusion equation is not required to estimate the timescale for any problem. It is always $t \sim L^2/D$ where L is the characteristic length of the system. A solution of the diffusion equation is needed to provide the detailed dependence of $T(x, t)$.

An interesting feature of the above problem is that the initial condition $t = 0$ produces the series

$$1 = \frac{4}{\pi} \sum_{n=0}^{\infty} \frac{1}{(2n+1)} \sin(k_n x) \tag{10.99}$$

The right-hand side of this equation appears to be a function of x, however, the left-hand side is a simple constant, so the right-hand side is also a constant. Also note that each term in the series vanishes at $x = 0$, yet the summation is not zero near this point. The latter assertion is a limiting process. The series vanishes at the point $x = 0$. But for any point $x = \varepsilon$, for very small ε, taking enough terms in the series will make a sum that gives one at that point. The series equals one in the neighborhood of $x = 0$, although not at the precise point of $x = 0$.

The solution by Laplace transform is more complicated. The transformed equation is

$$\bar{T}(x, p) = \int_0^{\infty} dt\, e^{-pt} T(x, t) \tag{10.100}$$

$$\left[D \frac{\partial^2}{\partial x^2} - p \right] \bar{T}(x, p) = -T_i \tag{10.101}$$

where the constant term on the right comes from integrating the time derivative. This differential equation has as its most general solution

$$\bar{T}(x, p) = \frac{T_i}{p} + Ae^{-x\sqrt{p/D}} + Be^{x\sqrt{p/D}} \tag{10.102}$$

The earlier solutions of the diffusion equation were for unbounded systems. In these cases, the last term was discarded ($B = 0$) since it diverges as $x \to \infty$. However, for bounded systems, this term must be retained, since it is needed to satisfy the two boundary conditions. It is also required to Laplace transform the boundary conditions

$$\bar{T}(0, p) = \frac{T_f}{p} = \frac{T_i}{p} + A + B \tag{10.103}$$

$$\left(\frac{\partial \bar{T}}{\partial x} \right)_L = 0 = \sqrt{\frac{p}{D}} \left[-Ae^{-L\sqrt{p/D}} + Be^{L\sqrt{p/D}} \right] \tag{10.104}$$

These two equations permit an algebraic solution for the unknown constants A and B

$$A = \frac{\Delta T}{p} \frac{1}{1 + \exp(-2L\sqrt{p/D})} \tag{10.105}$$

$$B = \frac{\Delta T}{p} \frac{1}{1 + \exp(2L\sqrt{p/D})} \tag{10.106}$$

$$T(x, t) = \int_{-i\infty}^{i\infty} \frac{dp}{2\pi i} \frac{e^{pt}}{p} \left[T_i + \Delta T \left(\frac{e^{-x\sqrt{p/D}}}{1 + \exp(-2L\sqrt{p/D})} + \frac{e^{x\sqrt{p/D}}}{1 + \exp(2L\sqrt{p/D})} \right) \right] \tag{10.107}$$

For $t > 0$ the contour integral is evaluated by closing it in a semicircle in the left-hand plane. There is a pole at $p = 0$ for both terms in the integrand, that gives the constant term of $T_f = T_i + \Delta T$. The factor of $\sqrt{p}$ suggests a branch point at the origin and a branch cut is drawn along the negative real axis. However, there is no need for a branch cut. When $p = -y$, where $0 \leqslant y \leqslant \infty$, then $\sqrt{p} = i\sqrt{y}$ above the branch cut, and $\sqrt{p} = -i\sqrt{y}$ below the branch cut. However, this particular integrand gives the same value for both these cases because of the symmetry of the two terms with regard to $\pm\sqrt{p}$. The branch cut is not needed. The two integrals above and below the cut cancel exactly. There are poles along the negative real axis and are at places where

$$1 + \exp[\pm 2L\sqrt{p/D}] = 0 \tag{10.108}$$

$$2L\sqrt{\frac{p}{D}} = \pm i\pi(2n + 1) \tag{10.109}$$

$$p_n = -Dk_n^2 \tag{10.110}$$

where k_n was introduced in the Fourier series solution. The residue at these poles are found by setting $p = p_n + \delta p$ and expanding in the small quantity δp

$$\sqrt{\frac{p}{D}} = \sqrt{\frac{p_n}{D}} \left[1 + \frac{\delta p}{2p_n} \right] \tag{10.111}$$

$$1 + \exp[\pm 2L\sqrt{p/D}] = \mp \delta p \frac{L}{\sqrt{p_n D}} = \pm i\delta p \frac{L}{Dk_n} \tag{10.112}$$

There are other contributions to the residue from the terms in the integrand. Collecting all these terms together, we find the solution

$$T(x, t) = T_f - \frac{4\Delta T}{\pi} \sum_{n=0}^{\infty} \frac{1}{(2n + 1)} e^{-tDk_n^2} \sin(k_n x) \tag{10.113}$$

that is exactly the same as before.

    If we compare the two methods of Fourier and Laplace transforms we can see that the Fourier is less work. One has to guess the correct sine or cosine function to use in the series, after which the remaining algebra is easy. The

Laplace transform method is more algebra, and the inverse transform can be daunting. However, at no place in the derivation is there guess work, we just plow through it. The Laplace transform is more straightforward.

10.3. Wave Equation

The wave equation in one dimension is written as

$$\left[v^2 \frac{\partial^2}{\partial x^2} - \frac{\partial^2}{\partial t^2} \right] \phi(x, t) = S(x, t) \tag{10.114}$$

where the right-hand side is the source. This equation will be solved for several different cases in one dimension. For problems that start at some time ($t = 0$), it is required to specify two initial conditions, $\phi(x, 0)$ and $\phi_t(x, 0)$, where the subscript t denotes the time derivative. Two conditions are required since the equation is second order.

Example 1. *Guitar notes:* Here the homogeneous wave equation is solved for a bounded system ($0 \leqslant x \leqslant L$), that has zero amplitude at both ends

$$\left[v^2 \frac{\partial^2}{\partial x^2} - \frac{\partial^2}{\partial t^2} \right] \phi(x, t) = 0 \tag{10.115}$$

$$\phi(0, t) = 0, \qquad \phi(L, t) = 0 \tag{10.116}$$

This problem gives the notes of a stringed instrument and solution is given in early courses in physics. Since the system is bounded, and clamped at both ends, the spatial part of the solution is $\sin(k_n x)$. The constraint that $\sin(k_n L) = 0$ is satisfied by the choice that $k_n L = n\pi$. If the temporal part of the solution is $\sin(\omega t)$, then the equations are

$$\phi(x, t) = \sum_n a_n \sin(k_n x) \sin(\omega_n t) \tag{10.117}$$

$$0 = \left[v^2 \frac{\partial^2}{\partial x^2} - \frac{\partial^2}{\partial t^2} \right] \phi(x, t) = -\sum_n a_n [v^2 k_n^2 - \omega_n^2] \sin(k_n x) \sin(\omega_n t) \tag{10.118}$$

that is satisfied by $\omega_n = vk_n = v\pi n/(L)$. The frequencies are multiples of the fundamental $\omega_1 = \pi v/(L)$. The coefficients a_n depend on how the string is plucked.

Example 2. *Green's function for guitar:* Solve the same problem of the bounded string clamped at both ends, but now it is plucked at the point x_0 at the time $t_0 = 0$. Green's function equation is

$$\left[v^2 \frac{\partial^2}{\partial x^2} - \frac{\partial^2}{\partial t^2} \right] G(x, x_0; t) = -\delta(x - x_0)\delta(t) \tag{10.119}$$

where the point x_0 is between the two clamped ends. The right-hand side of the equation is zero except at the point $t = 0$, $x = x_0$. Therefore, it obeys the homogeneous solution except at these points. Again choose the spatial function to be $\sin(k_n x)$ where $k_n = \pi n/L$, since then $\sin(k_n x) = 0$ at $x = 0$ and $x = L$. Start with a guess that

$$G(x, x_0; t) = \Theta(t) \sum_n \sin(k_n x)[A_n(x_0) \sin(vk_n t) + B_n(x_0) \cos(vk_n t)] \quad (10.120)$$

Both time options $\sin(vk_n t)$ and $\cos(vk_n t)$ have been included. Both satisfy the homogeneous equation and either or both could be part of the solution. The factor of $\Theta(t)$ ensures that no vibration occurs for $t < 0$. The derivative of this function gives a delta function of t, $d\Theta(t)/dt = \delta(t)$. The delta function $\delta(t)$ in Eq. (10.119) is obtained by differentiating this function. Examine the effect of differentiating on the temporal parts of this solution

$$\frac{\partial}{\partial t} \Theta(t)[A_n \sin(vk_n t) + B_n \cos(vk_n t)]$$

$$= \delta(t)B_n + \Theta(t)vk_n[A_n \cos(vk_n t) - B_n \sin(vk_n t)] \quad (10.121)$$

The first derivative gives a factor of $\delta(t)B_n$. When taking the second derivative this term gives a derivative of a delta function. Since this derivative is not in the differential equation, it is unwanted. It is eliminated by choosing $B_n = 0$. The right choice is $\sin(vk_n t)$. Taking one more derivative of this term gives

$$G(x, x_0; t) = \Theta(t) \sum_n A_n(x_0) \sin(k_n x) \sin(vk_n t) \quad (10.122)$$

$$\frac{\partial^2}{\partial t^2} G(x, x_0; t) = \delta(t) \sum_n A_n(x_0)vk_n \sin(k_n x)$$

$$- \Theta(t) \sum_n A_n(x_0)(vk_n)^2 \sin(k_n x) \sin(vk_n t) \quad (10.123)$$

There remains the problem of determining the constant $A_n(x_0)$ that depends on where the string was plucked (x_0). The first term on the right, in the above equation, must be $\delta(t)\delta(x - x_0)$, so that

$$\delta(x - x_0) = \sum_n A_n(x_0)vk_n \sin(k_n x) \quad (10.124)$$

The coefficient A_n is determined by taking the integral $\int dx \sin(k_m x)$ of both sides of the equation and using the orthogonality of the $\sin(k_n x)$ functions

$$\sin(k_m x_0) = \frac{L}{2} vk_m A_m \quad (10.125)$$

$$A_m = \frac{2}{Lvk_m} \sin(k_m x_0) \quad (10.126)$$

$$\delta(x - x_0) = \frac{2}{L} \sum_{n=1}^{\infty} \sin(k_n x) \sin(k_n x_0) \quad (10.127)$$

$$G(x, x_0; t) = \frac{2\Theta(t)}{vL} \sum_n \frac{1}{k_n} \sin(vk_n t) \sin(k_n x) \sin(k_n x_0) \quad (10.128)$$

Green's function is now completely determined. Note it is a symmetric function of (x, x_0). The series definition Eq. (10.127) of the delta function $\delta(x - x_0)$ was given earlier in Eq. (10.15).

Let us solve the problem again by using a Laplace transform. Multiply the initial Eq. (10.119) by $\exp(-pt)$ and integrate over all positive time. The initial conditions are $\phi(x, t = 0) = 0 = \phi_t(x, 0)$. This step produces the equations

$$\left[v^2 \frac{\partial^2}{\partial x^2} - p^2 \right] \bar{G}(x, x_0; p) = -\delta(x - x_0) \tag{10.129}$$

$$\bar{G}(x, x_0; p) = \sum_n C_n(x_0, p) \sin(k_n x) \tag{10.130}$$

$$\left[v^2 \frac{\partial^2}{\partial x^2} - p^2 \right] \bar{G}(x, x_0; p) = -\sum_n C_n [v^2 k_n^2 + p^2] \sin(k_n x) \tag{10.131}$$

$$= -\delta(x - x_0)$$

The first equation gives the transformed equation. The second equation gives the assumed form of the transformed solution. It still must vanish at the points $x = 0, L$ so that a factor of $\sin(k_n x)$ is required. In order to solve Eq. (10.131), multiply it by $\sin(k_m x)$ and integrate dx between $(0, L)$ to get

$$-\frac{L}{2} C_m [v^2 k_m^2 + p^2] = -\sin(k_m x_0) \tag{10.132}$$

$$C_m = \frac{2}{L} \frac{\sin(k_m x_0)}{p^2 + (vk_n)^2} \tag{10.133}$$

$$G(x, x_0; t) = \frac{2}{L} \int \frac{dp}{2\pi i} e^{pt} \sum_n \frac{\sin(k_n x) \sin(k_n x_0)}{p^2 + (vk_n)^2} \tag{10.134}$$

$$= \frac{2\Theta(t)}{L} \sum_n \frac{1}{vk_n} \sin(vk_n t) \sin(k_n x) \sin(k_n x_0)$$

that is the final result. The inverse Laplace transform is rather simple since there are no branch points and only simple poles at the points $p = \pm i\omega_n$, $\omega_n = vk_n$. The answer does have the form of $\Theta(t) \sin(vk_n t)$ that was deduced earlier.

The above solution by a Laplace transform was easy since we assumed the spatial dependence had the form of $\sin(k_n x)$, $k_n = \pi n/L$. Guessing is a very honorable way to solve differential equations, since the guess can be checked easily. However, assume that the spatial dependence is not yet known. How do we proceed? The solution to the homogeneous equation in Eq. (10.129) has the form of $\sinh(px/v)$ or $\cosh(px/v)$. The functions satisfying the boundary condition of vanishing at the two ends are

$$\bar{G}(x, x_0; p) = A \sinh(px/v), \quad \text{if } 0 \leqslant x < x_0$$

$$= B \sinh[(L - x)p/v], \quad \text{if } x_0 < x \leqslant L \tag{10.135}$$

These can be matched at the point $x = x_0$

$$A \sinh(px_0/v) = B \sinh[(L - x_0)p/v] \qquad (10.136)$$

A graph of this function shows it has a cusp at the point x_0. A cusp is exactly what is needed to get the delta function $\delta(x - x_0)$ on the right-hand side of Eq. (10.129). Take Eq. (10.129) and integrate it from $x_0 - \varepsilon$ to $x_0 + \varepsilon$ in the limit that $\varepsilon \to 0$. Neglect terms of $O(\varepsilon)$ that in practice is the term with the factor of p^2. On the right, the integral over the delta function gives minus one

$$-1 = v^2 \left[\left(\frac{\partial \bar{G}}{\partial x} \right)_{x_0 + \varepsilon} - \left(\frac{\partial \bar{G}}{\partial x} \right)_{x_0 - \varepsilon} \right] \qquad (10.137)$$

$$\left(\frac{\partial \bar{G}}{\partial x} \right)_{x_0 + \varepsilon} = -(p/v)B \cosh[(L - x_0)p/v] \qquad (10.138)$$

$$\left(\frac{\partial \bar{G}}{\partial x} \right)_{x_0 - \varepsilon} = (p/v)A \cosh(px_0/v) \qquad (10.139)$$

$$1 = pv\{B \cosh[(L - x_0)p/v] + A \cosh(px_0/v)\} \qquad (10.140)$$

The last equation shows that the change in slope of the function at $x = x_0$ is equal to the contribution from the delta function $\delta(x - x_0)$. This equation, along with Eq. (10.136) are sufficient to determine the constants A and B.

$$A = \frac{1}{pv} \frac{\sinh[(L - x_0)p/v]}{\sinh(pL/v)} \qquad (10.141)$$

$$B = \frac{1}{pv} \frac{\sinh(x_0 p/v)}{\sinh(pL/v)} \qquad (10.142)$$

The transformed Green's function $\bar{G}(x, x_0; p)$ is completely determined. The final task is to take the inverse Laplace transform, say $0 < x < x_0$

$$G(x, x_0; t) = \frac{1}{v} \int \frac{dp}{2\pi i} \frac{e^{pt}}{p} \sinh(px/v) \frac{\sinh[(L - x_0)p/v]}{\sinh(pL/v)} \qquad (10.143)$$

When $p \to 0$ there are two powers of p in the numerator and two in the denominator, so there is no pole at $p = 0$. The important poles are the zeros of $\sinh(pL/v)$ that occur at the points $p_n L/v = i\pi n$, $p_n = ik_n v$. These poles select the values of $k_n = \pi n/L$. The summation is over all positive and negative integers n except zero

$$G(x, x_0; t) = \frac{\Theta(t)}{ivL} \sum_n{}' \frac{e^{ik_n vt}}{k_n} \sin(k_n x) \sin(k_n x_0)$$

$$= \frac{2\Theta(t)}{vL} \sum_{n=1}^{\infty} \frac{\sin(k_n vt)}{k_n} \sin(k_n x) \sin(k_n x_0) \qquad (10.144)$$

In the second equation, the terms for $\pm n$ are combined to turn the exponential of time into a sine function. The final summation is over only positive values of n. This answer is the same as derived before. In this case there was no need to guess the proper Fourier components of the spatial part. They arose naturally from the poles of the transformed function.

Example 3. *Outward wave from a point source:* Green's function for the wave equation in one dimension is solved for an unbounded system $(-\infty < x < \infty)$,

$$\left[v^2 \frac{\partial^2}{\partial x^2} - \frac{\partial^2}{\partial t^2}\right] G(x, x_0; t) = -\delta(t)\delta(x - x_0) \tag{10.145}$$

This problem can be solved either by a Laplace or Fourier transform. We shall use a Fourier transform. Multiply the above equation by $\exp(-ikx)$ and integrate dx over all space

$$\bar{G}(k, x_0; t) = \int dx e^{-ikx} G(x, x_0; t) \tag{10.146}$$

$$\left[k^2 v^2 + \frac{\partial^2}{\partial t^2}\right] \bar{G}(k, x_0; t) = \delta(t) e^{-ikx_0} \tag{10.147}$$

$$\bar{G}(k, x_0; t) = \frac{\Theta(t)}{vk} \sin(kvt) e^{-ikx_0} \tag{10.148}$$

The first equation is the definition of the transformed Green's function and the second equation is the result of transforming the wave equation. There are no boundary conditions in this problem, since the boundaries are at infinity, where it is assumed nothing is happening. The third equation is the solution to the second equation and the solution is either $\Theta(t) \sin(kvt)$ or $\Theta(t) \cos(kvt)$. Only the sine solution gives a second derivative which has a delta function $\delta(t)$. There remains the task of taking the inverse Fourier transform

$$G(x, x_0; t) = \frac{\Theta(t)}{2\pi v} \int_{-\infty}^{\infty} \frac{dk}{k} \sin(kvt) e^{ik(x - x_0)} \tag{10.149}$$

The factor of $\sin(kvt)/k$ is an even function of k. Expand $(x' = x - x_0) \exp(ikx') = \cos(kx') + i \sin(kx')$. The $\sin(kx')$ is an odd function of k and this term averages to zero. The $\cos(kx')$ is combined with the $\sin(kvt)$ using

$$\sin(a)\cos(b) = \frac{1}{2}[\sin(a + b) + \sin(a - b)] \tag{10.150}$$

$$G(x, x_0; t) = \frac{\Theta(t)}{4\pi v} \int_{-\infty}^{\infty} \frac{dk}{k} \{\sin[k(vt - x')] + \sin[k(vt + x')]\}$$

$$= \frac{\Theta(t)}{4v}[\text{sgn}(vt - x') + \text{sgn}(vt + x')] \tag{10.151}$$

where the integral of $\sin(bk)/k$ is π times the sign of b. The sum of the two sign functions can also be written as

$$[\text{sgn}(vt - x') + \text{sgn}(vt + x')] = 2\Theta(vt - |x'|) \tag{10.152}$$

$$G(x, x_0; t) = \frac{\Theta(t)}{2v} \Theta(vt - |x - x_0|) \tag{10.153}$$

There is no signal until $vt > |x - x_0|$, where $t > 0$. The result for increasing time sequences is shown in Fig. 10.2(a) where there is a wave front travelling outward in both directions. Before the front $|x - x_0| > vt$ there is no disturbance. Behind the front $vt > |x - x_0|$ the amplitude has a constant value of $1/2v$.

Next solve the same problem using a Laplace transform. Apply a Laplace transform to the wave equation. The transform of the second time derivative is evaluated by an integration by parts

$$\int_0^\infty dt\, e^{-pt} \frac{\partial^2}{\partial t^2} G(x, t) = p^2 \bar{G}(x, p) - pG(x, 0) - \left(\frac{\partial G}{\partial t}\right)_{t=0} \tag{10.154}$$

$$\left[v^2 \frac{\partial^2}{\partial x^2} - p^2\right] \bar{G}(x, p) = -\delta(x - x_0) - pG(x, 0) - \left(\frac{\partial G}{\partial t}\right)_{t=0} \tag{10.155}$$

The two initial conditions are set equal to zero. The solution, which is well-behaved as $x \to \pm\infty$ is

$$\bar{G}(x, p) = A e^{-|x - x_0|p/v} \tag{10.156}$$

$$v^2 \frac{\partial^2}{\partial x^2} \bar{G}(x, p) = -2Apv\delta(x - x_0) + p^2 \bar{G} \tag{10.157}$$

(a)

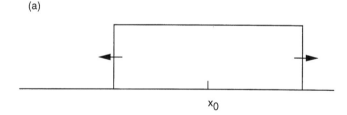

x_0

(b)

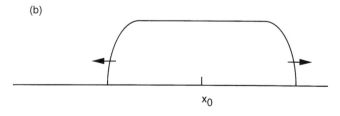

x_0

Figure 10.2 Green's function for 1D wave equation. For increasing time, a discontinuous front moves outward. (a) The front is a sharp step and (b) the front is a rounded edge.

The first term on the right must be $-\delta(x - x_0)$ so that $A = 1/(2pv)$ and the inverse transform is

$$G(x, t) = \frac{1}{2v} \int_{\varepsilon - \infty}^{\varepsilon + i\infty} \frac{dp}{2\pi i} \frac{1}{p} e^{p(t - |x - x_0|/v)} \tag{10.158}$$

The contour is closed in the RHP if $t < |x - x_0|/v$ and gives zero in this case. For $t > |x - x_0|/v$ the contour is closed in the LHP where there is a simple pole at $p = 0$. The result is identical to Eq. (10.153).

A problem closely related to the above Green's function is to find the displacement $\phi(x, t)$ of a wave that is plucked at $x = 0$, but the plucking is performed over a time interval. The equations to solve are

$$\left[v^2 \frac{\partial^2}{\partial x^2} - \frac{\partial^2}{\partial t^2} \right] \phi(x, t) = -\phi_0 \delta(x) f(t) \tag{10.159}$$

$$f(t) = \begin{cases} \sin(\omega t) & 0 < t < \dfrac{\pi}{\omega} \\ 0 & \text{otherwise} \end{cases} \tag{10.160}$$

This differential equation can be solved by a Laplace transform. However, it is easier to use the property of Green's function

$$\phi(x, t) = \phi_0 \int dt_0 \int dx_0 G(x - x_0; t - t_0) \delta(x_0) f(t_0)$$

$$= \frac{\phi_0}{2v} \int_0^{\pi/\omega} dt_0 \Theta(t - t_0) \Theta[v(t - t_0) - |x|] \sin(\omega t_0) \tag{10.161}$$

The last integral is evaluated assuming $t > \pi/\omega$ is past the initial transient so $t > t_0$ for all t. The second theta function requires that $t - |x|/v > t_0$. The integral has three regimes of time.

1. If $t > \pi/\omega + |x|/v$ so $t > t_0$ for all t and the integral is

$$\phi(x, t) = \frac{\phi_0}{2v\omega} [1 - \cos(\pi)] = \frac{\phi_0}{v\omega} \tag{10.162}$$

2. If $t < |x|/v$ then $\phi = 0$.
3. If $|x|/v < t < \pi/\omega + |x|/v$ the integral is

$$\phi = \frac{\phi_0}{2v} \int_0^{t - |x|/v} dt_0 \sin(\omega t_0) = \frac{\phi_0}{2v\omega} [1 - \cos(\omega t - |x|\omega/v)] \tag{10.163}$$

The wave front travels outward from the origin with velocity v, as shown in Fig. 10.2(b). It has a soft edge, in that the transition from no displacement ($\phi = 0$) to full displacement ($\phi = \phi_0/v\omega$) happens smoothly.

Example 4. *Image sources:* Earlier we solved the problem of Green's function for the wave equation for a wave bounded $0 \leqslant x$, $x_0 \leqslant L$ with the boundary

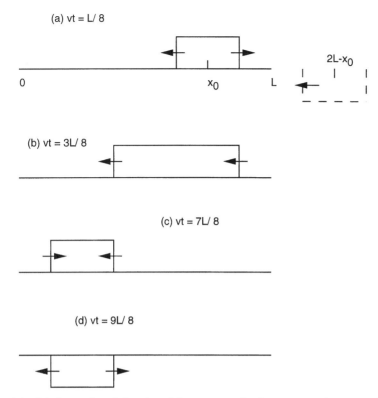

Figure 10.3 Solution to Green's function of the wave equation for a source point at $x_0 = 3L/4$ at different times, (a) $vt = L/8$, (b) $vt = 3L/8$, and (c) $vt = 7L/8$.

condition that $G = 0$ at both ends. Another way to find Green's function is to use the method of image sources. This method is illustrated in Fig. 10.3. Take $x_0 = 3L/4$. Part (a) shows the time of $vt = L/8$, where the two wave fronts are leaving the source point x_0. The wave going to the right will soon encounter the boundary point at $x = L$. In order to satisfy the boundary condition that $G = 0$ at that point, an image wave must be going leftward from the image source at $x = 2L - x_0$. This imaginary wave is shown dashed on the right. It also has a component going rightward, but this front never reaches the region of $0 \leqslant x \leqslant L$ and is ignored. At the time $vt = L/4$ the real wave going right and the image wave going left meet at the boundary and their sum is zero, that satisfies the boundary condition.

For later times, the image wave becomes the actual wave. In Fig. 10.3(b) at time $vt = 3L/8$ both wave fronts are going leftward The left-most edge will soon meet the boundary at $x = 0$. It is met by another image wave of negative amplitude coming from the image source at $x = -x_0$. They meet at point $x = 0$ at time $vt = 3L/4$ and cancel, satisfying the boundary condition. Then the image wave becomes the real wave.

Figure 10.3(c) shows the wave fronts at time $vt = 7L/8$ where the two fronts are now moving towards each other. They will meet at time $vt = L$. When they cross, the wave amplitude becomes negative. The rightward going wave has a rising amplitude, while the leftward going wave has a falling amplitude. These two fronts will each get to the boundary where they must be cancelled by other image waves with positive amplitudes.

The wave motion in the region $0 \leqslant x \leqslant L$ is a series of wave fronts that bounce back and forth. Some have positive amplitude and some have negative amplitude. The boundary conditions are satisfied by having an infinite series of image sources that alternate in sign. The general wave form including images are

$$G(x, x_0; t) = \frac{1}{2v} \left\{ \Theta(vt - |x - x_0|) + \sum_j (-1)^j \Theta[vt - f_j(x, x_0)] \right\}$$

where the summation is over the image sources. The first two image sources are the images from the two ends ($x = 0, L$) that are at the points ($-x_0, 2L - x_0$)

$$-\Theta(vt - |x + x_0|) - \Theta(vt - |x - 2L + x_0|) \tag{10.164}$$

The second set of images are those of these images. They have the opposite sign so are

$$\Theta(vt - |2L - x + x_0|) + \Theta(vt - |2L + x - x_0|) \tag{10.165}$$

the total response from the infinite set of image sources is

$$G(x, x_0; t) = \frac{1}{2v} \sum_{n=-\infty}^{\infty} \left\{ \Theta(vt - |2nL + x - x_0|) - \Theta(vt - |2nL - x - x_0|) \right\}$$

$$\tag{10.166}$$

The infinite series can be summed by using the integral representation for the theta function

$$\Theta(x) = \int \frac{dz}{2\pi i} \frac{e^{izx}}{z - i\varepsilon} \tag{10.167}$$

The integral goes along the real axis beneath the pole at $z = 0$. For $x > 0$ the contour is closed in the UHP and the integral gives one. For $x < 0$ the integral is closed in the LHP and this gives zero since no pole is enclosed. Then Green's function can be expressed as

$$G(x, x_0; t) = \frac{1}{2v} \int \frac{dz}{2\pi i} \frac{e^{izvt}}{z - i\varepsilon} \sum_{n=-\infty}^{\infty} \left[e^{-iz|2nL + x - x_0|} - e^{-iz|2nL + x + x_0|} \right]$$

$$= \frac{1}{2v} \int \frac{dz}{2\pi i} \frac{e^{izvt}}{z - i\varepsilon} \left[e^{-iz|x + x_0|} - e^{-iz|x - x_0|} \right.$$

$$\left. + \frac{2e^{-i2zL}}{1 - e^{-i2zL}} \left\{ \cos[z(x - x_0)] - \cos[z(x + x_0)] \right\} \right] \tag{10.168}$$

The summations over n have been done to get the factor of $(1 - e^{-2izL})$ in the denominators. The interesting aspect of the integrand is that there is no longer a pole at $z = 0$. The numerator also vanishes in this limit. There are poles where $1 - e^{-i2zL} = 0$ that are at the points $z = k_n = n\pi/L$ for all values of n except zero. The summation over these poles gives

$$G(x, x_0; t) = \frac{\Theta(t)}{2ivL} \sum_{n \neq 0} \frac{e^{ik_n vt}}{k_n} \{\cos[k_n(x - x_0)] - \cos[k_n(x + x_0)]\}$$

$$= \frac{2\Theta(t)}{L} \sum_{n=1}^{\infty} \frac{1}{vk_n} \sin(k_n vt) \sin(k_n x) \sin(k_n x_0) \qquad (10.169)$$

The final result is identical to the one derived earlier by Fourier and Laplace transforms. This formula describes the wave fronts bouncing back and forth between the ends of the system.

Example 5. *Soliton:* The next example of waves in one dimension assumes a string which stretches over the half space $(0 \leqslant x < \infty)$. The string is pulled tight and gives the waves a velocity v. At one end a single half pulse is initiated

$$\left[v^2 \frac{\partial^2}{\partial x^2} - \frac{\partial^2}{\partial t^2} \right] \phi(x, t) = 0 \qquad (10.170)$$

$$\phi(0, t) = \phi_0 \sin(\omega t), \quad \text{for } 0 \leqslant t \leqslant \frac{\pi}{\omega} \qquad (10.171)$$

$$\phi(x, 0) = 0 = \phi_t(x, 0) \qquad (10.172)$$

Problems with time dependent boundary conditions are solved best with a Laplace transform. Multiply the wave equation by $\exp(-pt)$ and integrate over all positive time. The integration of the derivative term gives constants that are the initial values and their initial derivative. Both are taken to be zero,

$$\bar{\phi}(x, p) = \int_0^\infty dt e^{-pt} \phi(x, t) \qquad (10.173)$$

$$\left[v^2 \frac{\partial^2}{\partial x^2} - p^2 \right] \bar{\phi}(x, p) = 0 \qquad (10.174)$$

$$\bar{\phi}(x, p) = A(p)e^{-px/v} + B(p)e^{xp/v} \qquad (10.175)$$

The first equation defines the transformed variable. The second equation is the transformed version of the wave equation. The last equation is the most general solution to the second equation. For problems that stretch to positive infinity the last term $\exp(xp/v)$ diverges which is avoided by setting $B(p) = 0$. The factor of $A(p)$ is determined by the boundary condition. The boundary condition must also be transformed

$$A(p) = \bar{\phi}(0, p) = \phi_0 \int_0^{\pi/\omega} dt e^{-pt} \sin(\omega t)$$

$$= \frac{\phi_0}{2i} [1 + e^{-p\pi/\omega}] \left[\frac{1}{p - i\omega} - \frac{1}{p + i\omega} \right]$$

$$= \phi_0 \frac{\omega}{p^2 + \omega^2} [1 + e^{-p\pi/\omega}] \qquad (10.176)$$

The final answer is obtained by the inverse Laplace transform. The path of integration is upward just to the right of the imaginary axis

$$\phi(x, t) = \phi_0 \omega \int_{-i\infty}^{i\infty} \frac{dp}{2\pi i} e^{p(t - x/v)} \frac{1 + e^{-p\pi/\omega}}{p^2 + \omega^2}$$

$$= \phi_0 \omega \int_{-i\infty}^{i\infty} \frac{dp}{2\pi i} \frac{e^{p(t - x/v)} + e^{p(t - x/v - \pi/\omega)}}{p^2 + \omega^2} \qquad (10.177)$$

The evaluation of this integral is subtle. There are three different results for three different time domains.

1. $t - x/v < 0$. In this case, close the contour of integration in the RHP. Since the function is analytic in this plane, the result is zero.

2. $t - x/v - \pi/\omega > 0$. In this case, close the contour of integration in the LHP. This contour circles the apparent poles at $p = \pm i\omega$. However, the residue at these poles is zero since

$$1 + e^{-(\pm i\omega)\pi/\omega} = 1 + e^{\mp i\pi} = 0 \qquad (10.178)$$

The wave amplitude is zero in this case too.

3. $vt > x > vt - \pi v/\omega$. In this case the two terms in Eq. (10.177) are treated differently. The second term has an exponent $(t - x/v - \pi/\omega)$ that is negative, so the contour is closed in the RHP. The integrand is analytic in the RHP, so this term is zero. The first term in Eq. (10.178) has an exponent $(t - x/v)$ that is positive. Here the contour of integration is closed in the LHP, where there are poles at $p = \pm i\omega$. In this case the residue is not zero, since the other term is not in this integral. The result is

$$\phi(x, t) = \phi_0 \sin[\omega(t - x/v)], \quad \text{for } vt > x > vt - \pi v/\omega \qquad (10.179)$$

This solution describes the single pulse traveling down the string. It has the behavior of a solitary wave, denoted as a *soliton*. It travels forever without damping. In practice, there is some damping and the wave amplitude decays slowly in time.

Problems

1. Solve the following equation in one dimension

$$\left[\frac{d^2}{dx^2} + k^2 \right] \phi(x) = \delta(x - x_0) \qquad (10.180)$$

where the values of $\phi(x)$ are confined to $0 \leqslant x \leqslant L$, and also $0 \leqslant x_0 \leqslant L$. The function vanishes $[\phi(0) = 0 = \phi(L)]$ on both boundaries.

2. Show that Eq. (10.57) is the solution to Green's function when it vanishes at the boundary.

3. Solve the diffusion equation for a semi-infinite system $(0 \leqslant x < \infty)$ with the initial condition that $T_i = 0$ and the boundary condition that $T(0, t) = T_f$ during the interval $0 \leqslant t \leqslant t_0$ and is T_i otherwise.

4. Solve the diffusion equation for a bar bounded on one end ($0 \leqslant x < \infty$). The bar is initially at temperature T_i for $t < 0$. Starting at $t = 0$ a constant energy current w_0 is injected into the bar at the $x = 0$ end. Find the subsequent temperature distribution of the bar.

5. Solve the diffusion equation in one dimension for a bar of length $0 \leqslant x \leqslant L$. The initial condition is $T(x, t < 0) = T_i$. For $t \geqslant 0$ the two ends are connected to different reservoirs, $T(0, t) = T_i + \Delta T$ and $T(L, t) = T_i - \Delta T$. Find $T(x, t)$ for all $t > 0$.

6. Solve the diffusion equation in one dimension for a bar of length $0 \leqslant x \leqslant 2L$. Initially the bar is at temperature T_i. Starting at $t = 0$, both ends of the bar are put in contact with a heat reservoir at temperature T_f. Derive an expression for $T(x, t)$ for $t > 0$.

7. Consider a bar of length L where $-L/2 \leqslant x \leqslant L/2$. Initially it is at temperature T_i and its two ends are maintained at that temperature due to a connection with a reservoir. At $t = 0$ a constant electrical current density J is imposed on the bar by voltage V.
   (a) Show that the proper differential equation (with Seebeck $S = 0$) is

$$C\frac{\partial T}{\partial t} - K\frac{\partial^2 T}{\partial x^2} = w_0 \Theta(t) \equiv \frac{JV}{L}\Theta(t) \tag{10.181}$$

   where w_0 is the Joule heating. Assume C, K, and w_0 are independent of temperature. The source term is the Joule heating.
   (b) Solve the differential equation.

8. Solve the diffusion equation in one dimension for a bar that extends $0 < x < \infty$. At the end $x = 0$ is the boundary condition that $T(0, t) = T_i + \Delta T \sin(\omega t)$ for all time. The solution is called a *diffusion wave*. Let the heating be due to the sun, so that $\omega = 2\pi/\text{day}$. Take the diffusion coefficient to be a typical number such as $D = 1$ cm^2/s. How far into the material does the thermal oscillation extend?

9. Solve for Green's function of the diffusion equation over the interval $0 \leqslant x, x_0 \leqslant L$ with the boundary conditions that $G = 0$ on the boundaries at $x = 0, L$.

$$\left[D\frac{\partial^2}{\partial x^2} - \frac{\partial}{\partial t}\right]G(x, x_0; t) = -\delta(t)\delta(x - x_0) \tag{10.182}$$

   It is easiest to use a Fourier series.

10. Solve the wave equation in one dimension for a system bounded at one end ($0 \leqslant x < \infty$). There is no motion for $t < 0$. Starting at $t = 0$ the end is moved $\phi(x = 0, t) = \phi_0 \sin(\omega t)$. Find $\phi(x, t)$ for $t > 0$.

11. Solve the following wave equation in one dimension ($-\infty < x < \infty$)

$$\left[v^2\frac{\partial^2}{\partial x^2} - \frac{\partial^2}{\partial t^2}\right]\phi(x, t) = -\delta(x)f(t) \tag{10.183}$$

$$f(t) = \alpha^2 t e^{-\alpha t} \tag{10.184}$$

where α is a constant.

12. Solve for Green's function the wave equation in one dimension for a system bounded at one end $(0 \leqslant x, x_0 < \infty)$,

$$\left[v^2 \frac{\partial^2}{\partial x^2} - \frac{\partial^2}{\partial t^2} \right] G(x, x_0; t) = -\delta(x - x_0)\delta(t) \tag{10.185}$$

with the initial condition that $G = 0$ for $t < 0$, and is clamped at the end, $0 = G(0, x_0, t) = G(x, 0, t)$.

13. The system is a circular wire loop of radius a. Waves travel with velocity v around the wire with a frequency of $\omega = v/a$. Solve Green's function equation for this system

$$\left[\omega^2 \frac{\partial^2}{\partial \theta^2} - \frac{\partial^2}{\partial t^2} \right] G(\theta, \theta_0; t) = -\delta(\theta - \theta_0)\delta(t) \tag{10.186}$$

where the position of the wave is denoted by the angle θ.

Two Dimensions

Laplace's equation, the diffusion equation, and the wave equation are solved in two dimensions for a variety of boundary conditions. The equations will be solved in two different coordinate systems: rectangular (x, y) and polar (ρ, θ). The latter case involves the properties of Bessel functions that are treated in an Appendix to this chapter.

11.1. Rectangular Coordinates

This section solves a variety of problems using cartesian coordinates (x, y). This choice of coordinates is preferred whenever the geometry of the differential equation is rectangular.

11.1.1. Laplace's Equation

Laplace's equation in two dimensions, in cartesian coordinates, is

$$\left[\frac{\partial^2}{\partial x^2} + \frac{\partial^2}{\partial y^2}\right]\phi(x, y) = 0 \tag{11.1}$$

The usual way to solve these problems is to assume that the individual eigenfunctions are of the product of two functions $\phi(x, y) = A(x)B(y)$, in which case the above equation can be written as

$$\frac{A''}{A} = -\frac{B''}{B} = \text{constant} \tag{11.2}$$

The left-hand side is only a function of x, while the right-hand side is only a function of y. How can they be equal? Only if they are equal to the same constant. The constant can either be positive or negative:

- The constant is positive $(=\alpha^2)$ and has the solutions

$$A(x) = A_c \cosh(\alpha x) + A_s \sinh(\alpha x) \tag{11.3}$$

$$B(y) = B_c \cos(\alpha y) + B_s \sin(\alpha y) \tag{11.4}$$

• The constant is negative $(= -\alpha^2)$ and has the solutions

$$A(x) = A_c \cos(\alpha x) + A_s \sin(\alpha x) \tag{11.5}$$

$$B(y) = B_c \cosh(\alpha y) + B_s \sinh(\alpha y) \tag{11.6}$$

One spatial direction has solutions in the form of sines and cosines, but the other spatial direction must have solutions in terms of hyperbolic sinh and cosh functions. This arrangement is dictated by Laplace's equation.

Example 1: Potential in a Rectangle. Consider a rectangle defined by the coordinates $0 \leqslant x \leqslant a$, $0 \leqslant y \leqslant b$. Find the electrostatic potential that obeys Laplace's equation, and has the boundary conditions that $\phi(0, y) = 0 = \phi(x, 0) = \phi(a, y)$, and the fourth side has $\phi(x, b) = \phi_0$. Figure 11.1 shows the geometry and boundary conditions. Since the boundary condition is zero at both $x = (0, a)$, it suggests eigenfunctions of the form

$$\phi \sim \sin(k_n x), \qquad k_n = \frac{\pi n}{a} \tag{11.7}$$

One cannot have a hyperbolic sinh in the x-direction since it has no zeros, and cannot be made to vanish at $x = a$. The choice of $\sin(k_n x)$ in that direction forces the y-direction to have a hyperbolic function. Choose $\sinh(k_n y)$ since it vanishes at $y = 0$

$$\phi(x,y) = \sum_n a_n \sin(k_n x)\,\sinh(k_n y) \tag{11.8}$$

This choice of eigenfunctions satisfies Laplaces equation, and gives the proper boundary condition of $\phi = 0$ on the three sides. On the top side

$$\phi(x, b) = \phi_0 = \sum_n a_n \sin(k_n x)\sinh(k_n b) \tag{11.9}$$

Multiply this equation by $\sin(k_m x)$ and integrate dx from 0 to a. Then $a_m = 0$

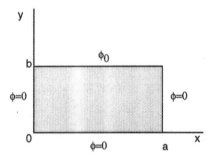

Figure 11.1   Rectangle with boundary conditions for Laplace's equation.

$$\frac{2\phi_0}{k_m} = \frac{a}{2} a_m \sinh(k_m b) \tag{11.10}$$

$$a_{2n+1} = \frac{4\phi_0}{\pi(2n+1)\sinh(k_n b)} \tag{11.11}$$

$$\phi(x, y) = \frac{4\phi_0}{\pi} \sum_{n=0}^{\infty} \frac{1}{2n+1} \sin(k_n x) \frac{\sinh(k_n y)}{\sinh[k_n b]} \tag{11.12}$$

where $k_n = \pi(2n+1)/a$. Although the summation over *n* extends to infinity, a reasonably accurate expression is achieved after 5–10 terms.

The boundary condition at $y = b$ for this problem can be summarized by the series

$$1 = \frac{4}{\pi} \sum_{n=0}^{\infty} \frac{\sin(k_n x)}{2n+1}, \qquad k_n = \frac{\pi}{a}(2n+1) \tag{11.13}$$

The right-hand side appears to be a function of the variable (*x*). However, the left-hand side is the constant one. Therefore, the summation on the right is a constant, and not a function of *x*. This series was encountered in the previous chapter.

A variation on the problem is to choose another side of the rectangle to have the nonzero value. Now choose that $\phi_0 = \phi(a, y)$ while the other three sides have zero boundaries. The solution is almost the same as above, after exchanging *x* and *y*, and *a* and *b*.

$$\phi(x, y) = \frac{4\phi_0}{\pi} \sum_{n=0}^{\infty} \frac{1}{2n+1} \sin(q_n y) \frac{\sinh(q_n x)}{\sinh(q_n a)} \tag{11.14}$$

where $q_n = \pi(2n+1)/b$.

Another similar problem is to have two boundaries at nonzero values, while the other two have zero potential. The solution for this problem is obtained by superposition. One finds a solution $\phi_1(x, y)$ of the type above for only one edge with nonzero potential. Then find a solution of the type above $\phi_2(x, y)$ that has the other edge nonzero while the other three are at zero potential. Then the final solution is to add the two: $\phi = \phi_1 + \phi_2$. This method satisfies Laplace's equation plus all of the boundary conditions.

Finally, consider the solution if all four edges are at the same potential ϕ_0. Then one adds four solutions together of the above type. After some algebra it is

$$\phi(x, y) = \phi_0 \left\{ \frac{4}{\pi} \sum_{n=0}^{\infty} \frac{1}{2n+1} \sin(k_n x) \frac{\cosh[k_n(y - b/2)]}{\cosh(k_n b/2)} \right.$$

$$\left. + \frac{4}{\pi} \sum_{n=0}^{\infty} \frac{1}{2n+1} \sin(q_n y) \frac{\cosh[q_n(x - a/2)]}{\cosh(q_n a/2)} \right\} \tag{11.15}$$

The solution for all four edges at the same potential has to be a constant $\phi(x, y) = \phi_0$, which means that the complex expression in the above large bracket must be identically one for any values of (x, y). It is; try it on the computer.

Example 2: Green's function for Laplace's Equation. The Green's function for Laplace's equation for a rectangle ($0 \leqslant x \leqslant a$, $0 \leqslant y \leqslant b$), with the boundary condition $G = 0$ on all four edges, is

$$\nabla^2 G(x, y; x_0, y_0) = -\delta(x - x_0)\delta(y - y_0) \tag{11.16}$$

$$G(x, y; x_0, y_0) = \frac{4}{ab} \sum_{nm} \frac{1}{k_n^2 + q_m^2} \sin(k_n x) \sin(k_n x_0) \sin(q_m y) \sin(q_m y_0)$$

$$\tag{11.17}$$

$$k_n = \frac{n\pi}{a}, \qquad q_m = \frac{m\pi}{b} \tag{11.18}$$

The solution is obvious by inspection. The sine functions are chosen since they vanish on the boundaries. The operator ∇^2 cancels the denominator of $k_n^2 + q_m^2$ which leaves the two factors of

$$\frac{2}{a} \sum_{n=1}^{\infty} \sin(k_n x) \sin(k_n x_0) = \delta(x - x_0) \tag{11.19}$$

$$\frac{2}{b} \sum_{m=1}^{\infty} \sin(q_m y) \sin(q_m y_0) = \delta(y - y_0). \tag{11.20}$$

This proves the result.

11.1.2. Diffusion Equation

In two dimensions the diffusion equation is

$$\left[\frac{\partial^2}{\partial x^2} + \frac{\partial^2}{\partial y^2} - \frac{1}{D}\frac{\partial}{\partial t}\right] T(x, y, t) = 0 \tag{11.21}$$

This equation is solved in rectangular coordinates for several problems.

Example 1: Diffusion from One Side. In this problem the initial condition is that at $t \leqslant 0$ the temperature is zero. Assume the system is a rectangle with sides $0 \leqslant x \leqslant a$, $0 \leqslant y \leqslant b$. At $t = 0$ one side, say the top, is put into contact with a reservoir at $T_f = T(x, b, t)$, while the other three sides remain in contact with reservoirs at $T = 0$. Find the temperature $T(x, y, t)$.

The first step is to determine the final temperature of the rectangle. At long times the diffusion ceases and the rectangle has its final temperature. In this instance the time derivative vanishes, and the diffusion equation becomes Laplace's equation. The solution to Laplace's equation for a rectangle with one edge at T_f and the other edges at $T = 0$ was solved in a previous subsection ($k_n = \pi(2n + 1)/a$).

$$T(x, y, \infty) = \frac{4T_f}{\pi} \sum_n \frac{1}{2n + 1} \sin(k_n x) \frac{\sinh(k_n y)}{\sinh(k_n b)} \tag{11.22}$$

$$T(x, y, t) = T(x, y, \infty) + \delta T(x, y, t) \tag{11.23}$$

It is convenient to write the solution as the final profile plus a transient contribution $\delta T(x, y, t)$. The transient solution has the following characteristics:

- It obeys the diffusion equation.
- It vanishes on all four boundaries. Note that $T(x, t, \infty)$ equals T_f on the top boundary, so that δT must equal zero there. If both terms were equal to T_f then the total function would be $2T_f$.
- It has an initial value of $-T(x, y, \infty)$, since the summation of the two solutions must be zero when the top surface comes into contact with the heat reservoir T_f at $t = 0$.

The feature that δT vanish on all four boundaries means that its eigenfunctions are of the form $\sin(k_n x) \sin(q_m y)$ where the wave vectors are $k_n = \pi(2n + 1)/a$, $q_m = \pi m/b$. That k_n has only odd integers follows from the fact that $T(x, y, \infty)$ has that feature. There is no such constraint on q_m.

$$\delta T(x, y, t) = \sum_{nm} a_{nm} \exp[-tD(k_n^2 + q_m^2)] \sin(k_n x) \sin(q_m y) \tag{11.24}$$

This choice for δT satisfies the boundary condition, and it obeys the diffusion equation. The constants a_{nm} are determined by the initial condition. At $t = 0$ set $\delta T(x, y, 0)$ equal to $-T(x, y, \infty)$

$$\sum_{nm} a_{nm} \sin(k_n x) \sin(q_m y) = -\frac{4T_f}{\pi} \sum_n \frac{1}{2n + 1} \sin(k_n x) \frac{\sinh(k_n y)}{\sinh(k_n b)} \tag{11.25}$$

First multiply the above equation by $\sin(k_{n'} x)$ and integrate dx. In both series the integral picks out the term with $n = n'$

$$\sum_m a_{n'm} \sin(q_m y) = -\frac{4T_f}{\pi} \frac{1}{2n' + 1} \frac{\sinh(k_{n'} y)}{\sinh(k_{n'} b)} \tag{11.26}$$

The next step is to multiply the entire equation by $\sin(q_{m'} y)$ and integrate dy. The integral on the right contains

$$\int_0^b dy \sin(q_m y) \sinh(k_n y) = -(-1)^m \frac{q_m}{k_n^2 + q_m^2} \sinh(k_n b) \tag{11.27}$$

$$a_{nm} = \frac{8T_f q_m (-1)^m}{\pi b(k_n^2 + q_m^2)(2n + 1)} \tag{11.28}$$

which completes the calculation.

Example 2: *Green's function for rectangle.* The Green's function for the diffusion equation obeys the differential equation

$$\left[\frac{\partial^2}{\partial x^2} + \frac{\partial^2}{\partial y^2} - \frac{1}{D}\frac{\partial}{\partial t}\right] G(x, x_0, y, y_0; t) = -\delta(x - x_0)\delta(y - y_0)\delta(t) \quad (11.29)$$

The system is a rectangle: $0 \leqslant x \leqslant a$, $0 \leqslant y \leqslant b$. The problem is undefined without boundary conditions. Instead of insisting that the function G vanish on the boundaries, impose the condition that no heat flows out of any boundary. This condition is $\hat{n} \cdot \nabla G = 0$ where $\hat{n}$ is the unit vector normal to the boundary line. It is also assumed that $G(x, x_0, y, y_0; t < 0) = 0$.

The function that has a zero value for the normal derivative on all four boundary lines is

$$\cos(k_n x)\cos(q_m y), \qquad k_n = \frac{\pi n}{a}, \qquad q_m = \frac{\pi m}{b} \qquad (11.30)$$

The derivative with respect to either x or y gives zero along $x = 0$ or $x = a$, and $y = 0$ or $y = b$. Therefore try a solution of the form

$$G(x, x_0, y, y_0; t) = D\Theta(t) \sum_{nm} a_{nm} \cos(k_n x) \cos(q_m y) \exp[-Dt(k_n^2 + q_m^2)] \tag{11.31}$$

The derivative ∇^2 gives a factor of $-(k_n^2 + q_m^2)$ in the argument of the series. The time derivative of the exponential factor produces the negative of this factor. The terms in the series each satisfy the diffusion equation. The factor of $\Theta(t)$ in front guarantees that there is no signal before the impulse at $t = 0$. The time derivative of $\Theta(t)$ gives $\delta(t)$ and the factors of D cancel. The above expression for $G(x, x_0, y, y_0; t)$ satisfies the equation

$$\left[\frac{\partial^2}{\partial x^2} + \frac{\partial^2}{\partial y^2} - \frac{1}{D}\frac{\partial}{\partial t}\right] G(x, x_0, y, y_0; t) = -\delta(t) \sum_{nm} a_{nm} \cos(k_n x)\cos(q_m y)$$
$$= -\delta(x - x_0)\delta(y - y_0)\delta(t) \quad (11.32)$$

Cancel the common factor of $\delta(t)$ and arrive at the expression

$$\delta(x - x_0)\delta(y - y_0) = \sum_{nm} a_{nm} \cos(k_n x)\cos(q_m y) \tag{11.33}$$

This equation serves to define the coefficients a_{nm}. An important point is that the summation over (n, m) both start at zero and go to positive infinity. For zero integer then $\cos(k_n x)$ is a constant, and a constant term is allowed. The coefficient a_{00} is found from (11.33) by integrating dx between $(0, a)$ and dy between $(0, b)$. The integral of a single cosine vanishes if $n \neq 0$

$$\int_0^a dx \cos(k_n x) = a\delta_{n=0} \tag{11.34}$$

and the integral over the delta function gives unity. This double integral gives $aba_{00} = 1$. Next multiply (11.33) by $\cos(k_{n'}x)$ and integrate dx and dy, that gives

$$\frac{ab}{2} a_{n'0} = \cos(k_{n'}x_0) \tag{11.35}$$

$$\frac{ab}{2} a_{0m'} = \cos(q_{m'}y_0) \tag{11.36}$$

$$\frac{ab}{4} a_{n'm'} = \cos(k_{n'}x_0)\cos(q_{m'}y_0) \tag{11.37}$$

$$a_{nm} = \frac{1}{ab}[2 - \delta_{n=0}][2 - \delta_{m=0}]\cos(k_n x_0)\cos(q_m y_0) \tag{11.38}$$

The various expression are obtained by doing the integral $dxdy$ with combinations of factors involving $\cos(q_m y)$ or $\cos(k_n x)$. The final expression for a_{nm} completes the solution. Note that a_{00} has a solution which is independent of time. The heat impulse at $t = 0$ spreads uniformly over the rectangle. Since it cannot escape the boundaries, it provides a net steady-state heating of the rectangle. Also note that the solution can be written as

$$T(x, y; t) = D\Theta(t)X(x, t)Y(y, t) \tag{11.39}$$

$$X(x, t) = \frac{1}{a}\left[1 + 2\sum_{n=1}^{\infty} e^{-tDk_n^2}\cos(k_n x)\cos(k_n x_0)\right] \tag{11.40}$$

$$Y(y, t) = \frac{1}{b}\left[1 + 2\sum_{m=1}^{\infty} e^{-tDq_m^2}\cos(q_m y)\cos(q_m y_0)\right] \tag{11.41}$$

The two dimensions behave independently.

The summation over integers produces a delta function in Eq. (11.33).

$$1 + 2\sum_{n=1}^{\infty} \cos(k_n x)\cos(k_n x_0) = \frac{1}{2}\sum_{n=-\infty}^{\infty} \left[e^{ik_n(x+x_0)} + e^{ik_n(x-x_0)}\right]$$

$$= a[\delta(x - x_0) + \delta(x + x_0)] \tag{11.42}$$

This result derives from the expression derived in the previous chapter

$$\sum_{n=-\infty}^{\infty} e^{i\theta n} = 2\pi\delta(\theta), \quad \sum_{m} e^{ik_n x} = 2\pi\delta(\pi x/a) = 2a\delta(x) \tag{11.43}$$

In the summations the angle $\theta = \pi(x \pm x_0)/a$ or $\pi(y \pm y_0)/b$. The delta functions $\delta(x + x_0)$ and $\delta(y + y_0)$ are also produced. They vanish since the integrands can never be zero if $(x, x_0) > 0$, $(y, y_0) > 0$.

11.1.3. Wave Equation

The wave equation in two dimensions in rectangular coordinates is

$$\left[\frac{\partial^2}{\partial x^2} + \frac{\partial^2}{\partial y^2} - \frac{1}{v^2}\frac{\partial^2}{\partial t^2}\right]\phi(x, y, t) = 0 \tag{11.44}$$

Two examples are solved using this equation.

Example 1: *Frequencies of a Rectangular Drum.* Consider a rectangle $(0 \leqslant x \leqslant a, 0 \leqslant y \leqslant b)$. Find the frequencies of vibration if all of the edges are clamped down, and do not vibrate. The condition that the edges do not vibrate suggests that the eigenfunctions are of the form $\sin(k_n x)\sin(q_m y)$, $k_n = \pi n/a$, $q_m = \pi m/b$. Then the solution has the general form

$$\phi(x, y, t) = \sum_{nm} \sin(k_n x)\sin(q_m y)\{a_{nm}\cos(\omega_{nm}t) + b_{nm}\sin(\omega_{nm}t)\} \tag{11.45}$$

$$\omega_{nm}^2 = v^2[k_n^2 + q_m^2] \tag{11.46}$$

The coefficients a_{nm}, b_{nm} depend upon how the drum is plunked. However, the frequencies of vibration are ω_{nm}.

Example 2: *Green's Function.* Green's function for the wave equation in two dimensions obeys the equation

$$\left[v^2\nabla^2 - \frac{\partial^2}{\partial t^2}\right]G(\mathbf{r}, \mathbf{r}_0; t) = -\delta(t)\delta(\mathbf{r} - \mathbf{r}_0) \tag{11.47}$$

Solve this equation using a Fourier transform, that gives the following equation and solution

$$\left[v^2 k^2 + \frac{\partial^2}{\partial t^2}\right]\bar{G}(\mathbf{k}, \mathbf{r}_0; t) = \delta(t)e^{-i\mathbf{k}\cdot\mathbf{r}_0} \tag{11.48}$$

$$\bar{G}(\mathbf{k}, \mathbf{r}_0; t) = \frac{\Theta(t)}{kv}\sin(kvt)e^{-i\mathbf{k}\cdot\mathbf{r}_0} \tag{11.49}$$

$$G(\mathbf{r}, \mathbf{r}_0; t) = \frac{\Theta(t)}{v}\int\frac{d^2k}{(2\pi)^2}\frac{\sin(kvt)}{k}e^{i\mathbf{k}\cdot(\mathbf{r}-\mathbf{r}_0)} \tag{11.50}$$

The integral is evaluated in polar coordinates: $d^2k = kdkd\theta$, where the angle θ is between the vectors $\mathbf{k}$ and $\mathbf{R} = \mathbf{r} - \mathbf{r}_0$. It is easiest to first do the integral over dk, where the integrand is assumed to converge to zero at infinity

$$G(\mathbf{r}, \mathbf{r}_0; t) = \frac{\Theta(t)}{8\pi^2 vi}\int_0^{2\pi}d\theta\int_0^\infty dke^{ikR\cos(\theta)}\left(e^{ikvt} - e^{-ikvt}\right)$$

$$= \frac{\Theta(t)}{8\pi^2 v}\int_0^{2\pi}d\theta\left(\frac{1}{vt + R\cos(\theta)} + \frac{1}{vt - R\cos(\theta)}\right)$$

$$= \frac{\Theta(vt - R)}{2\pi v\sqrt{(vt)^2 - R^2}} \tag{11.51}$$

The result is physically interesting. The impulse starts at the point $R = 0$ at $t = 0$ and spreads out radially. At a distance R from the origin, the wave does not arrive until the time $t_R = R/v$. Then the pulse goes by, and the trailing edge slowly decays. This behavior is very different than found in one dimension. In 1D the pulse went by and then there was no more disturbance. In 1D it is a soliton. In two dimensions it is not a soliton. Once the front has passed by, the wave continues to decay with decreasing amplitude in time.

Example 3: *Image Source.* A problem related to the above Green's function is to find the Green's function for the half-space $y > 0$ with the boundary condition that $G = 0$ when $y = 0$. The source point is assumed to be in the upper halfplane. For convenience set it at $\mathbf{r}_0 = (0, y_0)$. This problem can be solved in the same way as the above free space Green's function. However, the result can be written down by inspection using the concept of the image source. We put a source with the opposite sign at the image point $\mathbf{r}_0' = (0, -y_0)$ so that the solution is

$$G(\mathbf{r}, \mathbf{r}_0; t) = \frac{1}{2\pi v} \left[\frac{\Theta(v^2 t^2 - x^2 - (y - y_0)^2)}{\sqrt{(vt)^2 - x^2 - (y - y_0)^2}} - \frac{\Theta(v^2 t^2 - x^2 - (y + y_0)^2)}{\sqrt{(vt)^2 - x^2 - (y + y_0)^2}} \right] \tag{11.52}$$

The Green's function vanishes for $y = 0$ and obeys Eq. (11.47) for $y > 0$. It obeys a different equation for $y < 0$ but that is not part of our space!

11.2. Polar Coordinates

Two dimensional problems can also be solved in polar coordinates (ρ, θ).

$$\rho = \sqrt{x^2 + y^2} \tag{11.53}$$

$$\tan(\theta) = \frac{y}{x} \tag{11.54}$$

where ρ is the length of the vector while θ is the angle measured from the x-axis. The first step is to derive the expression for ∇^2 in polar coordinates. The easiest derivation is to examine how ∇^2 acts on a function of $F(\rho, \theta)$. The first step is to examine

$$\frac{\partial}{\partial x} F(\rho, \theta) = \frac{\partial F}{\partial \rho} \frac{\partial \rho}{\partial x} + \frac{\partial F}{\partial \theta} \frac{\partial \theta}{\partial x}$$

$$= \frac{x}{\rho} \frac{\partial F}{\partial \rho} - \frac{y}{\rho^2} \frac{\partial F}{\partial \theta} \tag{11.55}$$

Take one more derivative with respect to x and find

$$\frac{\partial^2}{\partial x^2} F = \frac{x^2}{\rho^2} \frac{\partial^2 F}{\partial \rho^2} + \left[\frac{1}{\rho} - \frac{x^2}{\rho^3} \right] \frac{\partial F}{\partial \rho} - \frac{2xy}{\rho^3} \frac{\partial^2 F}{\partial \rho \partial \theta}$$

$$+ \frac{2xy}{\rho^4} \frac{\partial F}{\partial \theta} + \frac{y^2}{\rho^4} \frac{\partial^2 F}{\partial \theta^2} \tag{11.56}$$

The next step is to take the double derivative with respect to y. The result can be obtained from the above formula by the substitutions $(x \rightarrow y, y \rightarrow -x)$ that gives

$$\frac{\partial^2}{\partial y^2} F = \frac{y^2}{\rho^2} \frac{\partial^2 F}{\partial \rho^2} + \left[\frac{1}{\rho} - \frac{y^2}{\rho^3} \right] \frac{\partial F}{\partial \rho} + \frac{2xy}{\rho^3} \frac{\partial^2 F}{\partial \rho \partial \theta}$$

$$- \frac{2xy}{\rho^4} \frac{\partial F}{\partial \theta} + \frac{x^2}{\rho^4} \frac{\partial^2 F}{\partial \theta^2} \tag{11.57}$$

$$\nabla^2 F(\rho, \theta) = \frac{\partial^2 F}{\partial \rho^2} + \frac{1}{\rho} \frac{\partial F}{\partial \rho} + \frac{1}{\rho^2} \frac{\partial^2 F}{\partial \theta^2}$$

$$= \frac{1}{\rho} \frac{\partial}{\partial \rho} \left(\rho \frac{\partial F}{\partial \rho} \right) + \frac{1}{\rho^2} \frac{\partial^2 F}{\partial \theta^2} \tag{11.58}$$

This important result is used in the remainder of this section.

11.2.1. Laplace's Equation

In polar coordinates in two dimensions, Laplace's equation is

$$0 = \nabla^2 \phi(\rho, \theta) = \left[\frac{\partial^2}{\partial \rho^2} + \frac{1}{\rho} \frac{\partial}{\partial \rho} + \frac{1}{\rho^2} \frac{\partial^2}{\partial \theta^2} \right] \phi(\rho, \theta) \tag{11.59}$$

The angular dependence always has the form of $\exp(in\theta)$ where n is an integer. The argument for an integer is that the system is periodic in the θ variable, so that $\phi(\rho, \theta + 2\pi) = \phi(\rho, \theta)$ since they are at the same point. The angular function $\exp(iv\theta)$ requires that v be an integer to maintain this periodicity. The solutions have the form

$$\phi(\rho, \theta) = \sum_n a_n R_n(\rho) e^{in\theta} \tag{11.60}$$

$$\nabla^2 \phi = \sum_n a_n e^{in\theta} \left[\frac{\partial^2}{\partial \rho^2} + \frac{1}{\rho} \frac{\partial}{\partial \rho} - \frac{n^2}{\rho^2} \right] R_n(\rho) = 0 \tag{11.61}$$

The expression in brackets is the differential equation satisfied by $R_n(\rho)$. Note that each term in this expression has the same power of ρ. In this special case, the solution has to be a power law of the form ρ^s where the exponent s needs

$$\left[\frac{\partial^2}{\partial\rho^2} + \frac{1}{\rho}\frac{\partial}{\partial\rho} - \frac{n^2}{\rho^2}\right]\rho^s = [s(s-1) + s - n^2]\rho^{s-2} = 0 \qquad (11.62)$$

$$s^2 = n^2, \qquad s = \pm n \qquad (11.63)$$

The exponents are $\pm n$. The most general solution to Laplaces equation in polar coordinates has the form

$$\phi(\rho, \theta) = \sum_{n=-\infty}^{\infty} e^{in\theta}[A_n\rho^n + B_n\rho^{-n}] + C\ln(\rho) \qquad (11.64)$$

The constants A_n, B_n are determined by the boundary conditions. Note that there are two solutions $\rho^{\pm n}$. Second-order differential equations always have two independent solutions. For solutions inside of a circle, then the solutions with ρ^{-n}, $n > 0$ diverge at the origin. They are usually omitted. Similarly, for solutions outside of a circle, the solutions ρ^n, $n > 0$ diverge at infinity and are usually omitted.

The case $n = 0$ is a special case since there appears to only be one case of $s = 0$. For $n = 0$ there is another solution of the form $\phi(\rho) = A_0 + C\ln(\rho)$. The second solution has the form of $\ln(\rho)$. One can check that this function satisfies Laplace's equation when $n = 0$.

Example 1: Electrostatics in a Circle. Figure 11.2 shows a circle of radius a that has the potential $\pm\phi_1$ or $\phi = 0$ on various parts of its boundary:

$$\phi(a, \theta) = \begin{cases} \phi_1 & -\theta_0 < \theta < \theta_0 \\ -\phi_1 & \pi - \theta_0 < \theta < \pi + \theta_0 \\ 0 & \text{elsewhere} \end{cases} \qquad (11.65)$$

The problem is to find the electrostatic potential inside of the circle with these boundary conditions. The angular function can be written as either $\cos(n\theta)$ or else $\sin(n\theta)$. Since the boundary conditions are symmetric in angle, then so is the solution. The $\sin(\theta)$ terms are set to zero. Therefore, the general solution

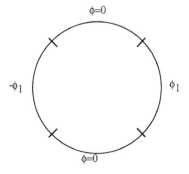

Figure 11.2. Boundary conditions on a circle.

has the form

$$\phi(\rho, \theta) = \sum_{n=0}^{\infty} a_n \left(\frac{\rho}{a}\right)^n \cos(n\theta) \tag{11.66}$$

The coefficients a_n are determined by the boundary conditions on the surface. Set $\rho = a$, multiply the above equation by $\cos(m\theta)$ and integrate over all angles. Since the cosine functions are orthogonal, the integral picks from the summation the term with $n = m$:

$$a_m = \int_0^{2\pi} \frac{d\theta}{\pi} \cos(m\theta)\phi(a, \theta) \tag{11.67}$$

This integral is evaluated using the boundary conditions specified above

$$a_m = \frac{\phi_1}{\pi} \left[\int_{-\theta_0}^{\theta_0} d\theta \cos(m\theta) - \int_{\pi-\theta_0}^{\pi+\theta_0} d\theta \cos(m\theta) \right]$$

$$= \frac{4\phi_1}{m\pi} \sin(m\theta_0)_{m=\text{odd}} \tag{11.68}$$

$$\phi(\rho, \theta) = \frac{4\phi_1}{\pi} \sum_{n=0}^{\infty} \frac{1}{2n+1} \left(\frac{\rho}{a}\right)^{2n+1} \cos[(2n+1)\theta] \sin[(2n+1)\theta_0] \tag{11.69}$$

which solves the problem.

Example 2: Green's Function of Laplace's Equation. The Green's function of Laplace's equation is the solution to

$$\nabla^2 G(\mathbf{r}, \mathbf{r}_0) = -\delta(\mathbf{r} - \mathbf{r}_0) \tag{11.70}$$

The Green's function is found for a circle of radius a, where $\mathbf{r}, \mathbf{r}_0$ are both inside of the circle. The boundary conditions are chosen so that $G = 0$ at the edge of a circle.

The first part of the discussion is to determine the delta function in polar coordinates. Given that $d^2r = \rho \, d\rho \, d\theta$ then the delta function obeys

$$f(\mathbf{r}_0) = f(\rho_0, \theta_0) = \int d^2r \delta(\mathbf{r} - \mathbf{r}_0) f(\mathbf{r})$$

$$= \int \rho \, d\rho \int d\theta \delta(\mathbf{r} - \mathbf{r}_0) f(\rho, \theta) \tag{11.71}$$

$$\delta(\mathbf{r} - \mathbf{r}_0) = \frac{1}{\rho} \delta(\rho - \rho_0)\delta(\theta - \theta_0) \tag{11.72}$$

The second part of the discussion is to determine how to generate the delta functions on the right of Eq. (11.70). The delta function of the radial variable will be found as the derivative of theta functions $\theta(\rho - \rho_0)$. The delta function of the angular variable is based on the identity

$$\sum_{n=-\infty}^{\infty} e^{in(\theta-\theta_0)} = 2\pi\delta(\theta-\theta_0) \tag{11.73}$$

Expand the factor of $\exp(in\theta)$ in sines and cosines. The sine terms cancel since they are odd functions of n

$$\sum_{n=-\infty}^{\infty} e^{in(\theta-\theta_0)} = 1 + 2\sum_{n=1}^{\infty} \cos[n(\theta-\theta_0)] = 2\pi\delta(\theta-\theta_0) \tag{11.74}$$

The delta function of angle can be expressed as a summation over cosines. The term with $n=0$ has a coefficient of one, but the coefficients of the cosine terms with $n \geqslant 1$ are two. The case with $n=0$ is special, and must always be treated separately.

As a function of angle, the impulse starts at the angle θ_0 and spreads evenly in both directions. It is an even function of $(\theta - \theta_0)$ and so the angular dependence goes according to $\cos[n(\theta - \theta_0)]$. Remember to treat the case $n=0$ as special, and write the solution,

$$G(\mathbf{r}, \mathbf{r}_0) = g_0(\rho, \rho_0) + \sum_{n=1}^{\infty} \cos[n(\theta-\theta_0)]g_n(\rho, \rho_0) \tag{11.75}$$

When $\rho \neq \rho_0$ the Green's function obeys the homogeneous equation Eq. (11.61) whose solutions are $\rho^{\pm n}$. For the case $n=0$ the two solutions are a constant, or else $\ln(\rho)$. The combinations which vanish at $\rho = a$ and do not diverge at $\rho = 0$ are

$$g_0(\rho, \rho_0) = \begin{cases} A_0 & 0 \leqslant \rho < \rho_0 \\ B_0 \ln(\rho/a) & \rho_0 < \rho \leqslant a \end{cases} \tag{11.76}$$

$$g_n = \begin{cases} A_n (\rho/a)^n & 0 \leqslant \rho < \rho_0 \\ B_n [(\rho/a)^n - (a/\rho)^n] & \rho_0 < \rho \leqslant a \end{cases} \tag{11.77}$$

The two expressions for $(\rho < \rho_0, \rho > \rho_0)$ are matched at ρ_0. Then the radial function is

$$A_0 = B_0 \ln(\rho_0/a) \tag{11.78}$$

$$A_n \left(\frac{\rho_0}{a}\right)^n = B_n \left[\left(\frac{\rho_0}{a}\right)^n - \left(\frac{a}{\rho_0}\right)^n\right] \tag{11.79}$$

$$g_0(\rho, \rho_0) = B_0[\Theta(\rho_0 - \rho)\ln(\rho_0/a) + \Theta(\rho - \rho_0)\ln(\rho/a)] \tag{11.80}$$

$$g_n(\rho, \rho_0) = B_n \left\{ \Theta(\rho_0 - \rho) \left(\frac{\rho}{\rho_0}\right)^n \left[\left(\frac{\rho_0}{a}\right)^n - \left(\frac{a}{\rho_0}\right)^n\right] \right.$$

$$\left. + \Theta(\rho - \rho_0) \left[\left(\frac{\rho}{a}\right)^n - \left(\frac{a}{\rho}\right)^n\right] \right\} \tag{11.81}$$

The next step is to operate on these expressions with the radial part of ∇^2. The derivatives of $\Theta(\rho - \rho_0)$, $\Theta(\rho_0 - \rho)$ give $\pm\delta(\rho - \rho_0)$. The coefficients of these delta functions cancel for the first derivatives, but not for the second.

$$\frac{1}{\rho}\frac{\partial}{\partial\rho}\left(\rho\frac{\partial}{\partial\rho}g_0\right) = \frac{B_0}{\rho}\,\delta(\rho - \rho_0) \tag{11.82}$$

$$\frac{1}{\rho}\frac{\partial}{\partial\rho}\left(\rho\frac{\partial}{\partial\rho}g_n\right) = \frac{n^2}{\rho^2}g_n + B_n\frac{2n}{\rho}\left(\frac{a}{\rho_0}\right)^n\delta(\rho - \rho_0) \tag{11.83}$$

In order to get the correct summation over the angles, set

$$B_n = \frac{B_0}{n}\left(\frac{\rho_0}{a}\right)^n \tag{11.84}$$

$$G(\mathbf{r}, \mathbf{r}_0) = B_0[\Theta(\rho_0 - \rho)\ln(\rho_0/a) + \Theta(\rho - \rho_0)\ln(\rho/a)]$$

$$+ B_0\sum_n\frac{\cos[n(\theta - \theta_0)]}{n}\left\{\Theta(\rho_0 - \rho)\left(\frac{\rho}{a}\right)^n\left[\left(\frac{\rho_0}{a}\right)^n - \left(\frac{a}{\rho_0}\right)^n\right]\right.$$

$$\left. + \Theta(\rho - \rho_0)\left(\frac{\rho_0}{a}\right)^n\left[\left(\frac{\rho}{a}\right)^n - \left(\frac{a}{\rho}\right)^n\right]\right\} \tag{11.85}$$

Operate on this expression by ∇^2 and find

$$\nabla^2 G(\mathbf{r}, \mathbf{r}_0) = \frac{B_0}{\rho}\delta(\rho - \rho_0)\left\{1 + 2\sum_n\cos[n(\theta - \theta_0)]\right\} \tag{11.86}$$

$$= 2\pi B_0\,\delta(\mathbf{r} - \mathbf{r}_0)$$

$$B_0 = -\frac{1}{2\pi} \tag{11.87}$$

The result of Eq. (11.85) can be summed into closed form. Recall that

$$\sum_{n=1}^{\infty}\frac{x^n}{n} = -\ln(1 - x) \tag{11.88}$$

$$\sum_n\frac{x^n}{n}\cos(n\theta) = \frac{1}{2}\sum_n\frac{x^n}{n}\left(e^{in\theta} + e^{-in\theta}\right)$$

$$= -\frac{1}{2}\{\ln[1 - xe^{i\theta}] + \ln[1 - xe^{-i\theta}]\}$$

$$= -\frac{1}{2}\ln[(1 - xe^{i\theta})(1 - xe^{-i\theta})]$$

$$= -\frac{1}{2}\ln[1 - 2x\cos(\theta) + x^2] \tag{11.89}$$

$$\ln(\rho_0/a) + \sum_n \frac{\cos[n(\theta - \theta_0)]}{n} \left(\frac{\rho}{a}\right)^n \left[\left(\frac{\rho_0}{a}\right)^n - \left(\frac{a}{\rho_0}\right)^n\right]$$

$$= \ln(\rho_0/a) + \frac{1}{2}\{\ln[1 - 2(\rho/\rho_0)\cos(\theta - \theta_0) + (\rho/\rho_0)^2]$$

$$- \ln[1 - 2(\rho\rho_0/a^2)\cos(\theta - \theta_0) + (\rho\rho_0/a^2)^2]\}$$

$$= \frac{1}{2}\ln\left[\frac{a^2[\rho_0^2 - 2\rho\rho_0\cos(\theta - \theta_0) + \rho^2]}{a^4 - 2\rho\rho_0 a^2\cos(\theta - \theta_0) + \rho^2\rho_0^2}\right]$$

$$G(\mathbf{r}, \mathbf{r}_0) = -\frac{1}{4\pi}\ln\left[\frac{a^2(\mathbf{r} - \mathbf{r}_0)^2}{(a^2\hat{\rho}_0 - \rho_0\mathbf{r})^2}\right] \tag{11.90}$$

Exactly the same result is obtained for the other case of $\Theta(\rho - \rho_0)$. In two dimensions, the solution to the Green's function for LaPlace's equation is $G \sim \ln(R)$, $\mathbf{R} = \mathbf{r} - \mathbf{r}_0$. The other factor, in the argument of the logarithm, is due to the potential of the image source, which maintains the boundary condition. Note that $G = 0$ if either ρ or ρ_0 equals to the radius a since then the argument of the logarithm is one.

11.2.2. Helmholtz Equation

The Helmholtz equation has the general form

$$0 = [\nabla^2 + k^2]\phi(\rho, \theta) \tag{11.91}$$

$$0 = \left[\frac{\partial^2}{\partial\rho^2} + \frac{1}{\rho}\frac{\partial}{\partial\rho} + \frac{1}{\rho^2}\frac{\partial^2}{\partial\theta^2} + k^2\right]\phi(\rho,\theta) \tag{11.92}$$

Again the angular part of the solution must have the form $\exp(in\theta)$ that gives for the radial part of the eigenfunctions

$$\phi(\rho, \theta) = \sum_n e^{in\theta} R_n(\rho) \tag{11.93}$$

$$0 = \left[\frac{\partial^2}{\partial\rho^2} + \frac{1}{\rho}\frac{\partial}{\partial\rho} - \frac{n^2}{\rho^2} + k^2\right]R_n(\rho) \tag{11.94}$$

$$R_n(\rho) = A_n J_n(k\rho) + B_n Y_n(k\rho) \tag{11.95}$$

Equation (11.94) *Bessels Equation* is important in mathematical physics. The two functions which satisfy this equation are the Bessel function $J_n(k\rho)$ and the Neumann function $Y_n(k\rho)$. They are both called Bessel's functions. Their properties are summarized in the appendix to this chapter. Many more properties are listed in standard references such as *Handbook of Mathematical Functions*[1]. The Neumann functions $Y_n(z)$ diverge at $z = 0$ while the Bessel functions are well behaved at the origin.

In solving Laplace's or Helmholtz's equation the symbol n must be an integer. But Bessel's equation is valid even for the case that n is not an integer.

Below we shall encounter Bessel functions where n is a half-integer, or even an irrational number. Another important point is observed by multiplying each term in Eq. (11.94) by ρ^2. Change variables to $z = k\rho$ and find

$$0 = \left[z^2 \frac{\partial^2}{\partial z^2} + z \frac{\partial}{\partial z} - n^2 + z^2 \right] R_n(z) \tag{11.96}$$

$$R_n(z) = A_n J_n(z) + B_n Y_n(z) \tag{11.97}$$

The equation depends only on $z = k\rho$.

Example 1: Normal Modes of a Drum. Now consider the normal modes of a circular drum. Start with the wave equation, and assume the solution oscillates in time with $\exp(-i\omega t)$

$$\left[v^2 \nabla^2 - \frac{\partial^2}{\partial t^2} \right] \phi(\mathbf{r}, t) = 0 \tag{11.98}$$

$$\phi(\mathbf{r}, t) = e^{-i\omega t} \phi(\rho, \theta) \tag{11.99}$$

$$\left[\nabla^2 + \frac{\omega^2}{v^2} \right] \phi(\rho, \theta) = 0 \tag{11.100}$$

The differential equation is Eq. (11.91), where $k = \omega/v$. The frequency is ω and the sound velocity is v. The solutions are regular at the origin, which eliminates the Neumann functions. The eigenfunctions must have the form

$$\phi(\rho, \theta, t) = \sum_{n\alpha} a_{n\alpha} e^{in\theta - i\omega_{n\alpha} t} J_n(k_{n\alpha} \rho) \tag{11.101}$$

The index α is explained below. The drum is drawn tightly at its rim at a radius of a. The boundary condition is that $\phi(a, \theta, t) = 0$. The Bessel function $J_n(z)$ oscillates with increasing values of z. It crosses zero in a nearly periodic fashion. The points where it vanishes are called, naturally, *zeros of Bessel functions*. Tables of such zeros are given in standard references. Index α labels the zeros. The parameter $\lambda_{n\alpha}$ are the values of z where $J_n(\lambda_{n\alpha}) = 0$. Typical values are shown in Table 11.1. The numbers in parenthesis for $\alpha = 4$ are estimated from the asymptotic expansion in the appendix. It shows that the Bessel functions vanishes according to the leading term in the asymptotic

Table 11.1 Some zeros of the Bessel function J_n for $n = 0, 1, 2$.

α	$\lambda_{0\alpha}$	$\lambda_{1\alpha}$	$\lambda_{2\alpha}$
1	2.405	3.832	5.136
2	5.520	7.016	8.417
3	8.654	10.174	11.620
4	11.792	13.324	14.796
	(11.781)	(13.352)	(14.923)

$$0 = \cos\left[\lambda_{n\alpha} - \frac{n\pi}{2} - \frac{\pi}{4}\right] \tag{11.102}$$

$$\lambda_{n\alpha} = \frac{\pi}{2}\left[2\alpha + n - \frac{1}{2}\right] \tag{11.103}$$

The latter formula is used to estimate the numbers in the table which are in parentheses. It is a fairly accurate estimate. The accuracy increases for increasing values of α. The zeros of the Bessel function occur at intervals which are separated approximately by π radians. The drum frequencies are given by the zeros of the Bessel function

$$k_{n\alpha}a = \frac{\omega_{n\alpha}}{v}a = \lambda_{n\alpha} \tag{11.104}$$

$$\omega_{n\alpha} = \frac{v}{a}\lambda_{n\alpha} \tag{11.105}$$

where the parameters $\lambda_{n\alpha}$ are shown in the table.

The eigenfunctions $\chi_{n\alpha}(\rho, \theta)$ of the wave equation for a circle, with the boundary condition that the amplitude vanishes at the edge $\rho = a$, are

$$\chi_{n\alpha}(\rho, \theta) = N_{n\alpha}e^{in\theta}J_n\left(\lambda_{n\alpha}\frac{\rho}{a}\right) \tag{11.106}$$

The normalization constant $N_{n\alpha}$ needs to be determined using

$$\int d^2\rho \chi_{n\alpha}^*(\rho, \theta)\chi_{m\beta}(\rho, \theta) = \delta_{nm}\delta_{\alpha\beta} \tag{11.107}$$

where $d^2\rho = \rho\, d\rho\, d\theta$. The $d\theta$ integral is completed first

$$\int_0^{2\pi} d\theta\, e^{i(m-n)\theta} = 2\pi\delta_{nm} \tag{11.108}$$

which forces $n = m$. The remaining integral is ($s = \rho/a$)

$$\int_0^1 sds J_n(\lambda_{n\alpha}s)J_n(\lambda_{n\beta}s) = \frac{1}{2}\delta_{\alpha\beta}[J_n'(\lambda_{n\alpha})]^2 \tag{11.109}$$

$$N_{n\alpha} = \frac{1}{a\sqrt{\pi}|J_n'(\lambda_{n\alpha})|} \tag{11.110}$$

The prime denotes derivative. Note that the two Bessel functions are only orthogonal for different values of α when they have the same value of n. They are forced to have the same value of n by the angular integral.

The above integral can be evaluated in a simple fashion. There are two steps: (1) To show that the integral vanishes if $\alpha \neq \beta$, and (2) To show it gives

Eq. (11.109) when $\alpha = \beta$. In doing the first step, let $k \equiv \lambda_{n\alpha}/a$, $p = \lambda_{n\beta}/a$. Write the Bessel's equation for each function

$$0 = \frac{d}{d\rho}\left(\rho \frac{dJ_n(k\rho)}{d\rho}\right) + \left[k^2\rho - \frac{n^2}{\rho}\right]J_n(k\rho) \tag{11.111}$$

$$0 = \frac{d}{d\rho}\left(\rho \frac{dJ_n(p\rho)}{d\rho}\right) + \left[p^2\rho - \frac{n^2}{\rho}\right]J_n(p\rho) \tag{11.112}$$

Multiply the first equation by $J_n(p\rho)$, the second equation by $J_n(k\rho)$, and subtract the two equations. The derivative terms can be combined to give the form

$$0 = \frac{d}{d\rho}\left[\rho J_n(p\rho)\frac{dJ_n(k\rho)}{d\rho} - \rho J_n(k\rho)\frac{dJ_n(p\rho)}{d\rho}\right] + (k^2 - p^2)\rho J_n(k\rho)J_n(p\rho) \tag{11.113}$$

Now integrate this equation between zero and a

$$(p^2 - k^2)\int_0^a d\rho\, \rho J_n(k\rho)J_n(p\rho) = a\left[J_n(pa)\frac{dJ_n(ka)}{da} - J_n(ka)\frac{dJ_n(pa)}{da}\right] \tag{11.114}$$

The right-hand side of this equation vanishes if $J_n(ka) = 0 = J_n(pa)$. If $\alpha \neq \beta$ then $k \neq p$ and the left-hand side vanishes only if the integral vanishes. These steps prove the integral in Eq. (11.109) vanishes if $\alpha \neq \beta$. Next, evaluate the integral for $\alpha = \beta$. Start with Bessel's equation which is written as ($z \equiv ka$)

$$0 = z^2 J_n''(z) + zJ_n'(z) + (z^2 - n^2)J_n(z) \tag{11.115}$$

Multiply this equation by J_n' and many of the terms can be combined into a derivative

$$0 = z^2 J_n'' J_n' + z(J_n')^2 + (z^2 - n^2)J_n' J_n \tag{11.116}$$

$$0 = \frac{1}{2}\frac{d}{dz}\left[z^2(J_n')^2 + (z^2 - n^2)J_n^2\right] - zJ_n^2 \tag{11.117}$$

Integrate this equation from $(0, \lambda)$ and obtain the identity

$$\int_0^\lambda dz\, zJ_n^2(z) = \frac{1}{2}\left[\lambda^2 J_n'(\lambda)^2 + (\lambda^2 - n^2)J_n(\lambda)^2\right] \tag{11.118}$$

The previous identity Eq. (11.109) applies when the boundary condition at the edge of the circle is $J_n(\lambda) = 0$. The above integral provides the normalization for other boundary conditions such as $J_n'(\lambda) = 0$.

Example 2: *Diffusion in a Circle.* Consider the following diffusion problem. The system is a circle of radius a. For $t < 0$ it has a temperature T_i. At $t \geqslant 0$ the rim of the circle is put in contact with a reservoir at the temperature T_f. Find the evolution of temperature $T(\mathbf{r}, t)$.

The entire rim is set at the same temperature. So the angular function $\exp(in\theta)$ must have $n = 0$ since there is no angular variation in this problem. Write the solution as the final temperature, at $t = \infty$, that is T_f, plus the transient

$$T(\rho, t) = T_f + \delta T(\rho, t) \tag{11.119}$$

$$\left[\nabla^2 - \frac{1}{D} \frac{\partial}{\partial t} \right] \delta T(\rho, t) = 0 \tag{11.120}$$

$$\delta T(\rho, t) = \sum_\alpha A_\alpha(t) J_0(\lambda_\alpha \rho / a) \tag{11.121}$$

The boundary condition for $t > 0$ is $T(a, t) = T_f$. Therefore, the transient has the boundary condition $\delta T(a, t) = 0$ since the factor of T_f already satisfies the boundary condition. Since $\delta T = 0$ at the edge of the circle, it is usefully expanded in terms of the functions $J_0(\lambda_{0\alpha} \rho / a)$ that are the Bessel functions that vanish at the surface. Putting this ansatz in the diffusion equation gives

$$\nabla^2 J_n(k\rho) = - k^2 J_n(k\rho) \tag{11.122}$$

$$0 = \left[\nabla^2 - \frac{1}{D} \frac{\partial}{\partial t} \right] \delta T(\rho, t) = - \sum_\alpha J_0(\lambda_\alpha \rho / a) \left[\left(\frac{\lambda_{0\alpha}}{a} \right)^2 + \frac{1}{D} \frac{\partial}{\partial t} \right] A_\alpha(t) \tag{11.123}$$

that has the solution

$$\delta T(\rho, t) = \sum_\alpha A_\alpha(0) e^{- Dt\lambda_{0\alpha}^2 / a^2} J_0(\lambda_{0\alpha} \rho / a) \tag{11.124}$$

Next apply the initial condition that $\delta T(\rho, t = 0) = T_i - T_f$.

$$T_i - T_f = \sum_\alpha A_\alpha(0) J_0(\lambda_{0\alpha} \rho / a) \tag{11.125}$$

This equation is solved by multiplying by $J_0(\lambda_{0\beta} \rho / a)$ and integrating $\rho \, d\rho$ from zero to a. Since the Bessel functions are orthogonal for different α, this process gives ($z = \lambda_{0\alpha} \rho / a$)

$$\frac{1}{2} A_\alpha(0) [J'_0(\lambda)]^2 = \frac{(T_i - T_f)}{\lambda^2} \int_0^\lambda dz z J_0(z) \tag{11.126}$$

The two recursion relations for Bessel functions can be subtracted to obtain the result which we specialize for $n = 1$

$$J_{n-1} = \frac{n}{z} J_n + J'_n = \frac{1}{z^n} \frac{d}{dz} [z^n J_n(z)] \tag{11.127}$$

$$z J_0(z) = J_1 + z J'_1 = \frac{d}{dz} [z J_1(z)] \tag{11.128}$$

$$\int_0^\lambda dz z J_0(z) = \lambda J_1(\lambda) \tag{11.129}$$

Using $J'_0 = -J_1$ gives

$$A_\alpha(0) = (T_i - T_f) \frac{2}{\lambda J_1(\lambda)} \qquad (11.130)$$

$$T(\rho, t) = T_f + 2(T_i - T_f) \sum_\alpha \frac{\exp[-tD\lambda^2/a^2]}{\lambda} \frac{J_0(\lambda\rho/a)}{J_1(\lambda)} \qquad (11.131)$$

where $\lambda \equiv \lambda_{0\alpha}$. The problem is solved. Note that at $t = 0$ the boundary conditions produce the identity

$$1 = \sum_\alpha \frac{2}{\lambda_{0\alpha}} \frac{J_0(\lambda_{0\alpha}\rho/a)}{J_1(\lambda_{0\alpha})} \qquad (11.132)$$

The right-hand side appears to be a function of ρ, while the left-hand side is the simple constant one. The summation on the right-hand side is actually independent of ρ. A similar summation was discussed earlier in one dimension, where the summation had $\sin(k_n x)$. In two dimensions, the Bessel functions are the equivalent to the sines and cosines of one dimension. They asymptotically behave as sines and cosines at large argument.

The same problem is worked using a Laplace transform on the time variable. The Laplace transform of the diffusion equation is

$$[D\nabla^2 - p]\bar{T}(\rho, p) = -T_i \qquad (11.133)$$

$$\bar{T}(\rho, p) = \frac{T_i}{p} + AI_0\left(\rho\sqrt{\frac{p}{D}}\right) \qquad (11.134)$$

The solution is the same for all angles so again $n = 0$. The radial equation is that for a Bessel function of imaginary argument, whose solutions are $I_0(k\rho)$ or $K_0(k\rho)$, where $k^2 = p/D$. The functions $K_0(z)$ diverge at small values of z and are not allowed for this problem. Therefore, the solution must be of the form $I_0(\rho\sqrt{p/D})$. The Laplace transform of the boundary condition gives $\bar{T}(a, p) = T_f/p$ that determines the constant A

$$\bar{T}(\rho, p) = \frac{T_i}{p} + \frac{T_f - T_i}{p} \frac{I_0(\rho\sqrt{p/D})}{I_0(a\sqrt{p/D})} \qquad (11.135)$$

$$\delta T(\rho, t) = (T_f - T_i) \int_{-i\infty}^{i\infty} \frac{dp}{2\pi i} \frac{e^{pt}}{p} \frac{I_0(\rho\sqrt{p/D})}{I_0(a\sqrt{p/D})} \qquad (11.136)$$

The inverse transform appears difficult but is easy. There is no response until $t > 0$ and then close the contour in the LHP. There is a simple pole at $p = 0$. There appears to be a branch cut along the negative real axis. However, setting $p = -y$, $\sqrt{p} = \pm i\sqrt{y}$ gives no cut since the Bessel functions $I_0(z)$ depend only on powers of z^2 that are the same above and below the cut. There is no branch cut. However, there are poles along the negative real axis at the points where

$$0 = I_0(ia\sqrt{y/D}) \equiv J_0(a\sqrt{y/D}) \qquad (11.137)$$

The Bessel function $I_0(z)$ is just J_0 when z becomes an imaginary number. The zeros of the Bessel functions occur at the points $\lambda_{0\alpha} = a\sqrt{y_\alpha/D}$ that gives

$$p_\alpha = -y_\alpha = -D\left(\frac{\lambda_{0\alpha}}{a}\right)^2 \tag{11.138}$$

The residue at the pole is found from the Taylor expansion of the argument of the Bessel function

$$J_0\left(\frac{a}{\sqrt{D}}\sqrt{y_\alpha + \delta y}\right) = J_0\left(a\sqrt{\frac{y_\alpha}{D}} + \frac{a\delta y}{2\sqrt{Dy_\alpha}}\right)$$

$$= J_0(\lambda) + \frac{a\delta y}{2\sqrt{Dy_\alpha}}J_0'(\lambda) + \cdots \tag{11.139}$$

The coefficient of δy determines the residue form the delta function. Collecting all of these factors gives

$$T(\rho, t) = T_i + (T_f - T_i)\left[1 - \sum_\alpha \frac{2}{\lambda}\frac{J_0(\rho\lambda/a)}{J_0'(\lambda)}e^{-tD(\lambda/a)^2}\right] \tag{11.140}$$

that is the same solution found by using a Fourier transform.

Example 3: Green's Functions in a Circle. The Green's functions for diffusion within a circle obeys the equation

$$\left[D\nabla^2 - \frac{\partial}{\partial t}\right]G(\mathbf{r}, \mathbf{r}_0; t) = -\delta(t)\delta(\mathbf{r} - \mathbf{r}_0) \tag{11.141}$$

Select the boundary condition that $G = 0$ on the edge of the circle. Based on previous problems, the solution is guessed to be similar to

$$G(\mathbf{r}, \mathbf{r}_0; t) = \Theta(t)\sum_{n\alpha} A_{n\alpha}\cos[n(\theta - \theta_0)]e^{-tD\lambda_{n\alpha}^2/a^2}J_n(\rho\lambda_{n\alpha}/a)J_n(\rho_0\lambda_{n\alpha}/a) \tag{11.142}$$

The summation over n extends from zero to infinity, and the case $n = 0$ is again special. Since the Green's function vanishes at the boundary, the radial functions must be the Bessel functions that vanish at $\rho = a$. Since the function is symmetric in (ρ, ρ_0), then the variable ρ_0 also has to be in the argument of a Bessel function. The exponential time dependence is familiar from previous solutions of the diffusion equation. The prefactor of $\Theta(t)$ gives the factor of $\delta(t)$ when differentiated. The only remaining question is to determine the constants $A_{n\alpha}$. In taking the time derivative of the right-hand side, there is one term that is the derivative of $\Theta(t)$ which gives $\delta(t)$. The coefficient of this term must be the delta function on the right-hand side

$$\delta(\mathbf{r} - \mathbf{r}_0) = \sum_{n\alpha} A_{n\alpha} \cos[n(\theta - \theta_0)] J_n(\rho\lambda_{n\alpha}/a) J_n(\rho_0\lambda_{n\alpha}/a) \qquad (11.143)$$

First integrate this equation $d\theta$ between $(0, 2\pi)$. On the right all terms vanish except $n = 0$,

$$\frac{\delta(\rho - \rho_0)}{\rho} = 2\pi \sum_\alpha A_{0\alpha} J_0(\lambda_{0\alpha}\rho/a) J_0(\lambda_{0\alpha}\rho_0/a) \qquad (11.144)$$

Next, multiply by $\rho d\rho J_0(\lambda_{0\beta}\rho/a)$ and integrate, which gives

$$J_0(\lambda_{0\beta}\rho_0/a) = \pi a^2 A_{0\beta} J_0(\lambda_{0\beta}\rho_0/a)|J_0'(\lambda_{0\beta})|^2 \qquad (11.145)$$

$$A_{0\beta} = \frac{1}{\pi a^2 |J_0'(\lambda_{0\beta})|^2} \qquad (11.146)$$

Next repeat the same steps, but initially multiply by $\cos[m(\theta - \theta_0)]$ before doing the integral over $d\theta$. Then again multiply by $\rho J_m(\lambda_{m\beta}\rho/a)$ and do the integral over ρ. One finds that

$$J_m(\rho_0\lambda_{m\alpha}/a) = \pi A_{m\alpha} \frac{a^2}{2} J_m(\rho_0\lambda_{m\alpha}/a)|J_m'(\lambda_{m\alpha})|^2 \qquad (11.147)$$

$$A_{m\alpha} = \frac{2}{\pi a^2 |J_m'(\lambda_{m\alpha})|^2} \qquad (11.148)$$

which completes the derivation. Again note that the coefficient for $n = 0$ is smaller by a factor of two when $n \geqslant 1$.

11.2.3. Hankel Transforms

A Hankel transform is similar to a Fourier transform, the difference being that Bessel functions are used instead of plane waves. The basic form of the transform is

$$F(k) = \int_0^\infty \rho d\rho J_0(k\rho) f(\rho) \qquad (11.149)$$

$$f(\rho) = \int_0^\infty k dk J_0(k\rho) F(k) \qquad (11.150)$$

The factor of $2/\pi$ is contained in the definition of the Bessel function. Actually, the same equations are valid for any $J_n(k\rho)$. In order to see whether these

$$f(\rho) = \int_0^\infty k\,dk\,J_0(k\rho) \int_0^\infty \rho'\,d\rho'\,J_0(k\rho')\,f(\rho')$$

$$= \int_0^\infty \rho'\,d\rho'\,f(\rho')\Lambda(\rho,\ \rho') \tag{11.151}$$

$$\Lambda(\rho,\ \rho') = \int_0^\infty k\,dk\,J_0(k\rho)J_0(k\rho') = \delta(\rho - \rho')/\rho \tag{11.152}$$

The last equality is required for the transform to be valid. This latter relationship can be proved by starting from the definition of the delta function in two dimensional cartesian space

$$\int \frac{d^2k}{(2\pi)^2}\,e^{i\mathbf{k}\cdot(\mathbf{r}-\mathbf{r}')} = \delta^2(\mathbf{r} - \mathbf{r}') = \frac{1}{\rho}\,\delta(\theta - \theta')\delta(\rho - \rho') \tag{11.153}$$

Rewrite the definition of the delta function Eq. (11.153) in polar coordinates, where: $\mathbf{k} = (k, \theta)$, $\mathbf{r} = (\rho, \theta_\rho)$, $\mathbf{r}' = (\rho', \theta_{\rho'})$.

$$\frac{1}{(2\pi)^2} \int_0^\infty k\,dk \int_0^{2\pi} d\theta\,\exp\{ik[\rho\cos(\theta - \theta_\rho) - \rho'\cos(\theta - \theta_{\rho'})]\}$$

$$= \frac{1}{\rho}\,\delta(\theta_\rho - \theta_{\rho'})\delta(\rho - \rho') \tag{11.154}$$

Since the Hankel transform is defined in terms of J_0, which are the functions associated with $n = 0$, the angular integrals can be eliminated. Integrate the above equation over $d\theta_\rho$ between 0 and 2π. Afterwards, do the integral $d\theta$. On the left, each integral produces a Bessel function

$$\int_0^{2\pi} d\theta_p\,e^{ik\rho\cos(\theta - \theta_p)} = 2\pi J_0(k\rho) \tag{11.155}$$

$$\int_0^{2\pi} d\theta\,e^{-ik\rho'\cos(\theta - \theta'_p)} = 2\pi J_0(k\rho') \tag{11.156}$$

$$\int_0^\infty k\,dk\,J_0(k\rho)J_0(k\rho') = \frac{1}{\rho}\,\delta(\rho - \rho') \tag{11.157}$$

The last equation is the relationship needed for the Hankel transform.

A short list of Hankel transforms is:

$$\int_0^\infty d\rho\,J_0(k\rho) = \frac{1}{k} \tag{11.158}$$

$$\int_0^\infty d\rho\,J_0(k\rho)\,\frac{\rho}{\sqrt{\rho^2 + a^2}} = \frac{1}{k}\,e^{-ka} \tag{11.159}$$

$$\int_0^\infty d\rho J_0(k\rho) \frac{\rho}{(\rho^2 + a^2)^{3/2}} = \frac{1}{a} e^{-ka} \tag{11.160}$$

$$\int_0^\infty d\rho J_0(k\rho) \frac{\rho}{\sqrt{\rho^2 + a^2}} \exp[-b\sqrt{\rho^2 + a^2}] = \frac{1}{\sqrt{k^2 + b^2}} \exp[-a\sqrt{k^2 + b^2}] \tag{11.161}$$

$$\int_0^\infty d\rho J_0(k\rho) \sin(a\rho) = \frac{\Theta(a - k)}{\sqrt{a^2 - k^2}} \tag{11.162}$$

$$\int_0^\infty d\rho J_0(k\rho) \cos(a\rho) = \frac{\Theta(k - a)}{\sqrt{k^2 - a^2}} \tag{11.163}$$

These integrals comprise many useful examples. Hankel transforms are useful since they are simply integrals of Bessel functions.

Some of these transforms are simple to evaluate. The integral with an exponential can be evaluated by using the integral definition of the Bessel function and interchanging the order of the two integrals

$$\int_0^\infty d\rho e^{-p\rho} J_0(k\rho) = \int_0^\infty d\rho e^{-p\rho} \int_0^\pi \frac{d\theta}{\pi} e^{ik\rho \cos(\theta)} = \int_0^\pi \frac{d\theta}{\pi} \int_0^\infty d\rho e^{-\rho[p - ik \cos(\theta)]}$$

$$= \int_0^\pi \frac{d\theta}{\pi[p - ik \cos(\theta)]} = \frac{1}{\sqrt{p^2 + k^2}} \tag{11.164}$$

The last integral was evaluated in Chapter 4.

Note that the Hankel transforms are symmetric. If $F(k)$ is the Hankel transform of $f(\rho)$ then $F(\rho)$ is the Hankel transform of $f(k)$.

Example 1: Green's functions for 2D Wave equation. The Green's function for the wave equation in two dimensions was earlier shown to be given by the integral

$$\bar{G}(\mathbf{k}, \mathbf{r}_0; t) = \frac{\Theta(t)}{vk} \sin(vkt) e^{-i\mathbf{k} \cdot \mathbf{r}_0} \tag{11.165}$$

$$G(\mathbf{r}, \mathbf{r}_0; t) = \frac{\Theta(t)}{2\pi v} \int_0^\infty dk \sin(vkt) \int_0^{2\pi} \frac{d\theta}{2\pi} e^{ikR \cos(\theta)}$$

$$= \frac{\Theta(t)}{2\pi v} \int_0^\infty dk \sin(vkt) J_0(kR)$$

$$= \frac{\Theta(vt - R)}{2\pi v \sqrt{(vt)^2 - R^2}} \tag{11.166}$$

The first equation gives the Fourier transform. The second expression gives the integral for the inverse transform. Carry out the angular integral first, this produces the Bessel function $J_0(kR)$. The last integral is carried out using the table of Hankel transforms. This procedure gives the same result as before.

Example 2: Plane wave scattering from a circle. Suppose a plane wave is incident along the positive x-direction. It encounters a circular target of radius a. The boundary conditions at the surface of the circle are the wave must vanish. The scattering produces an outgoing wave, which must be a Hankel function $H_m^{(1)}(k\rho)$. The incident wave is, with $x = \rho \cos(\theta)$

$$e^{ikx} = e^{ik\rho \cos(\theta)} = \sum_{m=-\infty}^{\infty} i^m J_m(k\rho)e^{im\theta} \tag{11.167}$$

where $k = \omega/v$ for a wave of frequency ω and velocity v. This imposing series is the generating function for the Bessel function. Note that it is actually required by the integral definition of the Bessel function. Multiplying both sides by $\cos(n\theta)$, and then integrating between 0 and 2π gives ($z = k\rho$)

$$\int_0^{2\pi} d\theta \cos(n\theta)e^{iz \cos(\theta)} = \pi[i^n J_n(z) + i^{-n} J_{-n}(z)] = 2\pi i^n J_n(z) \tag{11.168}$$

where $J_{-n} = (-1)^n J_n$. The latter equation is the integral definition of the Bessel function. It is equivalent to the expression for the generating function.

Back to the scattering problem. The outgoing wave Ψ_S must be expanded in a series similar to Eq. (11.167), except the appropriate Bessel function is the Hankel function. Each angular momentum state m has an unknown coefficient b_m

$$\psi(\rho, \theta) = e^{ikx} + \psi_s(\rho, \theta)$$

$$= \sum_{m=-\infty}^{\infty} i^m e^{im\theta}[J_m(k\rho) + b_m H_m^{(1)}(k\rho)] \tag{11.169}$$

The boundary condition that $\psi(a, \theta) = 0$ must be true for each value of m, which gives

$$b_m = -\frac{J_m(ka)}{H_m^{(1)}(ka)} \tag{11.170}$$

Using the asymptotic expansion for the Hankel function, the scattered wave at large distance has the form

$$\lim_{k\rho \gg 1} \psi_s \to \frac{f(\theta)}{\sqrt{k\rho}} e^{i(k\rho - \pi/4)} \tag{11.171}$$

$$f(\theta) = -\sqrt{\frac{2}{\pi}} \sum_m e^{im\theta} \frac{J_m(ka)}{H_m^{(1)}(ka)} \tag{11.172}$$

The factor $|f(\theta)|^2$ gives the cross section for scattering in two dimensions.

Appendix I: Vector Derivatives

Here are some derivatives in polar coordinates:

$$\nabla\phi(\mathbf{r}) = \hat{\rho}\,\frac{\partial\phi}{\partial\rho} + \frac{\hat{\theta}}{\rho}\,\frac{\partial\phi}{\partial\theta} \tag{11.173}$$

$$\nabla\cdot\mathbf{A}(\mathbf{r}) = \frac{1}{\rho}\,\frac{\partial}{\partial\rho}\,(\rho A_\rho) + \frac{1}{\rho}\,\frac{\partial A_\theta}{\partial\theta} \tag{11.174}$$

$$\nabla^2\phi = \frac{1}{\rho}\,\frac{\partial}{\partial\rho}\left(\rho\,\frac{\partial\phi}{\partial\rho}\right) + \frac{1}{\rho^2}\,\frac{\partial^2\phi}{\partial\theta^2} \tag{11.175}$$

$$\nabla^2\mathbf{A} = \hat{\rho}\left[\left(\nabla^2 - \frac{1}{\rho^2}\right)A_\rho - \frac{2}{\rho^2}\,\frac{\partial A_\theta}{\partial\theta}\right] + \hat{\theta}\left[\left(\nabla^2 - \frac{1}{\rho^2}\right)A_\theta + \frac{2}{\rho^2}\,\frac{\partial A_\rho}{\partial\theta}\right] \tag{11.176}$$

An expression such as $\nabla\cdot[K(\mathbf{r})\nabla\phi]$ is

$$\nabla\cdot[K(\mathbf{r})\nabla\phi] = \frac{1}{\rho}\,\frac{\partial}{\partial\rho}\left(\rho K(\rho,\theta)\,\frac{\partial\phi}{\partial\rho}\right) + \frac{1}{\rho^2}\,\frac{\partial}{\partial\theta}\left[K(\rho,\theta)\,\frac{\partial\phi}{\partial\theta}\right] \tag{11.177}$$

Appendix II: Bessel Functions

The solution to the Helmholtz equation in polar coordinates, in two dimensions, is

$$0 = [\nabla^2 + k^2]\phi(\mathbf{r}) \tag{11.178}$$

$$0 = \left(\frac{\partial^2}{\partial\rho^2} + \frac{1}{\rho}\,\frac{\partial}{\partial\rho} + \frac{1}{\rho^2}\,\frac{\partial^2}{\partial\theta^2} + k^2\right)\phi(\rho,\theta) \tag{11.179}$$

$$\phi(\rho,\theta) = e^{in\theta}R_n(\rho) \tag{11.180}$$

$$0 = \left(\frac{\partial^2}{\partial\rho^2} + \frac{1}{\rho}\,\frac{\partial}{\partial\rho} - \frac{n^2}{\rho^2} + k^2\right)\rho_n(\rho) \tag{11.181}$$

$$R_n(\rho) = AJ_n(k\rho) + B_nY_n(k\rho) \tag{11.182}$$

The second order differential equation has two solutions, which are called $J_n(z)$, $Y_n(z)$. The first (J_n) is a Bessel function. The second (Y_n) is a Neumann function. The symbol $N_n(z)$ is also commonly used for Neumann functions. Both are called Bessel functions. Here we list some of their properties

1. Series expansion

$$J_\nu(z) = \left(\frac{z}{2}\right)^\nu \sum_{n=0}^{\infty} \frac{(-z^2/4)^n}{n!\,\Gamma(n+1+\nu)} \tag{11.183}$$

$$J_0(z) = \sum_n \frac{(-z^2/4)^n}{(n!)^2} \tag{11.184}$$

The series converges for any value of z. The Neumann function $Y_\nu(z)$ does not have a convenient series expansion.

2. Asymptotic expansion for large z

$$J_\nu(z) \rightarrow \sqrt{\frac{2}{\pi z}} \cos\left[z - \frac{\nu\pi}{2} - \frac{\pi}{4}\right] [1 + O(1/z)] \qquad (11.185)$$

$$Y_\nu(z) \rightarrow \sqrt{\frac{2}{\pi z}} \sin\left[z - \frac{\nu\pi}{2} - \frac{\pi}{4}\right] [1 + O(1/z)] \qquad (11.186)$$

3. Integral representation

$$J_n(z) = \frac{1}{i^n \pi} \int_0^\pi d\theta \, e^{iz \cos(\theta)} \cos(n\theta) \qquad (11.187)$$

This representation makes it easy to show $J_{-n}(z) = (-1)^n J_n(z)$.

4. Recursion relations

$$J_{n+1}(z) + J_{n-1}(z) = \frac{2n}{z} J_n(z) \qquad (11.188)$$

$$J_{n+1}(z) - J_{n-1}(z) = -2\frac{d}{dz} J_n(z) \qquad (11.189)$$

They can be proved easily from the integral definition. For example, consider

$$J_{n+1}(z) - J_{n-1}(z) = \frac{1}{i^{n+1}\pi} \int_0^\pi d\theta e^{iz \cos(\theta)} \{\cos[(n+1)\theta] - i^2 \cos[(n-1)\theta]\}$$

$$= \frac{1}{i^{n+1}\pi} \int_0^\pi d\theta e^{iz \cos(\theta)} \{\cos[(n+1)\theta] + \cos[(n-1)\theta]\}$$

$$= \frac{1}{i^{n+1}\pi} \int_0^\pi d\theta e^{iz \cos(\theta)} \{2 \cos(n\theta) \cos(\theta)\}$$

$$= -2\frac{d}{dz} \frac{1}{i^n \pi} \int_0^\pi d\theta e^{iz \cos(\theta)} \cos(n\theta) = -2\frac{d}{dz} J_n(z)$$

$$(11.190)$$

5. Generating functions are series such as

$$\exp\left[\frac{1}{2} z(t - 1/t)\right] = \sum_{n=-\infty}^\infty t^n J_n(z) \qquad (11.191)$$

$$\exp[iz \cos(\theta)] = \sum_{n=-\infty}^\infty i^n e^{in\theta} J_n(z) \qquad (11.192)$$

The second equation is defined from the first using $t = ie^{i\theta}$. Generating functions are useful for summing series containing Bessel functions.

Besides these basic Bessels functions, there are several other important functions which are closely related.

Hankel functions are defined as

$$H_n^{(1)}(z) = J_n(z) + iY_n(z) \tag{11.193}$$

$$H_n^{(2)}(z) = J_n(z) - iY_n(z) \tag{11.194}$$

$$\lim_{z \to \infty} H_n^{(1,2)}(z) = \sqrt{\frac{2}{\pi z}} e^{\pm i\phi}[1 + O(1/z)] \tag{11.195}$$

$$\phi = z - \frac{\pi}{4}(2n + 1) \tag{11.196}$$

Note that $H_n^{(2)} = H_n^{(1)*}$. The Hankel functions correspond to travelling waves: $H_n^{(1)} \sim \exp[iz]$ is a wave that travels outward from the origin, while $H_n^{(2)} \sim \exp[-iz]$ is a wave that travels inward. They are the important functions for traveling wave problems.

Modified Bessel functions are useful for evanescent waves. In Bessel's equation change the sign of k^2. In one dimensions this gave eigenfunctions of the form $\sinh(kx)$, $\cosh(kx)$. In two dimensions they give the modified Bessel's functions $I_n(z)$ and $K_n(z)$

$$0 = \left(\frac{\partial^2}{\partial \rho^2} + \frac{1}{\rho}\frac{\partial}{\partial \rho} - \frac{n^2}{\rho^2} - k^2\right) \rho_n(\rho) \tag{11.197}$$

$$R_n(\rho) = AI_n(k\rho) + B_n K_n(k\rho) \tag{11.198}$$

The function $I_n(z)$ is wellbehaved at the origin ($z = 0$) and is given by a series that is absolutely convergent

$$I_n(z) = \left(\frac{z}{2}\right)^n \sum_{k=0}^{\infty} \frac{(z^2/4)^k}{k!(n + k)!} \tag{11.199}$$

This series is the same as for the function $J_n(z)$ except the minus sign is missing from the argument. The function $K_n(z)$ diverges at the origin, and does not have a convenient power series. Both functions have simple integral definitions, and asymptotic expansions

$$I_n(z) = \frac{1}{\pi}\int_0^\pi d\theta e^{z\cos(\theta)} \cos(n\theta) \tag{11.200}$$

$$K_n(z) = \int_0^\infty dt e^{-z\cosh(t)} \cosh(nt) \tag{11.201}$$

$$\lim_{z \to \infty} I_n(z) = \frac{e^z}{\sqrt{2\pi z}}[1 + O(1/z)] \tag{11.202}$$

$$\lim_{z \to \infty} K_n(z) = \sqrt{\frac{\pi}{2z}} e^{-z}[1 + O(1/z)] \qquad (11.203)$$

The functions I_n are regular at the origin but diverge at infinity. They are useful for solving problems where the system is inside a circle. The functions K_n diverge at the origin, but they converge exponentially at infinity. They are useful for problems which are outside of a circle.

Spherical Bessel Functions are based on half-integer index. They are defined with lowercase letters

$$j_n(z) = \sqrt{\frac{\pi}{2z}} J_{n+1/2}(z) \qquad (11.204)$$

$$y_n(z) = \sqrt{\frac{\pi}{2z}} Y_{n+1/2}(z) \qquad (11.205)$$

$$0 = \left[\frac{d^2}{dr^2} + \frac{2}{r}\frac{d}{dr} + k^2 - \frac{\ell(\ell+1)}{r^2}\right] R_\ell(kr) \qquad (11.206)$$

$$R_\ell(z) = A j_\ell(z) + B y_\ell(z) \qquad (11.207)$$

These functions are useful in three dimensions, where they are the solutions to the radial part of the wave equation (11.206) in spherical coordinates. Although the half-integer functions look complicated, they are actually quite simple. They are just sines and cosines times simple polynomials. Some functions of low index are:

$$j_0(z) = \frac{\sin z}{z} \qquad (11.208)$$

$$y_0(z) = -\frac{\cos(z)}{z} \qquad (11.209)$$

$$j_1(z) = \frac{\sin(z)}{z^2} - \frac{\cos(z)}{z} \qquad (11.210)$$

$$y_1(z) = -\frac{\cos(z)}{z^2} - \frac{\sin(z)}{z} \qquad (11.211)$$

$$j_2(z) = \left(\frac{3}{z^3} - \frac{1}{z}\right)\sin(z) - \frac{3}{z^2}\cos(z) \qquad (11.212)$$

$$y_2(z) = -\left(\frac{3}{z^3} - \frac{1}{z}\right)\cos(z) - \frac{3}{z^2}\sin(z) \qquad (11.213)$$

These functions appear to diverge as $z \to 0$. At small argument the functions $j_n \sim z^n$ are regular at the origin. The other functions $y_n \sim z^{-n-1}$ do indeed diverge.

References

1. *Handbook of Mathematical Functions*, ed. M. Abramowitz and I. A. Stegun (many publishers)
2. *Tables of Hankel Transforms*, A. Erdyli (McGraw-Hill, 1954)

Problems

1. Solve Laplace's equation in two dimensions for the half-space $x > 0$ assuming that the potential vanishes at $x = 0$, and there is a charge Q at $\mathbf{r}_0 = (d, 0)$ where $d > 0$

$$\nabla^2 \phi(\mathbf{r}) = -4\pi Q \delta(x - d)\delta(y) \qquad (11.214)$$

The problem is identical to a charge outside of a grounded metal conductor.

2. Consider the solution to Laplace's equation $\nabla^2 \phi(x, y) = 0$ for a rectangle defined by: $0 \leqslant x \leqslant a$, $0 \leqslant y \leqslant b$. The boundary conditions are that $\phi(0, y) = 0 = \phi(a, y)$ and $\phi(x, 0) = \phi_0 = \phi(x, b)$.

3. A rectangle is located $(0 \leqslant x \leqslant a, 0 \leqslant y \leqslant b)$. Find the steady-state temperature distribution if the boundaries have the following temperature $(q \equiv \pi/2b)$

$$x = 0 \qquad T(0, y) = -T_0 \sin(qy) \qquad (11.215)$$

$$x = a \qquad T(a, y) = +T_0 \sin(qy) \qquad (11.216)$$

$$y = 0 \qquad T(x, 0) = 0 \qquad (11.217)$$

$$y = b \qquad T(x, b) = -T_0[1 - 2x/a] \qquad (11.218)$$

4. Solve for the Green's function for Laplace's equation on a rectangle, with the boundary condition that $\hat{n} \cdot \nabla G = 0$ on all four boundaries. No current leaves the edges.

5. Consider solutions to the diffusion equation for a rectangle defined by $0 \leqslant x \leqslant a$, $0 \leqslant y \leqslant b$. Assume that $T(x, y, t < 0) = 0$. At $t = 0$ all four edges are put in contact with a heat reservoir at temperature T_f. Derive an expression for $T(x, y, t)$ for $t > 0$.

6. Solve the diffusion equation for a rectangle. Assume that $T = 0$ for $t < 0$. At $t = 0$ the top side is put in contact with a reservoir at temperature T_f. Assume the other three walls have the boundary condition that no heat can flow out of them. Solve for $T(x, y; t)$.

7. A rectangle has dimensions $(0 < x < a, 0 < y < b)$. The right-hand edge is held at a temperature $T(a, y) = T_0 \sin(\pi y/b)$, while the other three edges have $T = 0$. Find the steady-state temperature distribution in the rectangle.

8. The same rectangle as in the previous problem is initially at $T(x, y; t) = 0$, $t < 0$. At $t = 0$ the right-hand edge is put in contact with a reservoir with $T(a, y) = T_0 \sin(\pi y/b)$ while the other edges are held at $T = 0$. Solve the diffusion equation to find $T(x, y; t)$.

9. Solve the Green's function for diffusion in rectangular coordinates

$$\left[\nabla^2 - \frac{1}{D}\frac{\partial}{\partial t}\right] G(x, x_0, y, y_0; t) = -\delta(t)\delta(x - x_0)\delta(y - y_0) \qquad (11.219)$$

With the initial condition that $G = 0$, $t < 0$ and $T = 0$ on all boundaries.

10. A two dimensional waveguide is bounded $(0 \leqslant y \leqslant b)$ in the y-direction but is unbounded in the x-direction. The boundary conditions are $\phi(x, 0) = 0 = \phi(x, b)$. What is the frequency $\omega(p)$ and velocity v_p of a wave $\sim e^{ipx}$ traveling in the x-direction?

11. A hollow washer has inner radius a and outer radius of b. Assume the temperature is T_a on the inner surface and T_b on the outer surface. Find the distribution of temperatures everywhere assuming steady state diffusion.

12. Solve the same problem above but now $K(T) = K_0/(1 + \alpha T)$ where α is a small positive constant.

$$0 = \nabla \cdot [K(T)\nabla T(\mathbf{r})] \tag{11.220}$$

The problem can be solved exactly.

13. Assume a solid cylinder is heated uniformly inside, perhaps by the energy from nuclear fission. Solve the equation

$$K\nabla^2 T(\mathbf{r}) = -S_0 \tag{11.221}$$

where (K, S_0) are constants, and $T = T_0$ at the surface of the cylinder $\rho = a$.

14. A hollow washer has an inner radius of a, and an outer radius of b. The temperature on the inner radius is $T(a, \theta) = T_a \cos(\theta)$ and on the outer radius it is $T(b, \theta) = T_b \cos(\theta)$. Find the steady state temperature everywhere $T(\rho, \theta)$ for $a \leqslant \rho \leqslant b$.

15. Solve Laplace's equation $\nabla^2 \phi = 0$ in 2D for $\rho > a$, assuming that:

(a) $\phi(a, \theta) = \phi_1 \sin(\theta)$ $\tag{11.222}$

(b) $\phi(a, \theta) = \phi_1 \cos^2(\theta)$ $\tag{11.223}$

16. In two dimensions a potential $\phi(\mathbf{r}, t)$ satisfies Laplace's equation outside of a circle of radius a. The boundary condition on the surface is that the potential is ϕ_0 in the first and third quadrants, and $-\phi_0$ in the second and fourth quadrants. Find the solution.

17. Derive the recursion relations for the modified Bessel function

$$I_{n+1}(z) + I_{n-1}(z) = ? \tag{11.224}$$

$$I_{n+1}(z) - I_{n-1}(z) = ? \tag{11.225}$$

18. The Hankel function with $n = 0$ can be defined as

$$H_0^{(1)}(z) = -\frac{2i}{\pi} \int_0^\infty dt e^{iz\cosh(t)} \tag{11.226}$$

(a) Show this integral is a solution to Bessels equation with $n = 0$.
(b) Find the asymptotic behavior as $z \to \infty$.
(c) Find the asymptotic behavior as $z \to 0$.

19. Find the Green's function for the Helmholtz equation in unbounded two dimensional space.

$$[\nabla^2 + k^2]G(\mathbf{r}) = -\delta(\mathbf{r}) \tag{11.227}$$

Use outgoing wave boundary conditions.

20. The region is the inside of a circle of radius a. Find the steady state temperature distribution if the boundary of the circle is in contact with a reservoir that provides the temperature profile $T(a, \theta) = T_i + \Delta T \sin(\theta)$.

21. The region is the inside of a circle of radius a. The temperature is initially T_i everywhere. At $t = 0$ the boundary is put in contact with a reservoir with the same function as the prior problem: $T(a, \theta; t) = T_i + \Delta T \sin(\theta)$. Solve the diffusion equation to find the temperature $T(\rho, \theta; t)$ everywhere.

22. Solve for the Green's function of the wave equation inside a circle where the amplitude of the Green's function vanishes at the boundary.

Three Dimensions

Three dimensions provide some new type of problems that are different from those found in one or two dimensions. There are also some new functions that need to be introduced, such as Legendre polynomials and spherical Bessel functions. Problems will be solved in three coordinate systems, Cartesian, cylindrical, and spherical.

12.1. Cartesian Coordinates

Problems in Cartesian coordinates are those for a block of dimensions $0 < x < a, 0 < y < b, 0 < z < c$ and of volume $V_0 = abc$. Only a few problems will be solved, since the methods and solutions are very similar to those of two dimensions.

Example 1. *The potential in a box:* Consider a box of sides (a, b, c) with a potential Φ_0 on top, and $\phi = 0$ on the other five sides. Solve Laplace's equation $\nabla^2 \phi = 0$ to obtain the electrostatic potential everywhere in the box.

- In order that $\phi(0, y, z) = 0 = \phi(a, y, z)$, choose $\phi \propto \sin(k_n x)$ where $k_n = \pi n/a$. This function satisfies these boundary conditions.
- In order that $\phi(x, 0, z) = 0 = \phi(x, b, z)$, choose $\phi \propto \sin(q_m y)$ where $q_m = \pi m/b$. This function satisfies these boundary conditions.
- The equation satisfied by the z-coordinate is

$$0 = \left[\frac{\partial^2}{\partial z^2} - k_n^2 - q_m^2 \right] f(z) \tag{12.1}$$

$$f(z) = A \sinh(\lambda z) + B \cosh(\lambda z) \tag{12.2}$$

$$\lambda_{nm}^2 = k_n^2 + q_m^2 \tag{12.3}$$

The condition that $\phi(x, y, z = 0) = 0$ requires that $B = 0$.

The solution has the form

$$\phi(x, y, z) = \sum_{nm} A_{nm} \sin(k_n x) \sin(q_m y) \sinh(\lambda_{nm} z) \tag{12.4}$$

$$\Phi_0 = \sum_{nm} A_{nm} \sin(k_n x) \sin(q_m y) \sinh(\lambda_{nm} c) \tag{12.5}$$

The last equation is the boundary condition at the top of the box. It serves to determine the expansion constant A_{nm}. Multiply Eq. (12.5) by $\sin(k_{n'}x)\sin(q_{m'}y)$ and integrate over $dx\,dy$. The orthogonality of the sine functions picks out the term in the double summation with $n = n'$, $m = m'$ as the only nonzero term. The integral over the constant on the left gives zero when n' or m' are even, so they must be odd

$$\Phi_0 \frac{4}{k_{n'}q_{m'}} = \frac{ab}{4} A_{n'm'} \sinh(\lambda_{n'm'}c) \tag{12.6}$$

$$A_{nm} = \frac{16\Phi_0}{abk_n q_m} \frac{1}{\sinh(\lambda_{nm}c)} \tag{12.7}$$

$$\Phi(x, y, z) = \frac{16\Phi_0}{ab} \sum_{n\,\text{odd}} \sum_{m\,\text{odd}} \frac{1}{k_n q_m} \sin(k_n x) \sin(q_m y) \frac{\sinh(\lambda_{nm}z)}{\sinh(\lambda_{nm}c)} \tag{12.8}$$

which completes the derivation.

Example 2. *Green's function for diffusion:* Green's function in three dimensions for the diffusion equation is

$$\left[\nabla^2 - \frac{1}{D}\frac{\partial}{\partial t} \right] G(\mathbf{r}, \mathbf{r}_0; t) = -\delta(t)\delta^3(\mathbf{r} - \mathbf{r}_0) \tag{12.9}$$

The boundary conditions are chosen so that $G = 0$ on all six sides of the box. This problem naturally allows a product solution of the form

$$G(\mathbf{r}, \mathbf{r}_0; t) = D\Theta(t)X(x, x_0; t)Y(y, y_0; t)Z(z, z_0; t) \tag{12.10}$$

The various derivatives operating on this function are given below. A dot over a function denotes time derivative, while a subscript such as X_{xx} denotes a double derivative with respect to x,

$$\nabla^2 G = D\theta(t)[X_{xx}YZ + XY_{yy}Z + XYZ_{zz}] \tag{12.11}$$

$$\frac{1}{D}\frac{\partial}{\partial t}G = \delta(t)X(x, x_0; 0)Y(y, y_0; 0)Z(z, z_0; 0) + \Theta(t)[\dot{X}YZ + X\dot{Y}Z + XY\dot{Z}]$$

$$\tag{12.12}$$

The time derivative of the step function $\Theta(t)$ is a delta function $\delta(t)$. The delta function sets to zero the time arguments in the other functions. Putting together these various derivatives gives the formula

$$\left[\nabla^2 - \frac{1}{D}\frac{\partial}{\partial t} \right] G(\mathbf{r}, \mathbf{r}_0; t) = -\delta(t)X(x, x_0; 0)Y(y, y_0; 0)Z(z, z_0; 0)$$

$$+ \Theta(t)\left(YZ\left[X_{xx} - \frac{1}{D}\dot{X} \right] + XZ\left[Y_{yy} - \frac{1}{D}\dot{Y} \right] + XY\left[Z_{zz} - \frac{1}{D}\dot{Z} \right] \right)$$

The original diffusion equation is satisfied depending on

$$X_{xx} - \frac{1}{D}\dot{X} = 0, \qquad X(x, x_0; 0) = \delta(x - x_0) \qquad (12.13)$$

$$Y_{yy} - \frac{1}{D}\dot{Y} = 0, \qquad Y(y, y_0; 0) = \delta(y - y_0) \qquad (12.14)$$

$$Z_{zz} - \frac{1}{D}\dot{Z} = 0, \qquad Z(x, x_0; 0) = \delta(z - z_0) \qquad (12.15)$$

The three dimensional problem has been reduced to three one-dimensional problems that are alike. Futhermore, the one dimensional problems were solved in Chapter 10.

$$X(x, x_0; t) = \frac{2}{a}\sum_n e^{-tDk_n^2}\sin(k_n x)\sin(k_n x_0) \qquad (12.16)$$

$$Y(y, y_0; t) = \frac{2}{b}\sum_m e^{-tDq_m^2}\sin(q_m y)\sin(q_m y_0) \qquad (12.17)$$

$$Z(z, z_0; t) = \frac{2}{c}\sum_l e^{-tDp_l^2}\sin(p_l z)\sin(p_l z_0) \qquad (12.18)$$

where $k_n = \pi n/a$, $q_m = \pi m/b$, and $p_l = \pi l/c$.

12.2. Cylindrical Coordinates

For problems involving cylinders it is natural to use polar coordinates (ρ, θ) in the plane and to use z along the axis of the cylinder, in order that the spatial vector is $\mathbf{r} = (\rho, \theta, z)$. The expression for ∇^2 in this coordinate system is

$$\nabla^2 = \frac{\partial}{\partial\rho^2} + \frac{1}{\rho}\frac{\partial}{\partial\rho} + \frac{1}{\rho^2}\frac{\partial^2}{\partial\theta^2} + \frac{\partial^2}{\partial z^2} \qquad (12.19)$$

This expression will be used in a variety of problems. The eigenfunction relating to the variable θ must again be of the form $\exp(in\theta)$ where n is an integer. This choice is needed to satisfy the condition that $\phi(\rho, \theta + 2\pi, z) = \phi(\rho, \theta, z)$ which are the same point. The eigenfunction for ρ and z depend on the problem and boundary conditions.

Example 1. *Potential of a cylinder.* Solve for the electrostatic potential inside a cylinder of radius a. The z-direction spans the range $-d \leqslant z \leqslant d$ and the boundary conditions have the potential vanishing at both flat ends, or, $\phi(\rho, \theta, \pm d) = 0$. On the outside of the cylinder, the boundary conditions are

$$\phi(a, \theta, z) = \begin{cases} 0 & |z| > b \\ \Phi_0 & |z| < b \end{cases} \qquad (12.20)$$

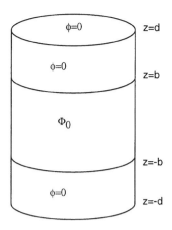

Figure 12.1

A sketch of the cylinder is shown in Fig. 12.1.The boundary conditions do not depend on the angle θ, nor do the solutions. The problem at hand is to solve

$$\left[\frac{\partial}{\partial\rho^2} + \frac{1}{\rho}\frac{\partial}{\partial\rho} + \frac{\partial^2}{\partial z^2}\right]\phi(\rho, z) = 0 \tag{12.21}$$

subject to the above boundary conditions.

In solving two dimensional Laplace's equation, it was remarked that if the functions in one direction, say x, were oscillatory such as $\sin(kx)$, $\cos(kx)$, then the solutions in the other direction were not oscillatory, $\sinh(ky)$, $\cosh(ky)$. The same idea applies to the present problem. If the dependence on z is oscillatory such as $\sin(kz)$, $\cos(kz)$ then the equation for the variable ρ is

$$\left[\frac{\partial}{\partial\rho^2} + \frac{1}{\rho}\frac{\partial}{\partial\rho} - k^2\right]\phi(\rho, z) = 0 \tag{12.22}$$

The solution to this equation are the modified Bessel functions $I_0(k\rho)$ and $K_0(k\rho)$ that are not oscillatory. Similarly, if the dependence on z is $\sinh(kz)$, $\cosh(kz)$, then the solutions for ρ are $J_0(k\rho)$, $Y_0(k\rho)$, that are oscillatory. An oscillatory solution in one direction means the other direction is not oscillatory.

In the present problem, the potential has to be zero at both ends, while not vanishing between. This boundary condition requires an oscillatory function of the form $\cos(k_\alpha z)$, $k_\alpha = \pi(2\alpha + 1)/2d$. The cosine function is chosen rather than the sine, since the boundary conditions at $\rho = a$ are symmetric. They are similar for $-z$ as for $+z$. Only a cosine function satisfies this requirement. The choice of k_α depends on $\cos(k_\alpha d) = 0$. Since there is an oscillatory function in the z-direction the radial solution is a modified Bessel function. Because K_0 diverges at the origin this function is not allowed. These arguments bring us to the form of the solution

$$\phi(\rho, z) = \sum_\alpha B_\alpha \cos(k_\alpha z)I_0(k_\alpha \rho) \tag{12.23}$$

Each term $\cos(k_\alpha z)I_0(k_\alpha \rho)$ satisfies Laplace's equation. The expansion constants B_α are determined by the boundary conditions outside the cylinder. Set $\rho = a$, and use the orthogonality of the functions $\cos(k_\alpha z)$. Multiply the equation by $\cos(k_m z)$ and integrate

$$\phi(a, z) = \sum_n B_\alpha \cos(k_\alpha z)I_0(k_\alpha a) \tag{12.24}$$

$$\int_{-d}^{d} dz \cos(k_\beta z) \left\{ \phi(a, z) = \sum_n B_n \cos(k_\alpha z)I_0(k_\alpha a) \right\} \tag{12.25}$$

$$dB_\alpha I_0(k_\alpha a) = \Phi_0 \int_{-b}^{b} dz \cos(k_\alpha z) = \frac{2\Phi_0}{k_\alpha} \sin(k_\alpha b) \tag{12.26}$$

$$B_\alpha = \frac{2\Phi_0}{dk_\alpha} \frac{\sin(k_\alpha b)}{I_0(k_\alpha a)} \tag{12.27}$$

$$\phi(\rho, z) = 2\Phi_0 \sum_{\alpha=0}^{\infty} \cos(k_\alpha z)\frac{\sin(k_\alpha b)}{k_\alpha d} \frac{I_0(k_\alpha \rho)}{I_0(k_\alpha a)} \tag{12.28}$$

this completes the solution.

Example 2. *Wave equation in a closed cylinder:* Consider a hollow cylinder that has a radius a and length L ($0 \leqslant z \leqslant L$). Find the allowed frequencies for scalar oscillations inside the cylindrical cavity,

$$\left[\frac{\partial^2}{\partial t^2} - v^2 \nabla^2 \right] \phi(\mathbf{r}, t) = 0 \tag{12.29}$$

The eigenfrequency is ω, and the Helmholtz equation is solved for the normal modes with $k = \omega/v$. The boundary condition is $\phi = 0$ on all boundaries. The eigenfunction has the form

$$\phi = e^{in\theta} J_n(\lambda_{n\alpha}\rho/a) \sin(k_m z)e^{-i\omega t} \tag{12.30}$$

Three dimensions require three independent eigenvalues, which are (α, m, n). The Bessel function is chosen for the solution in the ρ-variable since it vanishes at the walls of the cylinder. The function $\sin(k_m z)$ is chosen since it makes $\phi = 0$ at the ends $z = 0, L$ when $k_m = \pi m/L$. Putting this ansatz into the wave equation gives

$$\frac{\omega^2}{v^2} = k_m^2 + \left(\frac{\lambda_{n\alpha}}{a}\right)^2 \tag{12.31}$$

$$\omega_{n\alpha m} = v\left[k_m^2 + \left(\frac{\lambda_{n\alpha}}{a}\right)^2 \right]^{1/2} \tag{12.32}$$

which are the allowed frequencies. The results depend on the zeros of the Bessel functions given by $\lambda_{n\alpha}$.

Example 3. *Cylindrical wave guide:* Consider a scalar wave going down the interior of an infinitely long cylinder with a radius a. What is the velocity of the wave? The boundary condition is that the wave vanishes at the surface of the cylinder.

The solutions to this problem are similar to those of the previous problem. Now there is a traveling wave in the z direction, in order that the eigenfunctions have the form

$$\phi(\rho, \theta, z) \sim e^{in\theta + ipz - i\omega t} J_n(\lambda_{n\alpha}\rho/a) \tag{12.33}$$

$$\nabla^2\phi = -\frac{\omega^2}{v^2}\phi = -\left[p^2 + \left(\frac{\lambda_{n\alpha}}{a}\right)^2\right]\phi \tag{12.34}$$

$$\omega_{n\alpha}(p) = v\left[p^2 + \left(\frac{\lambda_{n\alpha}}{a}\right)^2\right]^{1/2} \tag{12.35}$$

$$v_{n\alpha}(p) = \frac{d\omega}{dp} = \frac{vp}{\sqrt{p^2 + (\lambda_{n\alpha}/a)^2}} \tag{12.36}$$

The velocity is linear in p at small p and goes to a constant v at large wave vector.

Example 4. *Circular boundary potential:* The system is the upper-half space, that includes the entire (x, y) plane and the positive values of $z(z \geqslant 0)$. Find the electrostatic potential in the UHP if the potential on the $z = 0$ plane is

$$\phi(\rho, 0) = \begin{cases} \phi_0, & \rho < a \\ 0, & \rho > a \end{cases} \tag{12.37}$$

The problem has circular symmetry so it is natural to use cylindrical coordinates. There is no dependence in the boundary conditions on the angle θ so the solution has $n = 0$. Laplace's equation in the remaining variables is

$$0 = \nabla^2\phi = \left[\frac{\partial^2}{\partial\rho^2} + \frac{1}{\rho}\frac{\partial}{\partial\rho} + \frac{\partial^2}{\partial z^2}\right]\phi(\rho, z) \tag{12.38}$$

The problem is a natural for the Hankel transform. Multiply the above equation by $\rho d\rho J_0(k\rho)$ and integrate over all positive ρ. The Hankel transform of the derivative operator is evaluated by a double integration by parts

$$\bar{\phi}(k, z) = \int_0^\infty \rho d\rho J_0(k\rho)\phi(\rho, z) \tag{12.39}$$

$$\int_0^\infty \rho d\rho J_0(k\rho)\left[\frac{1}{\rho}\frac{\partial}{\partial\rho}\left(\rho\frac{\partial\phi(\rho, z)}{\partial\rho}\right)\right] = -\int_0^\infty \rho d\rho \left(\frac{\partial J_0(k\rho)}{\partial\rho}\right)\left(\frac{\partial\phi(\rho, z)}{\partial\rho}\right)$$

$$= \int_0^\infty \rho d\rho \phi(\rho, z)\left[\frac{1}{\rho}\frac{\partial}{\partial\rho}\left(\rho\frac{\partial J_0(k\rho)}{\partial\rho}\right)\right]$$

$$= -k^2 \int_0^\infty \rho d\rho \phi(\rho, z) J_0(k\rho)$$

$$= -k^2\bar{\phi}(k, z) \tag{12.40}$$

The Hankel transform of the entire equation is

$$\left[\frac{\partial^2}{\partial z^2} - k^2\right]\bar{\phi}(k, z) = 0 \tag{12.41}$$

$$\bar{\phi}(k, z) = \bar{\phi}(k, 0)e^{-kz} \tag{12.42}$$

The familiar differential Eq. (12.41) has two solutions of the form $\exp(\pm kz)$. The solution $\exp(+kz)$ is discarded since it diverges at large z. The only remaining problem is to find the function $\bar{\phi}(k, 0)$. It is the Hankel transform of the boundary condition

$$\bar{\phi}(k, 0) = \int_0^\infty \rho d\rho J_0(k\rho)\phi(\rho, 0) \tag{12.43}$$

$$= \phi_0 \int_0^a \rho d\rho J_0(k\rho) = \frac{a\phi_0}{k} J_1(ka)$$

$$\phi(\rho, z) = a\phi_0 \int_0^\infty dk J_0(k\rho)J_1(ka)e^{-kz} \tag{12.44}$$

The final result is given in the last equation in terms of the inverse transform. Apparently this integral does not appear in any table in simple form and must be evaluated numerically. A simple answer is obtained in the limit of large $z \gg a$. The exponential factor guarantees that the important contributions to the integral are $k \sim z^{-1}$. Then the argument of the J_1 Bessel function is $ka = (kz)(a/z) \sim (a/z) \ll 1$. This Bessel function can be expanded in a series

$$J_1(z) = \frac{z}{2}\left[1 - \frac{z^2}{8} + O(z^4)\right] \tag{12.45}$$

$$\phi(\rho, z) = \frac{a^2}{2}\phi_0 \int_0^\infty k dk J_0(k\rho)e^{-kz}\left[1 - \frac{k^2a^2}{8} + \cdots +\right]$$

$$= \frac{a^2}{2}\phi_0\left[\frac{z}{r^3} - \frac{3a^2}{8}\frac{5z^3 - 3zr^2}{r^7} + \cdots +\right] \tag{12.46}$$

where $r = \sqrt{z^2 + \rho^2}$ and the integral is in [G and R 6.621(4)]. Write $z = r\cos(\theta)$ and the terms in the above series become

$$\phi(\rho, z) = \frac{\phi_0}{2}\left[\left(\frac{a}{r}\right)^2 P_1(\theta) - \frac{3}{4}\left(\frac{a}{r}\right)^4 P_3(\theta) + \cdots +\right] \tag{12.47}$$

The solution to Laplace's equation has the usual form of a summation over n for the function $P_n(\theta)r^{-n-1}$, where P_n are Legendre polynomials, $P_1(\theta) = \cos(\theta)$, $P_3 = [5\cos^3(\theta) - 3\cos(\theta)]/2$. The expansion is valid for $r \gg a$.

Example 5. *Diffusion in a cylinder:* Consider an infinitely long cylinder of radius a. Find Green's function for the diffusion equation in the cylinder, assuming that it vanishes at the surface of the cylinder. The points $(\mathbf{r}, \mathbf{r}_0)$ are within the cylinder

$$\left[D\nabla^2 - \frac{\partial}{\partial t} \right] G(\mathbf{r}, \mathbf{r}_0; t) = -\delta(t)\delta(\mathbf{r} - \mathbf{r}_0) \tag{12.48}$$

The radial functions are going to be $J_n(\lambda_{na}\rho/a)$ that vanish at the value $\rho = a$. The angular functions are going to be $\cos[n(\theta - \theta_0)]$. The dependence upon z is uncertain, so Fourier transform this variable to q. The transformed equation and its general solution are

$$-\delta(t)e^{-iqz_0}\delta^2(\boldsymbol{\rho} - \boldsymbol{\rho}_0) = \left[D(\nabla_\perp^2 - q^2) - \frac{\partial}{\partial t} \right] \bar{G}(\rho, \rho_0, q, z_0; t) \tag{12.49}$$

$$\bar{G}(\rho, \rho_0, q, z_0; t) = \Theta(t)e^{-iqz_0} \sum_{n\alpha} A_{n\alpha} \cos[n(\theta - \theta_0)]e^{-tD(q^2 + \lambda^2/a^2)}$$
$$\times J_n(\lambda\rho/a)J_n(\lambda\rho_0/a) \tag{12.50}$$

Insert this ansatz into the differential equation where there is a term with a factor of $\delta(t)$ equated to the delta function of radial position

$$\delta^2(\boldsymbol{\rho} - \boldsymbol{\rho}_0) = \sum_{n\alpha} A_{n\alpha} \cos[n(\theta - \theta_0)]J_n(\lambda\rho/a)J_n(\lambda\rho_0/a) \tag{12.51}$$

This identity determines the constant $A_{n\alpha}$, that can be found from two prior results for the delta functions of angle and radial variable

$$2\pi\delta(\theta - \theta_0) = 2 \sum_{n=0}^{\infty} \varepsilon_n \cos[n(\theta - \theta_0)] \tag{12.52}$$

$$\frac{\delta(\rho - \rho_0)}{\rho} = \frac{2}{a^2} \sum_\alpha \frac{J_n(\lambda\rho/a)J_n(\lambda\rho_0/a)}{|J_n'(\lambda)|^2} \tag{12.53}$$

$$A_{n\alpha} = \frac{2\varepsilon_n}{\pi a^2 |J_n'(\lambda)|^2} \tag{12.54}$$

$$\varepsilon_n = \begin{cases} 1 & n \geqslant 1 \\ 1/2 & n = 0 \end{cases} \tag{12.55}$$

The symbol λ always means λ_{na}. The summation over α gives $\delta(\rho - \rho_0)/\rho$ while the summation over n gives $\delta(\theta - \theta_0)$. The remaining step in the derivation is to evaluate the inverse Fourier transform to obtain the dependence on z

$$G(\mathbf{r}, \mathbf{r}_0; t) = \frac{2\Theta(t)}{\pi a^2} \int \frac{dq}{2\pi} e^{iq(z-z_0)} \sum_{n\alpha} \varepsilon_n e^{-tDq^2} \cos[n(\theta - \theta_0)]e^{-tD\lambda^2/a^2}$$
$$\times \frac{J_n(\lambda\rho/a)J_n(\lambda\rho_0/a)}{|J_n'(\lambda)|^2}$$
$$= \frac{\Theta(t)e^{-(z-z_0)^2/4Dt}}{\pi a^2\sqrt{\pi Dt}} \sum_{n\alpha} \varepsilon_n \cos[n(\theta - \theta_0)]e^{-tD\lambda^2/a^2} \tag{12.56}$$
$$\times \frac{J_n(\lambda\rho/a)J_n(\lambda\rho_0/a)}{|J_n'(\lambda)|^2}$$

Note that in the z direction there is ordinary one dimensional diffusion. In the radial direction there is the same answer as for diffusion in a circle. The answer factors into

$$G(\mathbf{r}, \mathbf{r}_0; t) = \Theta(t)G_1(z - z_0, t)G_2(\boldsymbol{\rho}, \boldsymbol{\rho}_0; t) \tag{12.57}$$

$$G_1(z, z_0; t) = \frac{e^{-(z-z_0)^2/4Dt}}{2\sqrt{\pi Dt}} \tag{12.58}$$

$$G_2(\boldsymbol{\rho}, \boldsymbol{\rho}_0; t) = \frac{2}{a^2} \sum_{n\alpha} \varepsilon_n \cos[n(\theta - \theta_0)]e^{-tD\lambda^2/a^2}$$
$$\times \frac{J_n(\lambda\rho/a)J_n(\lambda\rho_0/a)}{|J_n'(\lambda)|^2} \tag{12.59}$$

The two factor are the one dimensional Green's function for the z direction, and the two dimensional Green's function in the radial and angular directions. Each factor satisfies the diffusion equation for its variables.

12.3. Spherical Coordinates

Many problems in three dimensions are usefully solved in spherical coordinates, which are $\mathbf{r} = (r, \theta, \phi)$

$$x = r\sin(\theta)\cos(\phi) \tag{12.60}$$
$$y = r\sin(\theta)\sin(\phi) \tag{12.61}$$
$$z = r\cos(\theta) \tag{12.62}$$

$$\nabla^2\Phi = \left[\mathscr{L}_r + \frac{1}{r^2}\mathscr{L}(\theta, \phi)\right]\Phi(r, \theta, \phi) \tag{12.63}$$

$$\mathscr{L}_r\Phi = \frac{1}{r^2}\frac{\partial}{\partial r}\left(r^2\frac{\partial\Phi}{\partial r}\right) = \frac{1}{r}\frac{\partial^2}{\partial r^2}(r\Phi) = \left[\frac{\partial^2}{\partial r^2} + \frac{2}{r}\frac{\partial}{\partial r}\right]\Phi \tag{12.64}$$

$$\mathscr{L}(\theta, \phi)Y(\theta, \phi) = \frac{1}{\sin(\theta)}\frac{\partial}{\partial\theta}\left(\sin(\theta)\frac{\partial Y}{\partial\theta}\right) + \frac{1}{\sin^2\theta}\frac{\partial^2 Y}{\partial\phi^2} \tag{12.65}$$

Three versions of the derivative with respect to r are provided and they are all equivalent. Different books have different choices among these three versions.

The eigenfunctions are assumed to be the product of a function of the three separate variables. Write it as $\Phi = R(r)Y(\theta, \phi)$. Laplace's equation can be written as

$$0 = \frac{1}{\Phi}\nabla^2\Phi = \frac{1}{R}\mathscr{L}_r R + \frac{1}{r^2 Y}\mathscr{L}(\theta, \phi)Y(\theta, \phi) \tag{12.66}$$

$$0 = \frac{r^2}{R}\mathscr{L}_r R + \frac{1}{Y}\mathscr{L}(\theta, \phi)Y(\theta, \phi) \tag{12.67}$$

The two terms on the right of the equal sign must add to zero. The first term depends only on the variable r and the second term depends only on the variables (θ, ϕ). How can they add to zero? The only way is if they are each a constant. They obey the separate equations

$$0 = \left[\mathscr{L}_r - \frac{C}{r^2} \right] R \tag{12.68}$$

$$0 = [\mathscr{L}(\theta, \phi) + C] Y(\theta, \phi) \tag{12.69}$$

where the same constant C is in both equations. This separation of the problem into two separate equations is required when the eigenfunctions have the form of $\Phi = RY$.

The angular functions are discussed first. The azimuthal angle ϕ has the range of variables $0 \leqslant \phi \leqslant 2\pi$. The angular function must be periodic in this variable such that $Y(\theta, \phi + 2\pi) = Y(\theta, \phi)$ which forces the eigenfunction to have the form of $\exp(im\phi)$, where m is an integer. Writing $Y = e^{im\phi} f(\theta)$, the differential equation for the polar angle is

$$\mathscr{L}(\theta, \phi) e^{im\phi} f(\theta) = e^{im\phi} \left[\frac{1}{\sin(\theta)} \frac{\partial}{\partial \theta} \left(\sin(\theta) \frac{\partial f}{\partial \theta} \right) - \frac{m^2}{\sin^2\theta} f \right] \tag{12.70}$$

The operator on the right-hand side has eigenvalues $-l(l + 1)$ and eigenfunction in terms of Legendre polynomials P and Q

$$-l(l + 1) f = \frac{1}{\sin(\theta)} \frac{\partial}{\partial \theta} \left(\sin(\theta) \frac{\partial f}{\partial \theta} \right) - \frac{m^2}{\sin^2\theta} f \tag{12.71}$$

$$f = A P_l^{|m|}(\theta) + B Q_l^{|m|}(\theta) \tag{12.72}$$

with the constraint that l is a nonnegative integer and $-l \leqslant m \leqslant l$. For each l there are $(2l + 1)$ allowed values of m. Since the differential equation is second order, there must be two independent solutions that are P and Q. The solutions $Q_l^{|m|}(\theta)$ diverge for the allowed angles $0 \leqslant \theta \leqslant \pi$, so this solution is usually discarded. The usual case is to retain only the other function $P_l^{|m|}(\theta)$ which is quite well-behaved. The orthogonalized function for the variables (θ, ϕ) are called *spherical harmonics*

$$Y_{lm}(\theta, \phi) = N_{lm} e^{im\phi} P_l^{|m|}(\theta) \tag{12.73}$$

$$N_{lm}^2 = \frac{2l + 1}{4\pi} \frac{(l - |m|)!}{(l + |m|)!} \tag{12.74}$$

$$\mathscr{L}(\theta, \phi) Y_{lm}(\theta, \phi) = -l(l + 1) Y_{lm}(\theta, \phi) \tag{12.75}$$

Properties of these functions are reviewed in the appendix.

12.3.1. Laplace's Equation

Laplace's equation in spherical coordinates has the general solution $\phi(\mathbf{r}) = f(r) Y_{lm}(\theta, \phi)$ and the differential equation for the variable r is

$$\nabla^2 \phi(\mathbf{r}) = 0 \tag{12.76}$$

$$\left[\frac{1}{r} \frac{d^2}{dr^2} r - \frac{l(l + 1)}{r^2} \right] f(r) = 0 \tag{12.77}$$

The equation is simplified by multipling every term by r and then the differential equation is on the variable $rf(r)$

$$\left[\frac{d^2}{dr^2} - \frac{l(l+1)}{r^2}\right][rf(r)] = 0 \tag{12.78}$$

Since both term in the above equation have the same power of r the solution must be a power law such as $rf \propto r^s$. Inserting this assumption in the above equation produces the relation $s(s-1) = l(l+1)$ that has two solutions $s = -l, l+1$. So the most general solution to Laplace's equation in spherical coordinates is

$$\phi(\mathbf{r}) = \sum_{lm}\left[A_{lm}r^l + \frac{B_{lm}}{r^{l+1}}\right]Y_{lm}(\theta, \phi) \tag{12.79}$$

This equation is used in several examples.

Example 1. *Temperature in a shell:* The system is a spherical shell $a < r < b$. The temperature on the inner surface is $T(r = a) = T_a$ and on the outer surface is $T(r = b) = T_b$. Find the temperature $T(r)$ inside the shell.

The steady state temperature profile is given by the solutions to Laplace's equation. Since the boundary conditions have no dependence on angle, neither does the solution. Then $l = 0 = m$. The general solution must have the form

$$T(r) = A + \frac{B}{r} \tag{12.80}$$

Using the two boundary conditions at a and b gives two equations for the constants A and B.

$$T_a = A + \frac{B}{a} \tag{12.81}$$

$$T_b = A + \frac{B}{b} \tag{12.82}$$

Solving them gives the final solution

$$A = \frac{bT_b - aT_a}{b-a} \tag{12.83}$$

$$B = -\frac{ab}{b-a}(T_b - T_a) \tag{12.84}$$

$$T(r) = \frac{1}{r(b-a)}\left[a(b-r)T_a + b(r-a)T_b\right] \tag{12.85}$$

Example 2. *Potential in a sphere:* The system is the interior of a sphere of radius a. The potential on the surface is $\phi(a, \theta) = \phi_0 \cos(\theta)$. What is the potential inside the sphere?

The angular dependence is $\cos(\theta)$ which is an angular function with $l = 1$, $m = 0$. The radial solutions must be for $l = 1$. The solutions must be well-behaved at $r = 0$ which eliminates the term with B_1/r^2, so the radial part must go as $f = Ar$. Setting this equal to ϕ_0 on the surface gives $A = \phi_0/a$ and the solution

$$\phi(r, \theta) = \phi_0 \left(\frac{r}{a}\right) \cos(\theta) \tag{12.86}$$

Example 3. *Charge outside grounded sphere:* The problem is to find the electrostatic potential for a system with two sources of charge, (a) a grounded conductor of radius a centered at the origin and (b) a point charge Q at a distance $d > a$ from the origin. Another way to state the problem is that it is the solution to

$$\nabla^2 \Phi(\mathbf{r}) = -4\pi Q \delta(\mathbf{r} - \mathbf{r}_0) \tag{12.87}$$

where $\mathbf{r}_0 = d\hat{z}$ subject to the boundary condition that $\Phi(a, \theta) = 0$. The source charge Q has been put along the $\hat{z}$ axis and we can put it anywhere we choose. Selecting the z axis simplifies the analysis by eliminating any dependence on the azimuthal angle ϕ so only solutions with $m = 0$ are required.

There are two sources of potential, (a) charge Q and (b) sphere. The principle of superposition states that each makes a separate contribution to the potential, $\Phi = \Phi_Q + \Phi_S$. The solution from the charge Q is merely the potential from a point charge that is

$$\Phi_Q(\mathbf{r}) = \frac{Q}{|\mathbf{r} - \mathbf{r}_0|} = \frac{Q}{\sqrt{r^2 + d^2 - 2rd\cos\theta}} \tag{12.88}$$

Examine this solution near to the surface of the sphere. Expand it in spherical harmonics for $a < r < d$ using one of the generating series for spherical harmonics for $x < 1$ where $x = r/d$

$$\Phi_Q(r, \theta) = \frac{Q}{d} \frac{1}{\sqrt{1 + x^2 - 2x\cos\theta}} = \frac{Q}{d} \sum_n \left(\frac{r}{d}\right)^n P_n(\theta) \tag{12.89}$$

The potential Φ_S must cancel this series when $r = a$. It has the generic form for $m = 0$ of

$$\Phi_S = \frac{Q}{d} \sum_n \left[A_n r^n + \frac{B_n}{r^{n+1}}\right] P_n(\theta) \tag{12.90}$$

$$\Phi = \frac{Q}{d} \sum_n \left[A_n r^n + \frac{B_n}{r^{n+1}} + \left(\frac{r}{d}\right)^n\right] P_n(\theta) \tag{12.91}$$

The potential from the sphere must vanish as $r \to \infty$, which requires that $A_n = 0$. Since the potential at $r = a$ must vanish for all angles θ, then each coefficient of $P_n(\theta)$ must vanish separately. This requirement gives

$$B_n = -a^{n+1} \left(\frac{a}{d} \right)^n \tag{12.92}$$

$$\Phi(r, \theta) = \frac{Q}{d} \sum_n \left[\left(\frac{r}{d} \right)^n - \frac{a}{r} \left(\frac{a^2}{rd} \right)^n \right] P_n(\theta) \tag{12.93}$$

The above solution is valid for $a < r < d$. It can be expressed in terms of an image charge for the sphere. The first term above can be resummed back to the original formula that $\Phi_Q = Q/|\mathbf{r} - \mathbf{r}_0|$. The second term can be summed in a similar fashion. In the second term, from the sphere, let $y = a^2/(rd) < 1$ and this term can be summed to

$$\Phi_S = -\frac{aQ}{dr} \frac{1}{\sqrt{1 + y^2 - 2y \cos \theta}} = \frac{Q_I}{|\mathbf{r} - \mathbf{r}_I|} \tag{12.94}$$

$$Q_I = -\frac{aQ}{d}, \quad \mathbf{r}_I = \frac{a^2}{d} \hat{z} \tag{12.95}$$

The potential from the sphere can be represented, outside the sphere, as a potential from an image charge of Q_I located at the point $\mathbf{r}_I$. The charge has come from the ground to reside on the surface of the conductor and this charge distribution creates the potential Φ_S, which can be represented as an image charge.

12.3.2. Diffusion and Wave Equations

Here are some solutions to the diffusion and wave equations.

Example 1. *Heating outside a sphere:* Solve the diffusion equation to find the temperature distribution $T(r, t)$ for the space outside a sphere ($r > a$), where the temperature everywhere is $T(r, t < 0) = 0$ and at $t > 0$ the surface is put in contact with a reservoir that maintains its temperature at $T(a, t > 0) = T_0$.

The first step is to find the steady state temperature distribution in the space $r > a$ for very long periods. The steady state temperature is a solution to Laplace's equation $\nabla^2 T = 0$ with the boundary condition $T(r = a) = T_0$. Since the boundary conditions are independent of angle, so is the solution. The problem has $l = 0 = m$ and the steady state solution from Eq. (12.79) for $l = 0$ is

$$T(r) = T(\infty) + \frac{a}{r} T_0 \tag{12.96}$$

The space outside the sphere does not reach the same temperature T_0. Instead, it goes to the value $T(\infty)$ at $r = \infty$ which is set equal to zero. The steady state solution is similar to the problem in electrostatics which also has $\nabla^2 \phi = 0$. If there is a sphere of radius a with a charge Q then the potential is $\phi(r) = Q/r$. The potential at the surface is $\phi(a) = Q/a$, so that $\phi(r) = \phi(a)a/r$, which has the same form as the temperature solution.

The diffusion equation is best solved using a Laplace transform on the time variable. The transformed temperature is $\bar{T}(r, p)$ and obeys the equation

$$\left[\frac{1}{r}\frac{d^2}{dr^2}r - \frac{p}{D}\right]\bar{T}(r, p) = 0 \tag{12.97}$$

$$\left[\frac{d^2}{dr^2} - \frac{p}{D}\right][r\bar{T}(r, p)] = 0 \tag{12.98}$$

$$r\bar{T}(r, p) = a\bar{T}(a, p)\exp\left[-(r - a)\sqrt{\frac{p}{D}}\right] \tag{12.99}$$

The last needed information is $\bar{T}(a, p)$ which is the Laplace transform of the boundary condition

$$\bar{T}(a, p) = \frac{T_0}{p} \tag{12.100}$$

$$\bar{T}(r, p) = T_0\frac{a}{rp}\exp\left[-(r - a)\sqrt{\frac{p}{D}}\right] \tag{12.101}$$

$$T(r, t) = T_0\frac{a}{r}\int\frac{dp}{p2\pi i}e^{pt}\exp\left[-(r - a)\sqrt{\frac{p}{D}}\right] \tag{12.102}$$

The final answer is obtained by evaluating the contour integral which results from the inverse transform. There is a pole at $p = 0$ and a branch cut down the negative real axis. Exactly the same integral was evaluated in Chapter 11 for the problem of heat flow down a bar that is heated at one end. Using that earlier evaluation, the result is expressed in terms of a complimentary error function

$$T(r, t) = T_0\frac{a}{r}\text{erfc}\left[\frac{r - a}{\sqrt{4Dt}}\right] \tag{12.103}$$

The solution is quite similar to the one dimensional problem of heat flow down a bar that is heated at one end. The 1D problem has x instead of $r - a$, and lacked the prefactor of (a/r). The solution to (rT) has the same form as the 1D result, with the boundary condition (aT_0) at $r = a$. The analogy between 1D and 3D occurs because the radial derivative can be written as in Eq. (12.97) that makes it a simple double derivative on the function (rT). The two dimension problem cannot be written this way, and its solution is totally different.

Example 2. *Green's function for wave equation:* Green's function for the wave equation in three dimensional spherical coordinates is the solution to the equation

$$\left[\nabla^2 - \frac{1}{c^2}\frac{\partial^2}{\partial t^2}\right]G(\mathbf{r}, \mathbf{r}_0, t) = -\delta(t)\delta(\mathbf{r} - \mathbf{r}_0) \tag{12.104}$$

The differential equation is solved using the three dimensional Fourier transform on the **r** variable

$$\bar{G}(\mathbf{q}, \mathbf{r}_0, t) = \int d^3 r\, e^{-i\mathbf{q}\cdot\mathbf{r}} G(\mathbf{r}, \mathbf{r}_0, t) \tag{12.105}$$

$$\left[q^2 + \frac{1}{c^2} \frac{\partial^2}{\partial t^2} \right] \bar{G}(\mathbf{q}, \mathbf{r}_0, t) = \delta(t)\, \exp[-i\mathbf{q}\cdot\mathbf{r}_0] \tag{12.106}$$

The two solutions to the differential equation in time are $\sin(qct)$, $\cos(qct)$. The sine is the right answer since it gives the correct delta function for the second derivative

$$\bar{G}(\mathbf{q}, \mathbf{r}_0, t) = A\Theta(t)\, \sin(qct) \tag{12.107}$$

Taking two time derivatives of this expression gives

$$\frac{1}{c^2} \frac{\partial^2}{\partial t^2} \bar{G}(\mathbf{q}, \mathbf{r}_0, t) = -q^2 A\Theta(t) \sin(qct) + A \frac{q}{c} \delta(t) \tag{12.108}$$

Compare this result with the original equation (12.106) and conclude that

$$A = \frac{c}{q} \exp[-i\mathbf{q}\cdot\mathbf{r}_0] \tag{12.109}$$

$$G(\mathbf{r}, \mathbf{r}_0, t) = c\Theta(t) \int \frac{d^3 q}{(2\pi)^3} \frac{\sin(qct)}{q} \exp[i\mathbf{q}\cdot(\mathbf{r} - \mathbf{r}_0)] \tag{12.110}$$

The integrals are easy $(\mathbf{R} = \mathbf{r} - \mathbf{r}_0)$

$$\int d\Omega_q e^{i\mathbf{q}\cdot\mathbf{R}} = 2\pi \int_{-1}^{1} dv\, e^{iqRv} = 4\pi \frac{\sin(qR)}{qR} \tag{12.111}$$

$$\int_0^\infty dq \sin(qct) \sin(qR) = \frac{\pi}{2} [\delta(R - ct) - \delta(R + ct)] \tag{12.112}$$

$$G(\mathbf{r}, \mathbf{r}_0, t) = \frac{c\Theta(t)}{4\pi R} \delta(R - ct) \tag{12.113}$$

The delta function $\delta(R + ct)$ is not relevent since the argument is never zero when R and t are always positive.

The solution is a soliton. The wave pulse travels outward and is a delta function at the point $R = ct$. Similar behavior was found in one dimension where the solution was also a soliton. A different behavior is found in two dimensions where the solution is not a soliton but the pulse has trailing edge. The behavior of one and three dimensions are similar while that of two dimensions is different.

Example 3. *Wave from a moving source:* Consider the waves emanating from a source moving with the velocity u in the z direction. Assume the source has a smaller velocity than the wave $(u < v)$

$$\left[v^2 \nabla^2 - \frac{\partial^2}{\partial t^2} \right] \phi(\mathbf{r}, t) = -\delta(x)\delta(y)\delta(z - ut) \tag{12.114}$$

The problem is similar to Green's function, but it lacks the $\delta(t)$ on the right side of the equation since the source emits continuously.

The first step is to Fourier transform the space variables by multiplying the equation by $\exp(-i\mathbf{k} \cdot \mathbf{r})$ and integrating over all space

$$\left[v^2 k^2 + \frac{\partial^2}{\partial t^2} \right] \bar{\phi}(\mathbf{k}, t) = e^{-ik_z ut} \tag{12.115}$$

$$\bar{\phi}(\mathbf{k}, t) = \frac{e^{-ik_z ut}}{v^2 k^2 - k_z^2 u^2} + A \cos(kvt) + B \sin(kvt) \tag{12.116}$$

The solutions to the homogeneous equation are not needed so set $A = 0 = B$.

$$\begin{aligned} \phi(\mathbf{r}, t) &= \int \frac{d^3 k}{(2\pi)^3} \frac{e^{i\mathbf{k} \cdot \mathbf{r} - ik_z ut}}{v^2 k^2 - k_z^2 u^2} \\ &= \frac{1}{v^2} \int \frac{d^3 k}{(2\pi)^3} \frac{e^{i\mathbf{k} \cdot \mathbf{r} - ik_z ut}}{k_x^2 + k_y^2 + \gamma^2 k_z^2} \end{aligned} \tag{12.117}$$

where $\gamma^2 = 1 - u^2/v^2$. Rescale the coordinate system where $k_x' = k_x$, $k_y' = k_y$, $k_z' = \gamma k_z$, $x' = x$, $y' = y$, $z' = (z - ut)/\gamma$. Then the integral becomes

$$\begin{aligned} \phi(\mathbf{r}, t) &= \frac{1}{v^2 \gamma} \int \frac{d^3 k'}{(2\pi)^3} \frac{e^{i\mathbf{k}' \cdot \mathbf{r}'}}{(k')^2} = \frac{1}{4\pi v^2 \gamma r'} \\ &= \frac{1}{4\pi v^2} \frac{1}{\sqrt{\gamma^2(x^2 + y^2) + (z - ut)^2}} \end{aligned} \tag{12.118}$$

There is a compression of the wave by γ in the perpendicular direction. The supersonic case has $u > v$. Also note that if the source is not moving $(u = 0)$ the equation reverts to Laplace's equation with a solution $\phi = 1/(4\pi v^2 r)$.

Assume the wave has a frequency ω and the time dependence is $\cos(\omega t)$, then the wave equation becomes the Helmholtz equation with a wave vector $k = \omega/v$

$$[\nabla^2 + k^2]\phi(\mathbf{r}) = 0 \tag{12.119}$$

In spherical coordinates, the angular functions are spherical harmonics $Y_{lm}(\theta, \phi)$ giving a radial equation of the form

$$0 = \left[\frac{1}{r} \frac{d^2}{dr^2}(rR(r)) - \frac{l(l+1)}{r^2} R + k^2 R \right] \tag{12.120}$$

$$R_l(kr) = A_l j_l(kr) + B_l y_l(kr) \tag{12.121}$$

The solutions to the radial wave equation are called *spherical Bessel functions.* They are proportional to the regular Bessel and Neumann functions of half

$$j_l(z) = \sqrt{\frac{\pi}{2z}} J_{l+1/2}(z) \tag{12.122}$$

$$y_l(z) = \sqrt{\frac{\pi}{2z}} Y_{l+1/2}(z) \tag{12.123}$$

$$h_l^{(1)}(z) = j_l(z) + iy_l(z) \tag{12.124}$$

The last equation defines the spherical Hankel function that describes the outgoing wave solution. These functions appear to be complicated because of the half integer argument. The converse is true, they are simple functions that are sines and cosines multiplied by inverse polynomials

$$j_0(z) = \frac{\sin(z)}{z}, \quad y_0(z) = -\frac{\cos(z)}{z} \tag{12.125}$$

$$j_1(z) = \frac{\sin(z)}{z^2} - \frac{\cos(z)}{z}, \quad y_1(z) = -\frac{\cos(z)}{z^2} - \frac{\sin(z)}{z} \tag{12.126}$$

More properties of these functions are given in the Appendix.

Example 4. *Scattering by a sphere:* A plane wave is incident along the z axis on to a sphere of radius $r = a$. Assume the amplitude of the wave is zero at the surface of the sphere and has a frequency ω and a vector $k = \omega/c$. The solutions have the form

$$\psi = e^{ikz} + \psi_S \tag{12.127}$$

$$e^{ikz} = e^{ikr\cos(\theta)} = \sum_l (2l+1)i^l P_l(\theta) j_l(kr) \tag{12.128}$$

$$\psi_S = \sum_l (2l+1)i^l P_l(\theta) A_l h_l^{(1)}(kr) \tag{12.129}$$

$$\psi = \sum_l (2l+1)i^l P_l(\theta)[j_l(kr) + A_l h_l^{(1)}(kr)] \tag{12.130}$$

The scattered wave ψ_S must consist of outgoing waves that are the spherical Hankel functions $h_l^{(1)}$. The dependence on m was eliminated by having the incident particle come along the z axis. The boundary condition that the wave vanish at $r = a$ is true for all angles θ, which means that each coefficient of $P_l(\theta)$ must vanish separately

$$0 = j_l(ka) + A_l h_l^{(1)}(ka) \tag{12.131}$$

$$A_l = -\frac{j_l(ka)}{h_l^{(1)}(ka)} = -\frac{j_l}{j_l + iy_l} = -\frac{1}{1 + i\xi_l} \tag{12.132}$$

$$\xi_l(k) = \frac{y_l(ka)}{j_l(ka)} \tag{12.133}$$

$$\psi_S = -\sum_l (2l+1)i^l P_l(\theta) \frac{h_l^{(1)}(kr)}{1 + i\xi_l(k)} \tag{12.134}$$

The asymptotic form of the spherical Hankel function can be used to find the behavior of the scattered wave from a long distance, that is $(kr) \gg 1$

$$h_l^{(1)}(kr) \to \frac{e^{ikr}}{i^{l+1}kr}[1 + O(1/kr)] \tag{12.135}$$

$$\psi_S \to \frac{e^{ikr}}{r} f(k, \theta) \tag{12.136}$$

$$f(k, \theta) = \frac{i}{k} \sum_l (2l + 1) \frac{P_l(\theta)}{1 + i\xi_l(k)} \tag{12.137}$$

In scattering theory the quantity $|f(k, \theta)|^2$ is the differential cross section.

Example 5. *Green's function for the Klein–Gordon equation:* The equation is

$$\left[\nabla^2 - \frac{1}{v^2}\frac{\partial^2}{\partial t^2} - k^2\right] G(\mathbf{r}, \mathbf{r}_0; t) = -\delta(t)\delta(\mathbf{r} - \mathbf{r}_0) \tag{12.138}$$

Compared to the wave equation, the Klein–Gordon equation has an additional factor of the real positive constant k^2. The first step is to Fourier transform the equation $(\mathbf{r} \to \mathbf{q})$ by multiplying the equation by $\exp(-i\mathbf{r}\cdot\mathbf{q})$ and integrating over all space

$$\left[v^2(q^2 + k^2) + \frac{\partial^2}{\partial t^2}\right]\bar{G}(\mathbf{q}, \mathbf{r}_0; t) = v^2\delta(t)e^{-i\mathbf{q}\cdot\mathbf{r}_0} \tag{12.139}$$

$$\bar{G}(\mathbf{q}, \mathbf{r}_0; t) = v\Theta(t)\frac{\sin(vt\sqrt{q^2 + k^2})}{\sqrt{q^2 + k^2}}e^{-i\mathbf{q}\cdot\mathbf{r}_0} \tag{12.140}$$

It is easy to verify that this last expression is a solution to the differential equation. The only remaining task is to take the inverse Fourier transform. The angular integrals are easy since they give a spherical Bessel function $(\mathbf{R} = \mathbf{r} - \mathbf{r}_0)$

$$G(\mathbf{r}, \mathbf{r}_0; t) = v\Theta(t) \int \frac{d^3q}{(2\pi)^3} e^{i\mathbf{q}\cdot(\mathbf{r}-\mathbf{r}_0)} \frac{\sin(vt\sqrt{q^2 + k^2})}{\sqrt{q^2 + k^2}}$$

$$= \frac{v}{2\pi^2}\Theta(t) \int_0^\infty q^2\,dq\,j_0(qR)\frac{\sin(vt\sqrt{q^2 + k^2})}{\sqrt{q^2 + k^2}} \tag{12.141}$$

$$= \frac{v}{2\pi^2 R}\Theta(t) \int_0^\infty q\,dq\,\sin(qR)\frac{\sin(vt\sqrt{q^2 + k^2})}{\sqrt{q^2 + k^2}}$$

This integral was found in Table [G and R, 3.876(1)]

$$\int_0^\infty dq\,\cos(qR)\frac{\sin(L\sqrt{q^2 + k^2})}{\sqrt{q^2 + k^2}} = \frac{\pi}{2}J_0(k\sqrt{L^2 - R^2})\Theta(L - R)$$

with $L = vt$. Our integral is obtained by a simple differentiation with respect to R

$$\int_0^\infty q\, dq \sin(qR) \frac{\sin(vt\sqrt{q^2 + k^2})}{\sqrt{q^2 + k^2}}$$

$$= -\frac{\pi}{2} \frac{d}{dR} [J_0(k\sqrt{L^2 - R^2})\Theta(L - R)]$$

$$= \frac{\pi}{2}\left[\delta(L - R) - \frac{kR}{\sqrt{L^2 - R^2}} J_1(k\sqrt{L^2 - R^2})\Theta(L - R) \right] \quad (12.142)$$

$$G(\mathbf{r}, \mathbf{r}_0; t) = \Theta(t) \frac{v}{4\pi R}\left[\delta(L - R) - \frac{kR}{\sqrt{L^2 - R^2}} J_1(k\sqrt{L^2 - R^2})\Theta(L - R) \right]$$

$$(12.143)$$

The first term in brackets is a soliton type wave pulse that travels $R = L = vt$. This term was also found for Green's function of the wave equation. The second term is another solution that provides a trailing pulse behind the solitary wave.

12.4. Problems Inside a Sphere

Consider the class of problems that are confined inside a sphere. If the solution vanishes at the surface of the sphere ($r = a$), the natural functions are

$$\chi_{lm\alpha} = N_{lm\alpha} Y_l^m(\theta, \phi) j_l(\lambda_{l\alpha} r/a) \quad (12.144)$$

$$j_l(\lambda_{l\alpha}) = 0 \quad (12.145)$$

The coefficients $\lambda_{l\alpha}$ are defined as the points where the spherical Bessel functions vanish. For $l = 0$ these are at the points $\lambda_{0\alpha} = \alpha\pi$ but other values are found for nonzero values of l. In order to use these functions, we need to know the normalization, which is

$$\int_0^a r^2\, dr j_l(\lambda_{l\alpha} r/a) j_l(\lambda_{l\beta} r/a) = \frac{a^3}{2} \delta_{\alpha\beta} |j_l'(\lambda_{l\alpha})|^2 \quad (12.146)$$

A similar theorem was found in two dimensions. The present result is proved in the same fashion. Write the differential equation for the spherical Bessel function for two different values of λ, $k = \lambda_{l\alpha}/a$, $p = \lambda_{l\beta}/a$

$$0 = \frac{d}{dr}\left[r^2 \frac{dj_l(kr)}{dr} \right] + [k^2 r^2 - l(l + 1)]j_l(kr) \quad (12.147)$$

$$0 = \frac{d}{dr}\left[r^2 \frac{dj_l(pr)}{dr} \right] + [p^2 r^2 - l(l + 1)]j_l(pr) \quad (12.148)$$

Multiply the first equation by $j_l(pr)$, the second by $j_l(kr)$ then subtract the two. The derivative terms are combined to give

$$0 = \frac{d}{dr}\left[r^2 j_l(pr)\frac{d}{dr}j_l(kr) - r^2 j_l(kr)\frac{d}{dr}j_l(pr) \right] + r^2(k^2 - p^2)j_l(kr)j_l(pr) \quad (12.149)$$

Integrate this expression between zero and a and find

$$(p^2 - k^2)\int_0^a r^2 \, dr j_l(kr) j_l(pr) = a^2 \left[j_l(pa)\frac{d}{da}j_l(ka) - j_l(ka)\frac{d}{da}j_l(pa) \right] \quad (12.150)$$

The right-hand side vanishes if $j_l(ka) = 0 = j_l(pa)$. If $k \neq p$ then the left-hand side must vanish, this can only occur if the integral vanishes. The integral [Eq. (12.146)] vanishes if $\alpha \neq \beta$. Next, show if $\alpha = \beta$ the integral equals the right-hand side of Eq. (12.146). In Chapter 11 it was shown that if v is an integer then

$$\int_0^a r \, dr J_v(kr)^2 = \frac{a^2}{2}|J_v'(ka)|^2 \quad (12.151)$$

This result is valid for the case $J_v(ka) = 0$. When reviewing the prior derivation of this result, nowhere was it essential that v be an integer, in fact, the expression is valid for any value of v. Spherical Bessel functions have $v = l + 1/2$. Use the definition $j_l(z) = \sqrt{\pi/2z}\,J_{l+1/2}(z)$ to convert

$$\int_0^a r^2 \, dr j_l(ka)^2 = \frac{\pi}{2k}\int_0^a r \, dr J_{l+1/2}(kr)^2 = \frac{\pi a^2}{4k}|J_{l+1/2}'(ka)|^2 = \frac{a^3}{2}|j_l'(ka)|^2 \quad (12.152)$$

Note that $j_l(z) = 0$ implies that $J_{l+1/2}(z) = 0$. They differ by the factor of $\sqrt{z}$ which is not going to be zero. Similarly,

$$J_{l+1/2} = \sqrt{\frac{2z}{\pi}}\,j_l(z) \quad (12.153)$$

$$J_{l+1/2}' = \sqrt{\frac{2z}{\pi}}\,[j_l'(z) + j_l(z)/2z]$$
$$\quad (12.154)$$
$$= \sqrt{\frac{2z}{\pi}}\,j_l'(z)$$

the latter equality applies when $j_l(z) = 0$. These steps prove Eq. (12.146).

Example 1. *Oscillations of a sphere:* Find the allowed frequencies of a sphere of radius a. For a frequency ω the wave equation becomes a Helmholtz equation $[\nabla^2 + k^2]\phi = 0$ with $k = \omega/v$. The solutions in spherical coordinates are

$$\phi \sim e^{im\phi}P_l^{|m|}(\theta)j_l(kr) \quad (12.155)$$

First consider the boundary condition that the waves have zero amplitude on the surface of the sphere. Let $\bar{\lambda}_{l\alpha}$ be the zeros of the spherical Bessel functions. The boundary conditions on the surface of the sphere are

$$k = \frac{\bar{\lambda}_{l\alpha}}{a} \tag{12.156}$$

$$\omega_{l\alpha} = \bar{\lambda}_{l\alpha} \frac{v}{a} \tag{12.157}$$

The zeros of the spherical Bessel functions $\lambda_{0\alpha} = \alpha\pi$ are simple for $l = 0$. In the next section we solve a related problem for sound vibrations of a sphere.

Example 2. *Heating a sphere:* Consider a spherical object of radius a. Initially it is at temperature T_i. At $t = 0$ it is immersed in a heat reservoir at temperature T_f. Heat will diffuse into the sphere and reach a final temperature T_f. Solve the diffusion equation to determine $T(\mathbf{r}, t)$.

      Since the sphere is heated equally on all sides there is no angular dependence and the appropriate angular functions have $l = 0$, $m = 0$,

$$T(\mathbf{r}, t) = T_f + \delta T(r, t) \tag{12.158}$$

$$\delta T(r, t) = \sum_\alpha A_{0\alpha} e^{-tD(\lambda/a)^2} j_0(\lambda r/a) \tag{12.159}$$

The radial functions $j_0(kr)$ are used since they satisfy $\nabla^2 j_0(kr) = -k^2 j_0(kr)$. Here set $k \equiv \lambda/a$ where $\lambda \equiv \lambda_{0\alpha} = \alpha\pi$. The combination $\exp[-tDk^2]j_0(kr)$ satisfies the diffusion equation

$$\left[D\nabla^2 - \frac{\partial}{\partial t} \right] e^{-tDk^2} j_0(kr) = 0 \tag{12.160}$$

The coefficient $A_{0\alpha}$ is chosen to obey the initial condition at $t = 0$, which is

$$T_i - T_f = -\Delta T = \sum_\alpha A_{0\alpha} j_0(\lambda_{0\alpha} r/a) \tag{12.161}$$

Multiply this equation by $r^2 j_0(\lambda_{0\beta} r/a)$ and integrate dr from zero to a. Only the term in the series with $\alpha = \beta$ gives a nonzero result

$$A_{0\beta} \frac{a^3}{2} j_0'(\lambda_{0\beta})^2 = -\Delta T \int_0^a dr\, r^2 j_0(\lambda r/a) \tag{12.162}$$

$$\int_0^a r^2\, dr\, j_0(\lambda r/a) = \frac{a^3}{\lambda^3} \int_0^\lambda z\, dz\, \sin(z) = \frac{a^3}{\lambda^3} [\sin(z) - z\cos(z)]_0^\lambda = -(-1)^\alpha \frac{a^3}{\lambda^2} \tag{12.163}$$

Also, note that if $j_0(\lambda) = 0$ then $j_0'(\lambda) = \cos(\lambda)/\lambda - \sin(\lambda)/\lambda^2 = (-1)^\alpha/\lambda$. Collecting these results gives

$$A_\alpha = 2\Delta T(-1)^\alpha \tag{12.164}$$

$$\delta T(r, t) = 2\Delta T \sum_{\alpha=1}^{\infty} (-1)^\alpha j_0(\pi\alpha r/a)e^{-tD(\pi\alpha/a)^2} \tag{12.165}$$

$$1 = -2 \sum_{\alpha=1}^{\infty} (-1)^\alpha j_0(\pi\alpha r/a) \tag{12.166}$$

The last equation is an identity. The right-hand side appears to be a function of r, but is actually the constant one.

Example 3. *Green's function for diffusion:* Solve the usual diffusion equation

$$\left[D\nabla^2 - \frac{\partial}{\partial t} \right] G(\mathbf{r}, \mathbf{r}_0; t) = -\delta(t)\delta(\mathbf{r} - \mathbf{r}_0)$$

$$= \frac{\delta(t)}{r^2 \sin \theta} \delta(r - r_0)\delta(\theta - \theta_0)\delta(\phi - \phi_0) \tag{12.167}$$

where the three dimensional delta function is given on the right. Green's function is assumed to vanish on the surface of the sphere ($r = a$). The relevant radial functions will be $j_l(\lambda_{l\alpha}r/a)$. The most general angular function will be $Y_{lm}(\theta, \phi)$. However, the final answer will depend on the angle between the vectors $\mathbf{r}$ and $\mathbf{r}_0$, which is $\cos(\Omega) = \hat{r}\cdot\hat{r}_0$. Therefore, the relevant angular function is $P_l(\Omega)$, so the form for Green's function is

$$G(\mathbf{r}, \mathbf{r}_0; t) = \Theta(t) \sum_{l\alpha} A_{l\alpha} P_l(\Omega) j_l(\lambda_{l\alpha}r/a) j_l(\lambda_{l\alpha}r_0/a)e^{-tD(\lambda/a)^2}$$

Two Bessel functions are included, since the answer is symmetric in r and r_0. When this ansatz is included in the diffusion equation at $t = 0$ there is the identity

$$\delta(\mathbf{r} - \mathbf{r}_0) = \sum_{l\alpha} A_{l\alpha} P_l(\Omega) j_l(\lambda_{l\alpha}r/a) j_l(\lambda_{l\alpha}r_0/a) \tag{12.168}$$

The constant $A_{l\alpha}$ is found from two identities

$$\sum_{l=0}^{\infty} (2l + 1)P_l(\Omega) = 4\pi\delta(\phi - \phi_0)\delta(\cos \theta - \cos \theta_0) \tag{12.169}$$

$$\sum_{\alpha} \frac{2}{a^3} \frac{j_l(\lambda_{l\alpha}r/a)j_l(\lambda_{l\alpha}r_0/a)}{|j_l'(\lambda)|^2} = \frac{\delta(r - r_0)}{r^2}$$

$$A_{l\alpha} = \frac{2l + 1}{2\pi a^3} \frac{1}{|j_l'(\lambda)|^2} \tag{12.170}$$

The summation over α gives the delta function over the radial variable, while the summation over l gives the delta function over the angular variables. The summation over α was derived earlier, while the summation over l is the *completeness relation* for Legendre functions given in the Appendix.

So far, all the solutions to the wave equation in this book have been for scalar waves $\phi(\mathbf{r}, t)$. There are several examples of waves with vector fields, such as those for the electric and magnetic field. The student is familiar with this topic since it is covered in numerous physics courses. Instead, consider another vector field which are sound waves in solid materials.

Let $\mathbf{u}(\mathbf{r}, t)$ be the displacement of the material at point $(\mathbf{r}, t)$ in space time. The wave equation for sound waves is

$$\frac{\partial^2}{\partial t^2}\mathbf{u} = c_t^2 \nabla^2 \mathbf{u} + \Delta^2 \nabla(\nabla \cdot \mathbf{u}) \tag{12.171}$$

$$\Delta^2 = c_l^2 - c_t^2 \tag{12.172}$$

where c_t and c_l are transverse and longitudinal sound velocities. This equation is valid for materials that are completely isotropic for the propagation of sound. The waves have the same velocity in every direction. It applies to materials such as glasses and composites. Crystalline solids have a regular arrangement of atoms and a group symmetry. In crystals the sound velocity depends on direction and the wave equation is more complex than given above.

12.5.1. Bulk Waves

There is a theorem in vector calculus which states that any vector can be given exactly as the summation of a longitudinal $\mathbf{u}_l$ and transverse $\mathbf{u}_t$ vector, which satisfy the conditions

$$\mathbf{u} = \mathbf{u}_l + u_t \tag{12.173}$$

$$\nabla \cdot \mathbf{u}_t = 0, \quad \mathbf{u}_t = \nabla \times \mathbf{M}(\mathbf{r}, t) \tag{12.174}$$

$$\nabla \times \mathbf{u}_l = 0, \quad \mathbf{u}_l = \nabla\phi(\mathbf{r}, t) \tag{12.175}$$

where $\mathbf{M}(\mathbf{r}, t)$ and $\phi(\mathbf{r}, t)$ are vector and scalar functions. Assume a wave of wave vector $\mathbf{q}$ and frequency ω of the usual form

$$\mathbf{u}(\mathbf{r}, t) = \mathbf{u}_0 e^{i(\mathbf{q}\cdot\mathbf{r} - \omega t)} \tag{12.176}$$

$$\nabla \cdot \mathbf{u}_t = 0 \Rightarrow \mathbf{u}_0 \cdot \mathbf{q} = 0 \tag{12.177}$$

$$\nabla \times \mathbf{u}_l = 0 \Rightarrow \mathbf{u}_0 \times \mathbf{q} = 0 \tag{12.178}$$

These equations show:

- Transverse waves have the displacement $\mathbf{u}_0 \perp \mathbf{q}$ which is the usual definition of transverse. Putting this condition into the wave equation Eq. (12.171) gives for transverse waves the equation $\omega^2 = q^2 c_t^2$ so that c_t is the velocity of transverse waves.
- Longitudinal waves have $\mathbf{u}_0 \parallel \mathbf{q}$ resulting in

$$\nabla(\nabla \cdot \mathbf{u}) = -\mathbf{q}(\mathbf{q} \cdot \mathbf{u}_0)e^{i(\mathbf{q}\cdot\mathbf{r} - \omega t)} = -q^2 \mathbf{u}_0 e^{i(\mathbf{q}\cdot\mathbf{r} - \omega t)} \tag{12.179}$$

$$\omega^2 = q^2 c_l^2 \tag{12.180}$$

$$c_l^2 = c_t^2 + \Delta^2 \tag{12.181}$$

where c_l is the velocity of longitudinal sound. In a liquid or gas, $c_t = 0$ and sound is longitudinal.

The longitudinal and transverse sound propagate with different velocities. The two modes are uncoupled in bulk materials and are treated as totally independent. They have the separate wave equations

$$0 = \left[\frac{\partial^2}{\partial t^2} - c_t^2 \nabla^2 \right] u_t(\mathbf{r}, t) \tag{12.182}$$

$$0 = \left[\frac{\partial^2}{\partial t^2} - c_l^2 \nabla^2 \right] u_l(\mathbf{r}, t) \tag{12.183}$$

These cases are rather dull. The only mixing of the two modes occur at boundaries and interfaces. The next step is to learn about boundary conditions.

12.5.2. Boundary Conditions

First consider the case that the material has a boundary with air or vacuum. The sound wave cannot propagate across this boundary. For vacuum the statement is rigorous and for air some small fraction of the mode can be converted to sound waves in air. This fraction is quite small, so we assume it is zero for this simple discussion. The boundary is defined by a unit vector $\hat{n}$ normal to the surface and the boundary condition for sound is that the stress tensor σ_{ij} obeys the relationship at the boundary

$$0 = \sum_i n_i \sigma_{ij} \tag{12.184}$$

This equation is valid for all three values of j, which gives three possible boundary conditions. In order to discuss boundary conditions it is necessary to learn about stress and strain.

The *strain tensor* u_{ij} is defined as the symmetric derivative of the displacement $u_i(\mathbf{r}, t)$ that enters the wave equation

$$u_{ij} = \frac{1}{2} \left(\frac{\partial u_i}{\partial x_j} + \frac{\partial u_j}{\partial x_i} \right), \qquad u_{ii} = \frac{\partial u_i}{\partial x_i} \tag{122.185}$$

where (i, j) are the (x, y, z) coordinates. Obviously $u_{ij} = u_{ji}$. Some books omit the factor of $1/2$.

The *stress tensor* is σ_{ij}. For isotropic solids, which are considered here, the stress is proportional to the strain. These relationships are given in terms of the Poisson's ratio σ and Young's modulus E as

$$\sigma_{ii} = \frac{E}{(1 + \sigma)(1 - 2\sigma)} [(1 - 2\sigma)u_{ii} + \sigma \nabla \cdot \mathbf{u}] \tag{12.186}$$

$$\sigma_{ij} = \frac{E}{1 + \sigma} u_{ij}, \quad \text{for } i \neq j \tag{12.187}$$

$$F_i = \sum_j \frac{\partial}{\partial x_j} \sigma_{ij} \tag{12.188}$$

The quantity F_i is the force on the material in the direction x_i due to the displacement of the material. For a body in equilibrium, $\mathbf{F} = -g\rho\hat{z}$ is the equation that determines the elastic deformations, where g is the gravitational constant and ρ is the density of the material.

The two constants (σ, E) are related to the speed of sound

$$c_t^2 = \frac{E}{2\rho(1 + \sigma)} \tag{12.189}$$

$$c_l^2 = \frac{E(1 - \sigma)}{\rho(1 + \sigma)(1 - 2\sigma)} \tag{12.190}$$

$$\Delta^2 = \frac{E}{2\rho(1 + \sigma)(1 - 2\sigma)} \tag{12.191}$$

For example, using these relations one can show that the diagonal stress tensor is

$$\sigma_{ii} = \rho\{2c_t^2(u_{ii} - \mathbf{V} \cdot \mathbf{u}) + c_l^2 \mathbf{V} \cdot \mathbf{u}\} \tag{12.192}$$

which is a convenient form.

Example 1. *Stretching under gravity:* Consider a rod of length $L = 1$ m that hangs from the ceiling. How much does gravity stretch its length? Only the z dependence needs to be considered and only longitudinal stretching occurs. If $z = 0$ is the top of the rod,

$$0 = \frac{\partial\sigma_{zz}}{\partial z} + \rho g \tag{12.193}$$

$$0 = c_l^2 \frac{\partial^2}{\partial z^2} u_z + g \tag{12.194}$$

$$u_z = -\frac{gz^2}{2c_l^2} \tag{12.195}$$

If the speed of sound is $c_l = 3$ km/s, then $u_z = -0.5\ \mu$m at the end of the 1 m rod.

Example 2. *Vibrations of rods:* First consider the vibrations of rods. If an end is clamped and cannot move, then $\mathbf{u} = 0$. If the end is free we use the stress free boundary condition. If the rod is in the x direction and we consider only longitudinal oscillation, the only nonzero component of displacement is u_x. At a free end $du_x/dx = 0$. For rod of length $0 \leqslant x \leqslant L$, the different cases are:

- If the rod is clamped at both ends, then $u_x = u_0 \sin(k_n x)$, $k_n = n\pi/L$.
- If the rod is free at both ends, then $u_x = u_0 \cos[k_n(x - L/2)]$, $k_n = 2\pi n/L$.
- For a rod clamped at $x = 0$ and free at $x = L$, $u_x = u_0 \sin(k_n x)$, $k_n = \pi(2n + 1)/(2L)$.

In all cases the frequency is $\omega_n = c_l k_n$.

Example 3. *Vibrations of a drum:* Next consider the problem of the sound oscillations in a cylindrical drum. Sound is the displacement of atoms. Denote the vector $\mathbf{u}(\mathbf{r}, t)$ as the displacement of the atoms. The wave equation is now for a vector quantity

$$\left[\frac{\partial^2}{\partial t^2} - v^2 \nabla^2 \right] \mathbf{u}(\mathbf{r}, t) = 0 \tag{12.196}$$

The boundary conditions are that on the surface $\hat{n} \cdot \mathbf{u} = 0$, so the oscillation direction is not into the surface. The solutions to the wave equation for a vector field are different from those for a scalar field.

In early physics textbooks the sound oscillations in an organ pipe are considered to be one dimensional, so that $u_z = A \sin(k_n z)$ and $k_n = \pi n/L$. The allowed frequencies are $\omega_n = v k_n$. This solution is still valid for the present problem. Let the z axis be along the center of the cylinder. A valid solution to Eq. (12.196) is

$$\mathbf{u}(\mathbf{r}, t) = \hat{z} u_0 \sin(k_n z) \tag{12.197}$$

where $k_n = \pi n/L$. This solution obeys the differential equation and also the boundary conditions at the ends. There is no motion in the z direction on the ends of the cylinder. Along the walls with $\rho = a$ the gas slides along the walls, and there is no displacement perpendicular to the wall. The stresses $u_{\rho z} = \sigma_{\rho\rho} = 0$.

The reader may ask what happened to all the Bessel functions. There are additional solutions to the wave equation that involve radial oscillations. Their solution is rather easy because sound waves in air are always longitudinal. Sound waves in solids can be transverse, but not in air. Since only longtitudinal oscillations occur, then $\mathbf{u}_t = 0$.

The solution to the vector wave equation for longitudinal waves is simple. First, obtain the solution $\phi(\mathbf{r}, t)$ to the scalar wave equation of the form in Eq. (12.182). The solution to the vector wave equation is the gradient of this function. The solution involving only radial oscillations is

$$\phi = e^{in\theta - i\omega t} J_n(q\rho) \tag{12.198}$$

$$\mathbf{u}(\rho, \theta, t) = \nabla\phi = e^{i(n\theta - \omega t)} [\hat{\rho} u_\rho + \hat{\theta} u_\theta] \tag{12.199}$$

$$u_\rho(\rho) = q J_n'(q\rho) \tag{12.200}$$

$$u_\theta(\rho) = \frac{in}{\rho} J_n(q\rho) \tag{12.201}$$

$$\omega^2 = v^2 q^2 \tag{12.202}$$

Now apply the boundary conditions. At the top and bottom of the cylinder, the air slides along in a radial oscillation that is perpendicular to the ends. On the cylindrical surface, assume the radial oscillation is zero so that $u_\rho(a) = 0$ that is satisfied by $J_n'(q_\alpha a) = 0$. Now the constraint on the values of q_α is that the derivative of the Bessel function vanish. This condition is different from that found for scalar waves. The allowed frequencies are $\omega_{\alpha m} = v q_\alpha$. The purely radial solutions have no dependence on θ and are

$$u_\rho(\rho) = q_\alpha J_1(q_\alpha \rho) \tag{12.203}$$

where $\phi \propto J_0$ and $J'_0 = -J_1$. In this case the values of $q_\alpha = \lambda_{1\alpha}/a$ are determined by the zeros $[J_1(\lambda_{1\alpha}) = 0]$ of the Bessel function $J_1(z)$.

Finally, there are solutions combining the radial and z motions. They have eigenfunctions that are products of those found for the z and radial motion. The frequencies of the combined motions are given by $\omega^2 = v^2(q_\alpha^2 + k_m^2)$.

Example 4. *Oscillation of a sphere:* An interesting problem is solving the sound vibrations in a sphere. The scalar oscillations inside a sphere were solved earlier. Since sound is a displacement of air or matter the solution is a vector whose direction gives the direction of oscillation. In this case, the scalar wave equation is not suitable. To keep the discussion simple, only consider the vibrations that are spherically symmetric. The mode is the uniform oscillation of the surface of the sphere in the radial direction. For a spherical drum the boundary condition is $u_r = 0$ at the surface. For a solid object such as earth, the appropriate boundary condition is the radial stress $\sigma_{rr} = 0$ at the surface.

Assume the displacement $\mathbf{u}(\mathbf{r}, t) = \hat{r}u_r(r, t)$. The notation u_r denotes the radial component of the displacement which is the only component assumed to be nonzero. The vector direction is along the radial direction. The longitudinal sound velocity c_l is appropriate and assume the sound has frequency ω. The second line below is a vector identity

$$\left[c_l^2 \nabla^2 - \frac{\partial^2}{\partial t^2} \right] \mathbf{u}(\mathbf{r}, t) = 0 \tag{12.204}$$

$$\nabla^2(\hat{r}u_r) = \hat{r}\left[\nabla^2 u_r - \frac{2}{r^2} u_r \right] \tag{12.205}$$

$$= \hat{r}\left[\mathscr{L}_r - \frac{2}{r^2} \right] u_r, \quad \mathscr{L}_r = \frac{1}{r}\frac{d^2}{dr^2} r$$

$$c_l^2\left[\mathscr{L}_r - \frac{2}{r^2} \right] u_r + \omega^2 u_r = 0 \tag{12.206}$$

$$\left[\mathscr{L}_r + \frac{\omega^2}{c_l^2} - \frac{2}{r^2} \right] u_r = 0 \tag{12.207}$$

$$u_r = A j_1(kr) \tag{12.208}$$

$$k = \frac{\omega}{c_l} \tag{12.209}$$

The operator has the factor of $-2/r^2$ and is identical to the factor $-l(l + 1)/r^2$ when $l = 1$. So the solution is the spherical Bessel function for the case that $l = 1$ and not $l = 0$. This is different from the scalar wave equation in that the displacement is a vector. Even if it is the same in all directions its vector character gives a solution with $l = 1$.

For a spherical drum the displacement must vanish at the radius a of the drum. The allowed frequencies are those for which $j_1(\lambda_{1\alpha}) = 0$ and $\omega_\alpha = c_l\lambda_{1\alpha}/a$. For example, find the radius of the sphere whose fundamental sound is the A note at $\nu = 440$ Hz. The first zero of the $j_1(z)$ spherical bessel function is $\lambda_{11} = 4.4934$. The formula is $a = c_l\lambda_{11}/(2\pi\nu) = 0.56$ m, where the velocity of sound is $c_l = 343$ m/s.

Next, consider the free oscillations of a solid sphere. The boundary condition at the surface of the sphere is that the radial component of the stress must vanish. Without getting into the details of stress and strain, for the present case only radial oscillations which are isotropic are considered. The boundary condition is

$$0 = \frac{E}{(1 + \sigma)(1 - 2\sigma)}\left[(1 - \sigma)\frac{\partial u_r}{\partial r} + 2\sigma\frac{u_r}{r}\right] \tag{12.210}$$

$$0 = c_l^2 s\frac{\partial u_r}{\partial r} - 2(2c_t^2 - c_l^2)u_r \tag{12.211}$$

$$0 = z\frac{d}{dz}j_1(z) + 2(1 - 2s)j_1(z) \tag{12.212}$$

where $z = ka = \omega a/c_l$ and $s = c_t^2/c_l^2$. In the first line the boundary conditions are given in terms of Young's modulus E and the Poisson ration σ. In the second line the same formula is given in terms of the longitudinal (c_l) and transverse (c_t) speed of sound. Note that it has mixed boundary conditions that include both the eigenfunction and its derivative. The derivative is easy and another equation is obtained involving a tangent

$$j_1(z) = \frac{\sin(z)}{z^2} - \frac{\cos(z)}{z} \tag{12.213}$$

$$z\frac{d}{dz}j_1(z) = \sin(z)\left[1 - \frac{2}{z^2}\right] + 2\frac{\cos(z)}{z} \tag{12.214}$$

that gives for the boundary condition

$$0 = \sin(z)\left[1 - \frac{2}{z^2} + \frac{2}{z^2}(1 - 2s)\right] + \frac{\cos(z)}{z}[2 - 2(1 - 2s)] \tag{12.215}$$

$$0 = \tan(z)\left[1 - \frac{4s}{z^2}\right] + \frac{4s}{z} \tag{12.216}$$

$$\tan(z) = \frac{4sz}{4s - z^2} \tag{12.217}$$

The last equation is the final eigenvalue equation for the frequencies of the normal modes. Since $\tan(z)$ is periodic there are an infinite number of solutions. They depend on the ratio s of the square of the velocities. From the definitions

$$s = \frac{1 - 2\sigma}{2(1 - \sigma)}$$ (12.218)

Poisson's ratio σ varies between the limits $-1 < \sigma < 1/2$, which means that s varies between 0 and 3/4.

Appendix I: Vector Derivatives

Here are some derivatives in cylindrical coordinates, (r, θ, z)

$$\nabla\psi(\mathbf{r}) = \hat{r}\frac{\partial\psi}{\partial r} + \frac{\hat{\theta}}{r}\frac{\partial\psi}{\partial\theta} + \hat{z}\frac{\partial\psi}{\partial z}$$ (12.219)

$$\nabla \cdot \mathbf{A}(\mathbf{r}) = \frac{1}{r}\frac{\partial}{\partial r}(rA_r) + \frac{1}{r}\frac{\partial A_\theta}{\partial\theta} + \frac{\partial A_z}{\partial z}$$ (12.220)

$$\nabla \times \mathbf{A}(\mathbf{r}) = \hat{r}\left(\frac{1}{r}\frac{\partial A_z}{\partial\theta} - \frac{\partial A_\theta}{\partial z}\right) + \hat{\theta}\left(\frac{\partial A_r}{\partial z} - \frac{\partial A_z}{\partial r}\right) + \frac{\hat{z}}{r}\left(\frac{\partial(rA_\theta)}{\partial r} - \frac{\partial A_r}{\partial\theta}\right)$$ (12.221)

$$\nabla^2\psi(\mathbf{r}) = \frac{1}{r}\frac{\partial}{\partial r}\left(r\frac{\partial\psi}{\partial r}\right) + \frac{1}{r^2}\frac{\partial^2\psi}{\partial\theta^2} + \frac{\partial^2\psi}{\partial z^2}$$ (12.222)

$$\nabla^2\mathbf{A} = \hat{r}\left(\nabla^2 A_r - \frac{A_r}{r^2} - \frac{2}{r^2}\frac{\partial A_\theta}{\partial\theta}\right) + \hat{\theta}\left(\nabla^2 A_\theta - \frac{A_\theta}{r^2} + \frac{2}{r^2}\frac{\partial A_r}{\partial\theta}\right) + \hat{z}\nabla^2 A_z$$ (12.223)

Here are some derivatives in spherical coordinates,

$$\nabla\psi(\mathbf{r}) = \hat{r}\frac{\partial\psi}{\partial r} + \frac{\hat{\theta}}{r}\frac{\partial\psi}{\partial\theta} + \frac{\hat{\phi}}{r\sin(\theta)}\frac{\partial\psi}{\partial\phi}$$ (12.224)

$$\nabla \cdot \mathbf{A}(\mathbf{r}) = \frac{1}{r^2}\frac{\partial}{\partial r}(r^2 A_r) + \frac{1}{r\sin(\theta)}\left[\frac{\partial}{\partial\theta}(\sin\theta A_\theta) + \frac{\partial A_\phi}{\partial\phi}\right]$$ (12.225)

$$\nabla^2\psi = \frac{1}{r^2}\frac{\partial}{\partial r}\left(r^2\frac{\partial\psi}{\partial r}\right) + \frac{1}{r^2\sin\theta}\frac{\partial}{\partial\theta}\left(\sin\theta\frac{\partial\psi}{\partial\theta}\right) + \frac{1}{r^2\sin^2\theta}\frac{\partial^2\psi}{\partial\phi^2}$$ (12.226)

$$\nabla^2\mathbf{A} = \hat{r}\left\{\left[\nabla^2 - \frac{2}{r^2}\right]A_r - \frac{2}{r^2\sin(\theta)}\left[\frac{\partial}{\partial\theta}(\sin(\theta)A_\theta) + \frac{\partial A_\phi}{\partial\phi}\right]\right\}$$
$$+ \hat{\theta}\left\{\left[\nabla^2 - \frac{1}{r^2\sin^2(\theta)}\right]A_\theta + \frac{2}{r^2}\frac{\partial A_r}{\partial\theta} - \frac{2\cos(\theta)}{r^2\sin^2(\theta)}\frac{\partial A_\phi}{\partial\phi}\right\}$$ (12.227)
$$+ \hat{\phi}\left\{\left[\nabla^2 - \frac{1}{r^2\sin^2(\theta)}\right]A_\phi + \frac{2}{r^2\sin(\theta)}\frac{\partial A_r}{\partial\phi} + \frac{2\cos(\theta)}{r^2\sin^2(\theta)}\frac{\partial A_\theta}{\partial\phi}\right\}$$

An expression such as $\nabla \cdot [K(\mathbf{r})\nabla\psi]$ is

$$\nabla \cdot [K(\mathbf{r})\nabla\psi] = \frac{1}{r^2}\frac{\partial}{\partial r}\left[r^2 K(r,\theta,\phi)\frac{\partial\psi}{\partial r}\right] + \frac{1}{r^2\sin\theta}\frac{\partial}{\partial\theta}\left[\sin(\theta)K(\rho,\theta,\phi)\frac{\partial\psi}{\partial\theta}\right]$$
$$+ \frac{1}{r^2\sin^2(\theta)}\frac{\partial}{\partial\phi}\left[K(r,\theta,\phi)\frac{\partial\psi}{\partial\phi}\right] \qquad (12.228)$$

Appendix II: Legendre Polynomials

Legendre polynomials have two sets of notation that are interchangeable. As an example, consider the case $l = 1$, $m = 0$. The subscript is not written for $m = 0$. So this function is $P_1(\theta) = \cos(\theta)$. The other common notation is to write this as $P_1(\cos\theta) = \cos\theta$. Sometimes they are written as a function of θ and sometimes as a function of $\cos\theta$. Both notations are used, depending on what is convenient.

The functions for small values of l are,

$l = 0$

$$P_0 = 1 \qquad (12.229)$$

$$Q_0 = \frac{1}{2}\ln\left(\frac{1 + \cos\theta}{1 - \cos\theta}\right) \qquad (12.230)$$

$l = 1$

$$P_1 = \cos(\theta), \quad P_1^1 = \sin(\theta) \qquad (12.231)$$

$$Q_1 = \frac{\cos\theta}{2}\ln\left(\frac{1 + \cos\theta}{1 - \cos\theta}\right) - 1 \qquad (12.232)$$

$l = 2$

$$P_2 = \tfrac{1}{2}(3\cos^2\theta - 1), \quad P_2^1 = 3\cos\theta\sin\theta, \quad P_2^2 = 3\sin^2\theta \qquad (12.233)$$

$$Q_2 = \frac{P_2(\theta)}{2}\ln\left(\frac{1 + \cos\theta}{1 - \cos\theta}\right) - \frac{3}{2}\cos\theta \qquad (12.234)$$

Using the notation $P_l(z)$ they are,

$$P_0 = 1 \qquad (12.235)$$

$$P_1 = z \qquad (12.236)$$

$$P_2 = \tfrac{1}{2}[3z^2 - 1] \qquad (12.237)$$

$$P_3 = \tfrac{1}{2}[5z^3 - 3z] \qquad (12.238)$$

$$P_4 = \tfrac{1}{8}[35z^4 - 30z^2 + 3] \qquad (12.239)$$

$$P_5 = \tfrac{1}{8}[63z^5 - 70z^3 + 15z] \qquad (12.240)$$

$$P_6 = \tfrac{1}{16}[231z^6 - 315z^4 + 105z^2 - 5] \qquad (12.241)$$

The values of $m \neq 0$ can be found from the functions with $m = 0$ using

$$P_l^m = (\sin \theta)^m \frac{\partial^m P_l(\theta)}{\partial(\cos \theta)^m} \tag{12.242}$$

$$Q_l^m = (\sin \theta)^m \frac{\partial^m Q_l(\theta)}{\partial(\cos \theta)^m} \tag{12.243}$$

Integrals

Below are some integrals over Legendre polynomials. The integration variable $d\theta$ can be changed to dz where $z = \cos(\theta)$, $dz = -d\theta \sin(\theta)$. The integrals are written using both notations for the polynomials

$$\frac{2}{2l+1}\delta_{ll'} = \int_0^\pi d\theta \sin(\theta)P_l(\theta)P_{l'}(\theta) = \int_{-1}^1 dz P_l(z)P_{l'}(z) \tag{12.244}$$

$$\frac{2}{2l+1}\delta_{ll'}\frac{(l+m)!}{(l-m)!} = \int_0^\pi d\theta \sin(\theta)P_l^m(\theta)P_{l'}^m(\theta) = \int_{-1}^1 dz P_l^m(z)P_{l'}^m(z) \tag{12.245}$$

Remember that $0! = 1$.

Spherical Harmonics

The normalized angular functions are called *spherical harmonics*. They are normalized to give unity when a product of two of them are integrated over all of the solid angle

$$Y_{lm}(\theta, \phi) = N_{lm}e^{im\phi}P_l^{|m|}(\theta) \tag{12.246}$$

$$N_{lm} = \sqrt{\frac{2l+1}{4\pi}\frac{(l-|m|)!}{(l+|m|)!}} \tag{12.247}$$

$$\delta_{ll'}\delta_{mm'} = \int_0^{2\pi} d\phi \int_0^\pi d\theta \sin(\theta) Y_{lm}^*(\theta, \phi)Y_{l'm'}(\theta, \phi) \tag{12.248}$$

For any problem involving ∇^2 in spherical coordinates the spherical harmonics provide the angular functions. The radial functions $f(r)$ depend on the problem and the boundary conditions.

Recursion Relations

There are numerous recursion relations for Legendre polynomials. Two of the most useful ones are given below. The notation uses $z = \cos(\theta)$,

$$(n+1)P_{n+1} + nP_{n-1} = (2n+1)zP_n(z) \tag{12.249}$$

$$-(1-z^2)\frac{d}{dz}P_n(z) = n[zP_n - P_{n-1}] \tag{12.250}$$

Summation Formulas

Some useful summation formulas for Legendre functions are given below using the notation $P_1(z) = z$,

$$\sum_{l=0}^{\infty} (2l + 1)P_l(\mu)P_l(x) = 2\delta(\mu - x) \tag{12.251}$$

$$\sum_{l=0}^{\infty} (2l + 1)P_l(\mu)Q_l(x) = \frac{1}{x - \mu} \tag{12.252}$$

$$\sum_{l=0}^{\infty} x^l P_l(\mu) = \frac{1}{\sqrt{1 + x^2 - 2x\mu}} \tag{12.253}$$

The last identity is valid only when $|x| < 1$.

Addition Theorem

Consider two unit vectors $\hat{r}_1$ and $\hat{r}_2$ that have the angles in spherical coordinates of (θ_1, ϕ_1) and (θ_2, ϕ_2). The law of cosines states that the angle Θ_{12} between them is given by

$$\cos(\Theta_{12}) = \hat{r}_1 \cdot \hat{r}_2 = \cos(\theta_1)\cos(\theta_2) + \sin(\theta_1)\sin(\theta_2)\cos(\phi_1 - \phi_2) \tag{12.254}$$

To the left of the equal sign is $P_1(\Theta_{12})$. To the right of the equal sign are several terms that can be expressed as a summation over m

$$P_1(\Theta_{12}) = \frac{4\pi}{3} \sum_{m=-1}^{1} Y_{1m}(\theta_1, \phi_1)^* Y_{1m}(\theta_2, \phi_2)$$

$$= \frac{4\pi}{3}\left[\frac{3}{8\pi}\sin(\theta_1)\sin(\theta_2)e^{-i(\phi_1 - \phi_2)} + \frac{3}{4\pi}\cos(\theta_1)\cos(\theta_2) \right.$$

$$\left. + \frac{3}{8\pi}\sin(\theta_1)\sin(\theta_2)e^{i(\phi_1 - \phi_2)} \right] \tag{12.255}$$

$$= \cos(\theta_1)\cos(\theta_2) + \sin(\theta_1)\sin(\theta_2)\cos(\phi_1 - \phi_2)$$

This identity is the addition theorem for $l = 1$. The similar result for any value of l is

$$P_l(\Theta_{12}) = \frac{4\pi}{2l + 1} \sum_{m=-l}^{l} Y_{lm}(\theta_1, \phi_1)^* Y_{lm}(\theta_2, \phi_2) \tag{12.256}$$

which is the most general addition theorem.

The summation over all spherical harmonic functions gives the delta function of angles

$$\sum_{l=0}^{\infty} \sum_{m=-l}^{l} Y_{lm}(\theta_1, \phi_1) Y_{lm}^*(\theta_2, \phi_2) = \delta(\phi_1 - \phi_2)\delta(\cos\theta_1 - \cos\theta_2)$$

The addition theorem can be used to give the alternate expression

$$\sum_{l=0}^{\infty} (2l + 1)P_l(\Theta_{12}) = 4\pi\delta(\phi_1 - \phi_2)\delta(\cos\theta_1 - \cos\theta_2) \qquad (12.257)$$

The delta function on the cosine of the angles can be alternately written as

$$\delta(\cos\theta_1 - \cos\theta_2) = \frac{\delta(\theta_1 - \theta_2)}{\sin\theta_1} \qquad (12.258)$$

Appendix III: Spherical Bessel Functions

The radial part of the Helmholtz equation for three dimensional spherical coordinates has the form

$$0 = \left[\mathscr{L}_r + k^2 - \frac{l(l+1)}{r^2}\right]R_l(kr) \qquad (12.259)$$

$$\mathscr{L}_r\Phi(r) = \frac{1}{r}\frac{d^2}{dr^2}(r\Phi) = \frac{1}{r^2}\frac{d}{dr}\left(r^2\frac{d\Phi}{dr}\right) = \frac{d^2\Phi}{dr^2} + \frac{2}{r}\frac{d\Phi}{dr} \qquad (12.260)$$

The three forms for $\mathscr{L}_r$ are equivalent. The solutions to the differential equation are called *spherical Bessel functions*

$$R_l(z) = Aj_l(z) + By_l(z) \qquad (12.261)$$

$$j_l(z) = \sqrt{\frac{\pi}{2z}}J_{l+1/2}(z), \quad y_l(z) = \sqrt{\frac{\pi}{2z}}Y_{l+1/2}(z) \qquad (12.262)$$

$$h_l^{(1)}(z) = j_l(z) + iy_l(z), \quad h_l^{(2)}(z) = j_l(z) - iy_l(z) \qquad (12.263)$$

Lowercase letters denote the spherical Bessel and Neumann functions. They are the regular Bessel functions of half integer multiplied by a prefactor. The spherical Hankel functions are also defined and correspond to outgoing (1) and incoming (2) waves.

The spherical Bessel functions appear to be complicated because they involve regular Bessel functions of half integer. In fact, the opposite is true, they are simpler functions. Some functions of small l are,

$$j_0(z) = \frac{\sin(z)}{z}, \quad y_0(z) = -\frac{\cos(z)}{z} \tag{12.264}$$

$$j_1(z) = \frac{\sin(z)}{z^2} - \frac{\cos(z)}{z}, \quad y_1(z) = -\frac{\cos(z)}{z^2} - \frac{\sin(z)}{z} \tag{12.265}$$

$$j_2(z) = \left(\frac{3}{z^3} - \frac{1}{z}\right)\sin(z) - \frac{3}{z^2}\cos(z) \tag{12.266}$$

Also, note that $j_{-n} = (-1)^n y_{n-1}$. These functions have the property that as $z \to 0$ then $j_n \propto z^n$, $y_n \propto 1/z^{n+1}$. Again, the $j_n(z)$ functions are well-behaved at the origin, while the $y_n(z)$ functions diverge. The explicit formulas for $z \to 0$ are

$$j_n = \frac{z^n}{1 \cdot 3 \cdot 5 \cdots (2n + 1)}[1 - O(z^2)] \tag{12.267}$$

$$y_n = -\frac{1 \cdot 3 \cdot 5 \cdots (2n - 1)}{z^{n+1}}[1 - O(z^2)] \tag{12.268}$$

Some other properties of these functions are listed below:

1. *Integral definition*

$$j_n(z) = \frac{1}{2i^n} \int_0^\pi d\theta \, \sin(\theta) P_n(\theta) e^{iz\,\cos(\theta)}$$

$$= \frac{1}{2i^n} \int_{-1}^1 dv P_n(v) e^{izv} \tag{12.269}$$

2. *Asymptotic limits as $z \to \infty$*

$$j_n(z) \to \frac{1}{z} \sin\left(z - \frac{\pi n}{2}\right) \tag{12.270}$$

$$y_n(z) \to -\frac{1}{z} \cos\left(z - \frac{\pi n}{2}\right) \tag{12.271}$$

$$h_n^{(1)}(z) \to \frac{1}{iz} \exp\left(iz - i\frac{\pi n}{2}\right) = \frac{1}{zi^{n+1}} e^{iz} \tag{12.272}$$

3. *Recursion relations* apply to all $f_n = j_n, y_n, h_n^{(\alpha)}$

$$f_{n-1} + f_{n+1} = \frac{2n + 1}{z} f_n \tag{12.273}$$

$$n f_{n-1} - (n + 1) f_{n+1} = (2n + 1)\frac{d}{dx} f_n \tag{12.274}$$

$$\frac{1}{z} \cos\sqrt{z^2 - 2zt} = \sum_n \frac{t^n}{n!} j_{n-1}(z) \qquad (12.275)$$

$$e^{iz \cos(\theta)} = \sum_n (2n + 1)i^n j_n(z)P_n(\theta) \qquad (12.276)$$

The next three relationships require that $R = \sqrt{r^2 + \rho^2 - 2r\rho \cos \theta}$ and the assumption that $\rho > r$

$$\frac{\sin(kR)}{kR} = \sum_n (2n + 1) j_n(kr) j_n(k\rho)P_n(\theta) \qquad (12.277)$$

$$\frac{\cos(kR)}{kR} = -\sum_n (2n + 1) j_n(kr)y_n(k\rho)P_n(\theta) \qquad (12.278)$$

$$\frac{e^{ikR}}{kR} = \sum_n (2n + 1) j_n(kr)h_n^{(1)}(k\rho)P_n(\theta) \qquad (12.279)$$

References

1. A. E. H. Love, *A Treatise on the Mathematical Theory of Elasticity*. Dover, 1944.
2. L. D. Landau and E. M. Lifshitz, *Theory of Elasticity*. Addision-Wesley, 1959.
3. A. Erdélyi, ed., *Tables of Integral Transforms: Bateman Manuscript Project*, Vol. 1 and 2. McGraw-Hill, 1954.

Problems

1. Solve Laplace's equation to find the steady state temperature distribution inside a cylinder with a radius a and length h that has $T = 0$ on the end boundaries and a temperature distribution on the outside of the cylinder $T(a, z) = T_0 \sin(z\pi/h)$.

2. Solve Laplace's equation to find the steady state temperature distribution of a cylinder with a radius a and length h that has $T = 0$ on all boundaries except the top ($z = h$) that has T_0.

3. Treat the previous problem as a diffusion problem. The cylinder initially has $T = 0$ everywhere. Then at $t = 0$ the top is put in contact with a reservoir T_0 while the sides are in contact with a reservoir $T = 0$. Find $T(\rho, z, t)$.

4. A cylinder with radius a extends in infinity along the $\pm z$ directions. The surface is held at the potential $\phi_0 \sin(kz)$. Find the potential everywhere outside the cylinder.

5. Consider a semi-infinite circular hollow wave guide that extends $0 \leqslant z < \infty$. The field $\phi(\rho, z, t) = 0$ on the surface $\rho = a$. At the end $z = 0$ is the boundary condition $\phi(\rho, 0, t) = \phi_0 J_0(\lambda_{10}\rho/a) \sin(\omega t)$. Find the form of the potential $\phi(\rho, z, t)$ inside the guide for $z > 0$.

6. Find P_3^m for all values of m. Use the recursion relations to find P_7.

7. Use the integral definition of $j_n(z)$ to prove the recursion relation

$$j_n(z) = \frac{1}{2i^n} \int_0^\pi d\theta \, \sin(\theta) P_n(\theta) e^{iz \, \cos(\theta)}$$

$$nj_{n-1} - (n+1)j_{n+1} = (2n+1)\frac{dj_n}{dz}$$

8. Solve Laplace's equation $\nabla^2 T = 0$ for a spherical shell $(a < r < b)$ subject to the boundary conditions that

$$T(a, \theta, \phi) = T_a \sin(\theta) \cos(\phi) \qquad\qquad (12.280)$$

$$T(b, \theta, \phi) = T_b \sin(\theta) \cos(\phi) \qquad\qquad (12.281)$$

9. Solve Laplace's equation $\nabla^2 T = 0$ for a spherical shell $(a < r < b)$ subject to the boundary conditions that

$$T(a, \theta, \phi) = T_a \qquad\qquad (12.282)$$

$$T(b, \theta, \phi) = T_b \cos(\theta) \qquad\qquad (12.283)$$

10. Consider a hollow sphere with an inner radius of a and an outer radius of b. For $t < 0$ the temperature is zero. Beginning at $t = 0$ the inner radius is put into contact with a heat reservoir at T_a while the outer radius is at T_b. Solve the diffusion equation to find $T(r, t)$ for $t > 0$.

11. Consider a hollow sphere with an inner radius of a and outer radius of b. For $t < 0$ the temperature is zero. Beginning at $t = 0$ the outer radius is put into contact with a heat reservoir at T_b while the inner radius has no heat flow, $\partial T/\partial r = 0$. Solve the diffusion equation to find $T(r, t)$ for $t > 0$.

12. Solve Green's function for the scalar wave equation inside a sphere with a radius a. Assume that it vanishes at the surface of the sphere.

13. Solve the vector wave equation for a torsional wave going down a solid rod.
    (a) Show that the following function satisfies the vector wave equation

$$\mathbf{u}(\mathbf{r}, t) = u_0 \rho \hat{\theta} \exp[i(kz - \omega t)] \qquad\qquad (12.284)$$

    (b) Find the relationship between frequency and wave vector.
    (c) Show that this wave obeys the stress free boundary conditions at $\rho = a$.

13

Odds and Ends

This chapter is concerned with four overlapping topics. The first is to discuss the solution of differential equations using series. This method is quite powerful and is very useful on new equations for which the solutions are unknown. The second topic is to discuss orthogonal functions such as Laguerre polynomials and Hermite polynomials. The third topic is the Sturm–Liouville theory, that generalizes the properties of orthogonal polynomials. The final topic is singular integral equations.

Before discussing series, it is important to emphasize that all solutions to differential equations are actually series. For example, the Helmholtz equation in two dimensions is solved in terms of Bessel functions $J_n(z)$, $N_n(z)$. They are of course, just series. They are series which occur often enough that they have been given a name. Sines and cosines are other series which occur often enough to have been given names. Many differential equations can be solved using series, as is shown below.

13.1. Hypergeometric Functions

Hypergeometric functions are series that have a particular structure. They are found in solving many differential equations. Some of their properties are reviewed.

The first case is called *confluent hypergeometric functions* and is denoted as $F(a, b, z)$ and $U(a, b, z)$. They both are the solution to the differential equation

$$\left[z \frac{d^2}{dz^2} + (b - z) \frac{d}{dz} - a \right] F(a, b, z) = 0 \qquad (13.1)$$

By agreement $F(a, b, 0) = 1$ at $z = 0$. A series expansion is used to find this function. Assume a form

$$F = \sum_{n=0}^{\infty} A_n z^n \qquad (13.2)$$

where the coefficients A_n need to be determined. Insert this series into the differential equation

$$\frac{d}{dz}F = \sum_n A_n n z^{n-1} \tag{13.3}$$

$$\frac{d^2}{dz^2}F = \sum_n A_n n(n-1)z^{n-2} \tag{13.4}$$

The differential equation is

$$\begin{aligned}
0 &= \sum_n A_n[n(n-1)z^{n-1} + (b-z)nz^{n-1} - az^n] \\
&= \sum_n A_n[z^{n-1}n(n-1+b) - z^n(n+a)]
\end{aligned} \tag{13.5}$$

In the above series, rearrange terms so that each one has the same coefficient of z^n. In the term with z^{n-1} redefine $n' = n - 1$, $n = n' + 1$.

$$0 = \sum_n z^n[A_{n+1}(n+1)(n+b) - A_n(n+a)] \tag{13.6}$$

The summation is zero for any value of z. The only way this can happen is if each coefficient of z^n vanishes separately. This requirement produces the two term recursion relation

$$A_{n+1}(n+1)(n+b) = A_n(n+a) \tag{13.7}$$

Evaluate this equation successively for $n = 0, 1, \ldots$ which gives

$$A_1 = A_0 \frac{a}{b} \tag{13.8}$$

$$A_2 = A_1 \frac{(a+1)}{2(b+1)} = A_0 \frac{a(a+1)}{2!b(b+1)} \tag{13.9}$$

$$A_n = A_0 \frac{(a)_n}{n!(b)_n} \tag{13.10}$$

$$(a)_n \equiv a(a+1)(a+2)\cdots(a+n-1) = \frac{\Gamma(a+n)}{\Gamma(a)} \tag{13.11}$$

All expansion coefficients can be expressed as A_0 times a polynomial in z. Furthermore, the requirement $F(a, b, z = 0) = 1$ forces $A_0 = 1$. The series for the confluent hypergeometric function is

$$F(a, b, z) = \sum_{n=0}^{\infty} \frac{z^n}{n!} \frac{(a)_n}{(b)_n} \tag{13.12}$$

The other solution $U(a, b, z)$ cannot be expanded as a series in a simple way.

At large values of z there is the asymptotic expansion

$$\lim_{z \to \infty} F(a, b, z) = \frac{\Gamma(b)}{\Gamma(a)} e^z z^{a-b} [1 + O(1/z)] \qquad (13.13)$$

This result is used often. The *Kummer transformation* gives an important exact identity

$$F(a, b, z) = e^z F(b - a, b, - z) \qquad (13.14)$$

There are also many recursion relations between different functions.

Another important feature of this function is if a is a negative integer then the series has a finite number of terms. As an example

$$F(-2, b, z) = 1 + \frac{z}{1!} \frac{(-2)}{b} + \frac{z^2}{2!} \frac{(-2)(-1)}{b(b+1)} \qquad (13.15)$$

Further terms are zero since they have $(2 - 2) = 0$. That is, $(-m)_n = 0$ if $n > m$. Of course, in this case the asymptotic expansion is not an exponential but is a polynomial of $O(z^m)$.

We now list some simple functions that result from particular choices for (a, b),

$$F(a, a, z) = e^z \qquad (13.16)$$

$$F(1, 2, z) = \frac{e^{2z} - 1}{2z} \qquad (13.17)$$

$$F(1/2, 3/2, -x^2) = \frac{\sqrt{\pi}}{2x} \mathrm{erf}(x) \qquad (13.18)$$

$$F(v + 1/2, 2v + 1, 2z) = \Gamma(1 + v) \left(\frac{2}{z}\right)^v e^z I_v(z) \qquad (13.19)$$

$$F(-n, 1 + \alpha, x) = \frac{n!}{(\alpha + 1)_n} L_n^{(\alpha)}(x) \qquad (13.20)$$

where the special functions are, error, modified Bessel, and Laguerre. Many other such identities can be found in standard references.

A *hypergeometric function* is the series

$$F(a, b; c; z) \equiv {}_2F_1(a, b; c; z)$$

$$= \sum_{n=0}^{\infty} \frac{z^n}{n!} \frac{(a)_n (b)_n}{(c)_n} \qquad (13.21)$$

In the notation $_pF_q$ the first subscript p gives the number of terms $(a)_n$ in the numerator and q gives the number of such terms in the denominator of the coefficients in the series. This hypergeometric function obeys the differential equation $[w = {}_2F_1(a, b; c; z)]$

$$\left[z(1 - z) \frac{d^2}{dz^2} + [c - (a + b + 1)z] \frac{d}{dz} - ab \right] w = 0 \tag{13.22}$$

that can be shown by direct substitution into the differential equation.

$$F(z) = \sum_n A_n z^n \tag{13.23}$$

$$0 = \left[z(1 - z) \frac{d^2}{dz^2} + [c - (a + b + 1)z] \frac{d}{dz} - ab \right] F(z)$$

$$= \sum_n A_n \{ n(n - 1)(z^{n-1} - z^n) + n[cz^{n-1} - (a + b + 1)z^n] - abz^n \}$$

$$\tag{13.24}$$

$$= \sum_n A_n \{ z^{n-1} n(n - 1 + c) - z^n [n(n - 1) + n(a + b + 1) + ab] \}$$

$$= \sum_n z^n [A_{n+1}(n + 1)(n + c) - A_n(n + a)(n + b)]$$

The last equation is derived by changing $n' = n - 1$ in the first term of the prior equation. Since the equation vanishes for all values of z, the coefficient of each power of z^n must vanish separately. This step gives the two term recursion relation

$$A_{n+1} = A_n \frac{(n + a)(n + b)}{(n + 1)(n + c)} \tag{13.25}$$

The recursion relation gives the terms in the series for the hypergeometric function.

The hypergeometric functions have many interesting relationships. For example, a single derivative gives

$$\frac{d}{dz} F(a, b; c; z) = \frac{ab}{c} F(a + 1, b + 1; c + 1; z) \tag{13.26}$$

There are many recursion relations among hypergeometric functions. One example is

$$(b - a)F(a, b; c; z) + aF(a + 1, b; c; z) = bF(a, b + 1; c; z) \tag{13.27}$$

When $a, b, c,$ or z have a special value then the hypergeometric function is often a simple function, such as

$$F(a, b; c; 1) = \frac{\Gamma(c)\Gamma(c - a - b)}{\Gamma(c - a)\Gamma(c - b)} \tag{13.28}$$

$$F(1, 1; 2; z) = -\frac{1}{z} \ln(1 - z) \tag{13.29}$$

$$F\left(\frac{1}{2}, 1; \frac{3}{2}; z^2\right) = \frac{1}{2z} \ln\left(\frac{1+z}{1-z}\right) \tag{13.30}$$

$$F\left(\frac{1}{2}, 1; \frac{3}{2}; -z^2\right) = \frac{1}{z} \tan^{-1}(z) \tag{13.31}$$

$$F(a, b; b; z) = \frac{1}{(1-z)^a} \tag{13.32}$$

Many more such relationships are given in books on functions.

13.1.1. Continued Fractions

Some series solutions have two term recursion relations such as $a_{n+1} = a_n r_n$ where a_n are the coefficients of the series expansion and r_n is some known function of n. A two term recursion relation can always be solved in terms of a hypergeometric function.

The next level of complexity is to have the series obey a three term recursion relation such as

$$0 = \alpha_{n+1} a_{n+1} + \beta_n a_n + \gamma_{n-1} a_{n-1} \tag{13.33}$$

where α_n, β_n, and γ_n are known functions. The unknown coefficients a_n are determined by continued fractions. First, rewrite the above expression as

$$a_n\left[\beta_n + \alpha_{n+1} \frac{a_{n+1}}{a_n}\right] = -\gamma_{n-1} a_{n-1} \tag{13.34}$$

$$\frac{a_n}{a_{n-1}} = \frac{-\gamma_{n-1}}{\beta_n + \alpha_{n+1}\dfrac{a_{n+1}}{a_n}} \tag{13.35}$$

$$= -\frac{\gamma_{n-1}/\beta_n}{1 + \dfrac{\alpha_{n+1}}{\beta_n}\dfrac{a_{n+1}}{a_n}}$$

The above expression is a type of two term recursion relation. It can be iterated

$$r_n = \frac{\alpha_{n+1}\gamma_n}{\beta_n\beta_{n+1}} \tag{13.36}$$

$$\frac{a_n}{a_{n-1}} = \frac{-\gamma_{n-1}/\beta_n}{1 - \dfrac{r_n}{1 - \dfrac{r_{n+1}}{\ddots}}} \tag{13.37}$$

The notation is obviously cumbersome. Another notation is

$$\frac{a_n}{a_{n-1}} = \frac{-\gamma_{n-1}/\beta_n}{1 - r_n/1 - r_{n+1}/1 - r_{n+2}/\cdots} \tag{13.38}$$

The general form of a continued fraction is

$$F = \frac{a_1}{b_1 + a_2/b_2 + a_3/b_3 + a_4/b_4 + \cdots}$$

$$= \frac{a_1}{b_1+} \frac{a_2}{b_2+} \frac{a_3}{b_3+} \cdots$$

(13.39)

The second set of notation is common. A continued fraction can be periodic if the coefficients (a_n, b_n) repeat in a regular pattern. A period-1 continued fraction has constants alike $(a_n = a, b_n = b)$. Then we can evaluate it trivially

$$F = \frac{a}{b + F}$$

(13.40)

$$F(F + b) = a$$

(13.41)

$$F = -\frac{b}{2} \pm \sqrt{a + b^2/4}$$

(13.42)

A real result requires that $4a + b^2 > 0$. A period-2 continued fraction has four different constants (a_1, b_1, a_2, b_2) and $a_{j+2m} = a_j$. It can be evaluated exactly in a similar fashion. For a general series that is not periodic, the coefficients can be evaluated on the computer.

Many simple functions obey three term recursion relations and can be used to generate a continued fraction. As an example, start with the recursion relation for Bessel functions

$$\frac{2n}{z} J_n(z) - J_{n+1}(z) = J_{n-1}(z)$$

(13.43)

$$\frac{2n}{z} J_n(z) \left[1 - \frac{z}{2n} \frac{J_{n+1}}{J_n} \right] = J_{n-1}(z)$$

(13.44)

$$\frac{J_n}{J_{n-1}} = \frac{z/(2n)}{1 - \frac{z}{2n} \frac{J_{n+1}}{J_n}}$$

(13.45)

$$\frac{J_n}{J_{n-1}} = \frac{z/(2n)}{1-(z/2)^2 \dfrac{1}{n(n+1)} \bigg/ 1-(z/2)^2 \dfrac{1}{(n+1)(n+2)} \bigg/ 1-(z/2)^2 \dfrac{1}{(n+2)(n+3)} \bigg/ 1 - \cdots}$$

(13.46)

If we set $z = 2$ this gives the continued fraction for $r_n = 1/[n(n + 1)]$. When $n = 1$

$$\frac{J_1(z)}{J_0(z)} = \frac{z/2}{1 - \frac{1}{2}(z/2)^2/1 - \frac{1}{6}(z/2)^2/1 - \cdots}$$

(13.47)

A different continued fraction can be derived for each function that has a three term recursion relation.

Many differential equations have a three or four term recursion relation if they are solved directly. It is often possible to reduce the problem to a two term recursion relation by making various substitutions. A two term recursion relation can always be solved in terms of a hypergeometric function. The substitutions are found by examining the behavior of the solution at its end points.

Example 1. *One dimensional Coulomb potential:* As an example, consider the equation

$$0 = \left[\frac{d^2}{dz^2} - \frac{C}{z} - a \right] \phi(z) \tag{13.48}$$

where $0 \leqslant z < \infty$. The constant C is known and a is the eigenvalue. We want the solution that is regular at point $z = 0$. This differential equation will generate a three term recursion relation if solved directly by the series method. Instead, examine the behavior of the solution as $z \to \infty$. Then term C/z is negligible compared with term a, so at very large values of z we effectively have the equation

$$0 = \left[\frac{d^2}{dz^2} - a \right] \phi(z) \tag{13.49}$$

$$\phi(z) \to A e^{-z\sqrt{a}} + B e^{+z\sqrt{a}} \tag{13.50}$$

Since we want the solution to be well-behaved at large z, set $B = 0$. Since we know the solution goes as an exponential at large z, consider the form of the solution

$$\phi(z) = e^{-z\sqrt{a}} G(z) \tag{13.51}$$

The problem has been altered to find the function $G(z)$. If it is found exactly, then so has $\phi(z)$. This assumed form is inserted into the original differential equation in order to generate the equation obeyed by $G(z)$. First, take some derivatives where a prime denotes a derivative

$$\frac{d}{dz} [e^{-z\sqrt{a}} G(z)] = e^{-z\sqrt{a}} [G' - \sqrt{a}\, G(z)] \tag{13.52}$$

$$\frac{d^2}{dz^2} [e^{-z\sqrt{a}} G(z)] = e^{-z\sqrt{a}} [G'' - 2\sqrt{a}\, G' + aG] \tag{13.53}$$

The original differential equation is now

$$0 = e^{-z\sqrt{a}} \left[G'' - 2\sqrt{a}\, G' - \frac{C}{z} G \right] \tag{13.54}$$

$$G(z) = \sum_n A_n z^n \tag{13.55}$$

$$0 = \sum_n A_n[n(n-1)z^{n-2} - z^{n-1}(2\sqrt{a}\,n + C)] \tag{13.56}$$

$$A_{n+1} = A_n \frac{2\sqrt{a}\,n + C}{n(n+1)} = 2\sqrt{a}\,A_n \frac{C' + n}{n(n+1)} \tag{13.57}$$

$$C' = \frac{C}{2\sqrt{a}} \tag{13.58}$$

The differential equation for $G(z)$ has only a two term recursion relation, whose solution is given by a confluent hypergeometric function. Note that there is no solution for $n = 0$, so $A_0 = 0$ and the first nonzero term is A_1. The solution is

$$G(z) = A_1[F(C' + 1, 2, z') - 1] \tag{13.59}$$

$$z' = 2\sqrt{a}\,z \tag{13.60}$$

$$\phi(z) = A_1 e^{-z\sqrt{a}}[F(C' + 1, 2, z') - 1] \tag{13.61}$$

At large values of z the confluent hypergeometric function, $F(z') \sim e^{z'} = e^{2\sqrt{a}z}$, makes $\phi(z)$ diverge. The only way to keep $\phi(z)$ from diverging in this limit is to require that the series truncate after m terms. This truncation is achieved by requiring that $C' + 1 = -m$ and gives

$$\frac{C}{2\sqrt{a}} = -(m + 1) \tag{13.62}$$

The above identity can only be satisfied if $C < 0$. Otherwise there are no solutions that are well-behaved at infinity or at the origin. We can satisfy one or the other of these requirements, but in order to satisfy both, $C < 0$ and the eigenvalue is

$$a = \frac{C^2}{4(m+1)^2} \tag{13.63}$$

In physics this problem corresponds to the bound states of the one dimensional Coulomb potential. A physical example of this problem is a particle bound by its own image potential outside the surface of a conductor, for example a good metal. The requirement that $C < 0$ is that the particle must be attracted to its image rather than repulsed. There are no bound states if it is repulsed. For bound states, $F(-m, 2, z')$ is just the Laguerre polynomial $L_m^{(1)}(z')$ discussed in the next section.

Example 2. *Hydrogen atom:* Another problem where a three term recursion relation is turned into a two term relation is the equation in spherical coordinates

$$\left[\nabla^2 + \frac{2}{r} - \varepsilon\right]\phi(\mathbf{r}) = 0 \tag{13.64}$$

where ε is the eigenvalue determined by the boundary condition that the solution converge in the limit $r \to \infty$. This differential equation is the dimensionless form of the wave functions of an electron bound to a proton – the hydrogen atom. The variable r is the distance in terms of Bohr radii, while ε is the bound state energy in terms of Rydbergs. Here, it is just a differential equation whose solutions are required to be well-behaved at $r = 0$ and $r \to \infty$.

Assume the solution has angular functions given by the spherical harmonics $Y_{lm}(\theta, \phi)$, and then we only need to solve the radial equations

$$\phi(\mathbf{r}) = Y_{lm}(\theta, \phi) R_l(r) \tag{13.65}$$

$$0 = \left[\mathcal{L}_r - \frac{l(l+1)}{r^2} + \frac{2}{r} - \varepsilon \right] R_l(r) \tag{13.66}$$

Multiply both equations by r and define $\chi_l(r) = r R_l(r)$ which obeys the equation

$$0 = \left[\frac{d^2}{dr^2} - \frac{l(l+1)}{r^2} + \frac{2}{r} - \varepsilon \right] \chi_l(r) \tag{13.67}$$

This equation has a three term recursion relation if solved directly by series. Instead, examine some possible limiting solutions. At large values of r the largest nonderivative term is ε, so the asymptotic solution is $\phi \sim \exp(-r\sqrt{\varepsilon})$. The other end point is at $r \to 0$. Then, the largest nonderivative term is $-l(l+1)/r^2$ and the solution is

$$0 = \left[\frac{d^2}{dr^2} - \frac{l(l+1)}{r^2} \right] \chi(r) \tag{13.68}$$

$$\lim_{r \to 0} \chi(r) \to r^{l+1} \tag{13.69}$$

These two limits suggest the substitution

$$\chi_l(r) = r^{l+1} e^{-r\sqrt{\varepsilon}} G(r) \tag{13.70}$$

which defines the new unknown series $G(r)$. This form of the solution is inserted into the differential equation [Eq. (13.67)]. First, evaluate the second derivative

$$\frac{d}{dr}[r^{l+1} e^{-r\sqrt{\varepsilon}} G(r)] = e^{-r\sqrt{\varepsilon}} [r^{l+1} G'(r) - \sqrt{\varepsilon} r^{l+1} G + (l+1) r^l G] \tag{13.71}$$

$$\frac{d^2}{dr^2}[r^{l+1} e^{-r\sqrt{\varepsilon}} G(r)] = e^{-r\sqrt{\varepsilon}} r^{l+1}$$

$$\times \left\{ G'' - 2\left[\sqrt{\varepsilon} - \frac{l+1}{r} \right] G' + \left[\varepsilon - \frac{2\sqrt{\varepsilon}(l+1)}{r} + \frac{l(l+1)}{r^2} \right] G \right\} \tag{13.72}$$

The original equation [Eq. (13.67)] gives a differential equation for $G(r)$

$$0 = G'' - 2\left[\sqrt{\varepsilon} - \frac{l+1}{r}\right]G' + \frac{2G}{r}[1 - \sqrt{\varepsilon}(l+1)] \tag{13.73}$$

Multiply the entire equation by $r/2\sqrt{\varepsilon}$, then define the new variable $z = 2r\sqrt{\varepsilon}$, and the equation is now

$$0 = z\frac{d^2}{dz^2}G + [2(l+1) - z]\frac{d}{dz}G - \left(l+1 - \frac{1}{\sqrt{\varepsilon}}\right)G \tag{13.74}$$

This has the classic form for a confluent geometric function, where $a = l+1 - 1/\sqrt{\varepsilon}$ and $b = 2(l+1)$. Bound state require that the series be truncated or $a = -n_r$. The integer n_r is called *the radial quantum number* while $n \equiv l+1+n_r$ is the *principle quantum number*,

$$R_l(r) = r^l e^{-r/n} F(-n_r, 2(l+1), 2r/n) \tag{13.75}$$

$$a = l + 1 - \frac{1}{\sqrt{\varepsilon}} = n_r \tag{13.76}$$

$$\varepsilon = \frac{1}{n^2} \tag{13.77}$$

$$z = 2r\sqrt{\varepsilon} = \frac{2r}{n} \tag{13.78}$$

Again the confluent hypergeometric series is related to a Laguerre polynomial. These results are well known solutions for quantum states of an electron in the hydrogen atom.

Example 3. *Harmonic oscillator:* This example considers the solution of the differential equation

$$0 = \left[\frac{d^2}{dz^2} - z^2 + a\right]\phi(z) \tag{13.79}$$

Here the space is $-\infty < z < \infty$ and we want the solutions that are regular at the origin and vanish as $z^2 \to \infty$. At large values of $|z|$ the largest nonderivative term is $-z^2$. Asymptotically the equation becomes

$$0 = \left[\frac{d^2}{dz^2} - z^2\right]\phi(z) \tag{13.80}$$

$$\phi(z) = e^{-z^2/2} \tag{13.81}$$

$$\frac{d^2}{dz^2}e^{-z^2/2} = e^{-z^2/2}[z^2 - 1] \tag{13.82}$$

The factor of $\exp(-z^2/2)$ gives the correct asymptotic solution. Note that its derivative also has another term of $O(1)$ which is acceptable. So try a solution of the form

$$\phi(z) = e^{-z^2/2}G(z) \qquad (13.83)$$

$$\frac{d^2}{dz^2}[e^{-z^2/2}G(z)] = e^{-z^2/2}[G'' - 2zG' + (z^2 - 1)G] \qquad (13.84)$$

If we put this form for $\phi(z)$ into the original equation [Eq. (13.79)] we get a differential equation for $G(z)$

$$0 = \frac{d^2}{dz^2}G - 2z\frac{d}{dz}G - (1-a)G \qquad (13.85)$$

This equation can be solved by a series and gives a two term recursion relation

$$G(z) = \sum_n A_n z^n \qquad (13.86)$$

$$0 = \sum_n A_n[n(n-1)z^{n-2} - z^n(2n+1-a)] \qquad (13.87)$$

$$A_{n+2} = A_n \frac{2n+1-a}{(n+1)(n+2)} \qquad (13.88)$$

$$G(z) = A_0\left[1 + \frac{z^2}{2!}(1-a) + \frac{z^4}{4!}(1-a)(5-a) + \cdots + \right]$$
$$+ A_1\left[z + \frac{z^3}{3!}(3-a) + \frac{z^5}{5!}(3-a)(7-a) + \cdots + \right] \qquad (13.89)$$

Bound states are obtained by truncating the series. For symmetric solutions $G(-z) = G(+z)$ then $A_1 = 0$, $a = 2m+1$ gives a series of m terms. For antisymmetric states $G(-z) = -G(z)$ then $A_0 = 0$, $a = 2m+3$ gives the proper solution. These polynomials are related to the Hermite polynomials discussed in the next section.

13.2. Orthogonal Polynomials

A number of orthogonal polynomials have already been treated in previous chapters, such as $\sin(k_n x)$, $\cos(k_n x)$, $P_n(\theta)$, $Y_{lm}(\theta, \phi)$, $J_n(z)$, and $j_l(z)$. Several more are discussed in this section, parabolic cylinder functions, Laguerre polynomials, and Hermite polynomials. These we have found to be the most useful. Other orthogonal functions or polynomials are omitted because they are only used in very special types of problems such as Chebyshev, Jacobi, or Gegenbauer polynomials.

13.2.1. Parabolic Cylinder Functions

Consider the differential equation

$$0 = \left[\frac{d^2}{dz^2} - \frac{z^2}{4} - a \right] y(z) \tag{13.90}$$

The solutions to this equation are given in terms of parabolic cylinder functions. Since it is a second order equation there are two independent solutions. The solutions which are regular at the origin can be given in terms of confluent hypergeometric functions. There is a solution y_1 that is even in z, and another solution y_2 that is an odd function of z;

$$y_1(z) = e^{-z^2/4} F(1/4 + a/2, 1/2; z^2/2)$$
$$= e^{z^2/4} F(1/4 - a/2, 1/2, -z^2/2) \tag{13.91}$$

$$y_2(z) = z e^{-z^2/4} F(3/4 + a/2, 3/2; z^2/2) \tag{13.92}$$

The second form of y_1 is derived from the first by a Kummer transformation. Physics students recognize Eq. (13.90) as the dimensionless form of Schrödinger's equation for the one dimensional harmonic oscillator

$$0 = \left[-\frac{\hbar^2}{2m} \frac{d^2}{dx^2} + \frac{K}{2} x^2 - E \right] \phi(x) \tag{13.93}$$

where $\hbar$ is Planck's constant, m is the mass of the particle, K is the spring constant, and E is the eigenvalue. The dimensionless form is found by using the unit of length $l = [\hbar^2/(mK)]^{1/4}$. Then, make the variable change $z = \sqrt{2}\, x/l$, $a = -E\sqrt{m/K\hbar^2}$. Physics students know the solution to the harmonic oscillator is given in terms of Hermite polynomials. What is the relationship between parabolic cylinder functions and Hermite polynomials?

If the differential equation [Eq. (13.90)] acts over the finite range $z_1 \leqslant z \leqslant z_2$ the solutions are given in terms of parabolic cylinder functions. In this case the constant a can have any real value. However, if the differential equation applies over all space $-\infty < z < \infty$ the constant $a = -(n + 1/2)$ is the only permissible eigenvalue. The parabolic cylinder functions become proportional to Hermite polynomials for $a = -(n + 1/2)$.

13.2.2. Hermite Polynomials

Hermite polynomials $H_n(x)$ are solutions to the differential equations

$$0 = \left[\frac{d^2}{dx^2} - 2x \frac{d}{dx} + 2n \right] H_n(x) \tag{13.94}$$

$$0 = \left[\frac{d^2}{dx^2} - x^2 + 2n + 1 \right] f_n(x), \quad f_n(x) = e^{-x^2/2} H_n(x) \tag{13.95}$$

The polynomials are another confluent hypergeometric function. The series has a different form depending on whether the integer n is even ($n = 2m$) or odd

$$H_{2m}(x) = (-1)^m \frac{2m)!}{m!} F(-m, 1/2, x^2) \tag{13.96}$$

$$H_{2m+1}(x) = (-1)^m 2x \frac{(2m+1)!}{m!} F(-m, 3/2, x^2) \tag{13.97}$$

$$H_0 = 1 \tag{13.98}$$

$$H_1 = 2x \tag{13.99}$$

$$H_2 = -2(1 - 2x^2) \tag{13.100}$$

$$H_3 = -4x(3 - 2x^2) \tag{13.101}$$

$$H_4 = 4[3 - 12x^2 + 4x^4] \tag{13.102}$$

A typical recursion relation is

$$2xH_n(x) = H_{n+1}(x) + 2n\, H_{n-1}(x) \tag{13.103}$$

The final two results are the orthogonality integral and the generating function

$$\int_{-\infty}^{\infty} dx e^{-x^2} H_n(x) H_m(x) = \delta_{nm} \sqrt{\pi}\, 2^n n! \tag{13.104}$$

$$e^{2xz - z^2} = \sum_{n=0}^{\infty} \frac{z^n}{n!} H_n(x) \tag{13.105}$$

Generating functions are useful for evaluating integrals involving polynomials. As an example, the orthgonality integral Eq. (13.104) is derived using generating functions. Start with two generating functions and take the integral of their product

$$e^{2xz - z^2} = \sum_{n=0}^{\infty} \frac{z^n}{n!} H_n(x) \tag{13.106}$$

$$e^{2xy - y^2} = \sum_{m=0}^{\infty} \frac{y^m}{m!} H_m(x) \tag{13.107}$$

$$\sum_{nm} \frac{z^n y^m}{n! m!} I_{nm} = e^{-y^2 - z^2} \int_{-\infty}^{\infty} dx e^{-x^2 + 2x(y+z)} \tag{13.108}$$

$$I_{nm} = \int_{-\infty}^{\infty} dx e^{-x^2} H_n(x) H_m(x) \tag{13.109}$$

$$e^{-y^2 - z^2} \int_{-\infty}^{\infty} dx e^{-x^2 + 2x(y+z)} = e^{-y^2 - z^2 + (y+z)^2} \int_{-\infty}^{\infty} dx e^{-(x-y-z)^2} = \sqrt{\pi}\, e^{2yz} \tag{13.110}$$

$$\sum_{nm} \frac{z^n y^m}{n! m!} I_{nm} = \sqrt{\pi}\, e^{2yz} = \sqrt{\pi} \sum_l \frac{(2yz)^l}{l!} \tag{13.111}$$

Compare the two series in the last equation. On the right, the exponents for y and z are identical, so they must be identical on the left. This shows that $I_{nm} = \delta_{nm} \tilde{I}_n$. When $n = m = l$ we get the identity

$$\frac{\tilde{I}_n}{(n!)^2} = \sqrt{\pi} \frac{2^n}{n!} \tag{13.112}$$

$$\tilde{I}_n = \sqrt{\pi} \, 2^n n! \tag{13.113}$$

which proves Eq. (13.104). The same technique can be used to evaluate many integrals that contain Hermite polynomials.

13.2.3. Laguerre Polynomial

The Laguerre polynomials are familiar to physics students as the eigenfunctions of Schrödinger's equation for the hydrogen atom. They also are encountered in manipulating certain operators in the statistical mechanics of the harmonic oscillator.

The ordinary Laguerre polynomial is denoted as $L_n(x)$ and the *associated Laguerre polynomial* is denoted as $L_n^{(\alpha)}(x)$. Setting $\alpha = 0$ recovers $L_n(x)$. They obey the differential equation

$$0 = \left[x \frac{d^2}{dx^2} + (1 + \alpha - x) \frac{d}{dx} + n \right] L_n^{(\alpha)}(x) \tag{13.114}$$

This differential equation is recognized as the classic form for confluent hypergeometric functions

$$L_n^{(\alpha)}(x) = \binom{n + \alpha}{n} F(-n, 1 + \alpha; x) \tag{13.115}$$

$$L_0 = 1 \tag{13.116}$$

$$L_1 = 1 - x \tag{13.117}$$

$$L_2 = 1 - 2x + \frac{x^2}{2} \tag{13.118}$$

All sources agree on the definition of $L_n(x)$. For the associated polynomials there are two different versions that differ by the prefactor in Eq. (13.115). Some books omit this prefactor. The normalization integral and generating function are

$$\int_0^\infty dx x^\alpha e^{-x} L_n^{(\alpha)}(x) L_m^{(\alpha)}(x) = \delta_{nm}$$

$$\frac{\exp\left[\dfrac{xz}{z-1}\right]}{(1-z)^{1+\alpha}} = \sum_{n=0}^\infty z^n L_n^{(\alpha)}(x) \tag{13.119}$$

The three term recursion formula is

$$(n + 1)L_{n+1}^{(\alpha)}(x) = (2n + 1 + \alpha - x)L_n^{(\alpha)}(x) - (n + \alpha)L_{n-1}^{(\alpha)}(x) \quad (13.120)$$

The Laguerre polynomials have a number of interesting series identities such as

$$\sum_{m=0}^{n} L_m^{(\alpha)}(x)L_{n-m}^{(\beta)}(y) = L_n^{(\alpha+\beta+1)}(x + y) \quad (13.121)$$

$$\sum_{m=0}^{n} L_m^{(\alpha)}(x) = L_n^{(1+\alpha)}(x) \quad (13.122)$$

The final equation is the relation to the differential equation for the wavefunction of the hydrogen atom. Below is an equation and its solution

$$0 = \left[\frac{d^2}{dz^2} + \frac{2n + \alpha + 1}{z} + \frac{1 - \alpha^2}{z^2} - \frac{1}{4}\right] f_n(z) \quad (13.123)$$

$$f_n(z) = e^{-z/2} z^{\alpha/2} L_n^{(\alpha)}(z) \quad (13.124)$$

The first term in the differential equation is the kinetic energy operator. The second term is the Coulomb potential $(1/r)$ while the third term $(1/r^2)$ is the angular momentum part of the kinetic energy. The term $(-1/4)$ is the eigenvalue. The differential equation is the dimensionless form of Schrödinger's equation for an electron in the hydrogen atom.

13.3. Sturm–Liouville Theory

The *Liouville equation* is

$$\frac{d}{dz}\left(p(z)\frac{d\phi(z)}{dz}\right) + [q(z) + \lambda r(z)]\phi(z) = 0 \quad (13.125)$$

This section is devoted to elucidating the properties of this differential equation. The functions $p(z)$, $q(z)$, and $r(z)$ are assumed to be known. The functions $p(z)$ and $r(z)$ are required to be positive. The eigenvalue is λ. With the proper choice of (p, q, r) the Liouville equation represents every one dimensional differential equation discussed so far in this book. This assertion is explained below.

The equation will be solved over the region $z_1 \leqslant z \leqslant z_2$, where $z_{1,2}$ are the boundaries. The function $p(z)$ cannot have a zero in this interval. If it does, the solutions become ill-defined.

First, we prove that this choice is the most general linear homogeneous second order differential equation. This assertion is surprising since it does not appear to be true. The most general homogeneous second order differential equation is

$$\left[a_2(z)\frac{d^2}{dz^2} + a_1(z)\frac{d}{dz} + a_0(z)\right]\phi(z) = 0 \quad (13.126)$$

where $a_j(z)$ are arbitrary functions. The Liouville equation does not have that generality, since we identify, $a_2 = p$, $a_1 = p'$, and $a_0 = (q + \lambda r)$. It requires that $a_1 = a'_2$, where the prime denotes derivative. That is not true in general. However, Eq. (13.126) can be made to have the form Eq. (13.125) by multiplying the former by the function $w(z)$

$$\left[w(z)a_2(z) \frac{d^2}{dz^2} + w(z)a_1(z) \frac{d}{dz} + w(z)a_0(z) \right]\phi(z) = 0 \qquad (13.127)$$

The function $w(z)$ is chosen so that:

$$p(z) = w(z)a_2(z) \qquad (13.128)$$

$$\frac{d}{dz} p(z) = \frac{d}{dz} [wa_2] = wa_1 = wa_2 \left(\frac{a_1}{a_2} \right) \qquad (13.129)$$

The last differential equation has the solution

$$p(z) = w(z)a_2(z) = \exp\left[\int^z dz' \frac{a_1(z')}{a_2(z')} \right] \qquad (13.130)$$

$$w(z)a_0(z) = q(z) + \lambda r(z) \qquad (13.131)$$

This choice of $w(z)$ gives $p = wa_2$, $p' = wa_1$. The most general homogeneous second order differential equation can be cast into the form of a Liouville equation. Table 13.1 shows how the standard equation for various polynomials and functions can be cast into this form. In each case the variable is written as z, although for many functions other symbols are common. Some of these choices deserve an explanation.

1. Legendre functions have $z = \cos(\theta)$.
2. Bessel functions equation is usually written as $(z = \rho)$

$$\frac{1}{z}\frac{d}{dz}\left(z \frac{d}{dz} J_n(kz) \right) + \left(k^2 - \frac{n^2}{z^2} \right) J_n(kz) = 0 \qquad (13.132)$$

Table 13.1 Liouville Equation Parameters for Standard Equations

Name	Symbol	$p(z)$	$q(z)$	$r(z)$	λ
Trigometric	$\sin(kz)$, $\cos(kz)$	1	0	1	k^2
Hyberbolic	$\sinh(kz)$, $\cosh(kz)$	1	0	1	$-k^2$
Legendre	$P_l^m(z)$, $Q_l^m(z)$	$1 - z^2$	$-\dfrac{m^2}{1 - z^2}$	1	$l(l + 1)$
Bessel	$J_n(kz)$, $N_n(kz)$	z	$-\dfrac{n^2}{z}$	z	k^2
Spherical	$j_l(kz)$, $y_l(kz)$	z^2	$-l(l + 1)$	z^2	k^2
Laguerre	$L_n(z)$	ze^{-z}	0	e^{-z}	n
Hermite	$H_n(z)$	e^{-z^2}	0	e^{-z^2}	n

The Liouville form is found by multiplying every term in this equation by z.

3. Spherical Bessel functions have the usual equation ($z \equiv r$)

$$\frac{1}{z^2}\frac{d}{dz}\left(z^2\frac{d}{dz}j_n(kz)\right) + \left(k^2 - \frac{n(n+1)}{z^2}\right)j_n(kz) = 0 \qquad (13.133)$$

and the Liouville form is found by multiplying by z^2.

4. The Laguerre polynomial is listed for the case that $\alpha = 0$, since $L_n^{(0)}(z) \equiv L_n(z)$. Its differential equation is Eq. (13.114). This equation is converted to the Liouville form by multiplying every term by $\exp(-z)$.

5. The differential equation for the Hermite polynomial is Eq. (13.94) and is converted to the Liouville form by multiplying the equation by $\exp(-z^2)$.

The Liouville equation is a generic differential equation which applies to all of the polynomials and special functions used in our solutions. The properties of the Liouville equation are shared by all of these polynomials.

Wronskian

For each value of the eigenvalue λ, the second order equation has two solutions, $\phi_\lambda^{(1)}(z)$ and $\phi_\lambda^{(2)}(z)$. They each obey the same equation

$$0 = \frac{d}{dz}\left(p(z)\frac{d\phi_\lambda^{(1)}(z)}{dz}\right) + [q(z) + \lambda r(z)]\phi_\lambda^{(1)}(z) \qquad (13.134)$$

$$0 = \frac{d}{dz}\left(p(z)\frac{d\phi_\lambda^{(2)}(z)}{dz}\right) + [q(z) + \lambda r(z)]\phi_\lambda^{(2)}(z) \qquad (13.135)$$

Multiply the top equation by $\phi_\lambda^{(2)}(z)$ and the bottom equation by $\phi_\lambda^{(1)}(z)$ and then subtract the two equations,

$$0 = \phi_\lambda^{(2)}\frac{d}{dz}\left(p(z)\frac{d}{dz}\phi_\lambda^{(1)}\right) - \phi_\lambda^{(1)}\frac{d}{dz}\left(p(z)\frac{d}{dz}\phi_\lambda^{(2)}(z)\right) \qquad (13.136)$$

The terms with $(q + \lambda r)$ cancel during the subtraction. The remaining terms can be written as an exact differential,

$$0 = \frac{d}{dz}[p(z)W(\phi_\lambda^{(1)}, \phi_\lambda^{(2)})] \qquad (13.137)$$

$$W(\phi_\lambda^{(1)}, \phi_\lambda^{(2)}) = \phi_\lambda^{(2)}\frac{d}{dz}\phi_\lambda^{(1)} - \phi_\lambda^{(1)}\frac{d}{dz}\phi_\lambda^{(2)} \qquad (13.138)$$

The quantity W is called the *Wronskian*. Since the derivative of (pW) is zero, then this function is a constant,

$$W(\phi_\lambda^{(1)}, \phi_\lambda^{(2)}) = \frac{C}{p(z)} \qquad (13.139)$$

where the constant C depends on how the functions are normalized. This expression is quite powerful.

In some cases it is trivial. If the functions are $\sin(kz)$ or $\cos(kz)$ it states

$$W(\sin, \cos) = \cos(kz) \frac{d}{dz} \sin(kz) - \sin(kz) \frac{d}{dz} \cos(kz)$$

$$= k[\cos^2(kz) + \sin^2(kz)] = k \quad (13.140)$$

which is the correct result with $p = 1$ and $C = k$. A less obvious identity for Bessel functions is

$$W(J_n, N_n) = N_n(z) \frac{d}{dz} J_n(z) - J_n(z) \frac{d}{dz} N_n(z) = \frac{2}{\pi z} \quad (13.141)$$

which has $C = 2/\pi$ and $p = z$. The above identity for Bessel functions is valid for any positive value of z and for any integer n. The Wronskian for spherical Bessel functions is

$$W(j_n, y_n) = \frac{1}{z^2} \quad (13.142)$$

Boundary Conditions

The eigenfunctions exist over the interval $z_1 \leqslant z \leqslant z_2$, where $z_{1,2}$ are the end points. The boundary conditions are applied at these two points but the general boundary condition could depend on the function or its derivative. A linear combination is taken of these two options, where the coefficients α_i, β_i are constants

$$\alpha_1 \left(\frac{d}{dz} \psi_\lambda(z) \right)_{z_1} + \beta_1 \psi_\lambda(z_1) = 0 \quad (13.143)$$

$$\alpha_2 \left(\frac{d}{dz} \psi_\lambda(z) \right)_{z_2} + \beta_2 \psi_\lambda(z_2) = 0 \quad (13.144)$$

$$\psi_\lambda = A\phi_\lambda^{(1)} + B\phi_\lambda^{(2)} \quad (13.145)$$

The solution is a linear combination of $\phi_\lambda^{(1)}$ and $\phi_\lambda^{(2)}$. The two constants A and B are determined by the two boundary conditions at the end points.

Orthogonality for Different λ

Consider two solutions $\psi_{\lambda_1}(z)$ and $\psi_{\lambda_2}(z)$, where it is assumed that $\lambda_1 \neq \lambda_2$ These solutions obey the same boundary conditions, as determined by the above procedure. Each satisfies an equation

$$0 = \frac{d}{dz} \left(p(z) \frac{d\psi_{\lambda_1}(z)}{dz} \right) + [q(z) + \lambda_1 r(z)] \psi_{\lambda_1}(z) \quad (13.146)$$

$$0 = \frac{d}{dz} \left(p(z) \frac{d\psi_{\lambda_2}(z)}{dz} \right) + [q(z) + \lambda_2 r(z)] \psi_{\lambda_2}(z) \quad (13.147)$$

Multiply Eq. (13.146) by ψ_{λ_2} and Eq. (113.147) by ψ_{λ_1}, then subtract the two equations. The first term is again a perfect differential

$$0 = \frac{d}{dz}[p(z)W(\psi_{\lambda_1}, \psi_{\lambda_2})] + r\psi_{\lambda_1}\psi_{\lambda_2}(\lambda_1 - \lambda_2) \tag{13.148}$$

Integrate this equation over dz between the two end points. Since the first term is a perfect differential it is evaluated at the end points

$$0 = p(z_2)W(z_2) - p(z_1)W(z_1) + (\lambda_1 - \lambda_2)\int_{z_1}^{z_2} dz r(z)\psi_{\lambda_1}(z)\psi_{\lambda_2}(z) \tag{13.149}$$

The two Wronskians vanish at the end points. For example, at the lower end point

$$W(z_1) \equiv \psi_{\lambda_2}(z_1)\left(\frac{d}{dz}\psi_{\lambda_1}\right)_{z_1} - \psi_{\lambda_1}(z_1)\left(\frac{d}{dz}\psi_{\lambda_2}\right)_{z_1} \tag{13.150}$$

Multiply each term by α_1 and add or subtract the terms in β_1

$$\alpha_1 W(z_1) = \psi_{\lambda_2}(z_1)\left[\alpha_1\left(\frac{d}{dz}\psi_{\lambda_1}\right)_{z_1} + \beta_1\psi_{\lambda_1}(z_1) - \beta_1\psi_{\lambda_1}(z_1)\right]$$
$$- \psi_{\lambda_1}(z_1)\left[\alpha_1\left(\frac{d}{dz}\psi_{\lambda_2}\right)_{z_1} + \beta_1\psi_{\lambda_2}(z_1) - \beta_1\psi_{\lambda_2}(z_1)\right]$$
$$= -\beta_1\psi_{\lambda_1}\psi_{\lambda_2}(1 - 1) = 0 \tag{13.151}$$

where the boundary conditions have been used to remove the terms with $\alpha_1\psi' + \beta_1\psi$. The same method is used to prove that $W(z_2) = 0$. There remains the integral in Eq. (13.149) that must vanish if $\lambda_1 \neq \lambda_2$

$$\int_{z_1}^{z_2} dz r(z)\psi_{\lambda_1}(z)\psi_{\lambda_2}(z) = \delta_{1=2}I_1 \tag{13.152}$$

This integral is the basic orthogonality expression. Note that the integral over the functions is weighted by the factor of $r(z)$ in the integrand. That factor is familiar, since previously we have encountered expressions such as

$$\int_0^\infty \rho d\rho J_n(k\rho)J_n(k'\rho) = \frac{1}{k}\delta(k - k') \tag{13.153}$$

The integrand has a factor of $\rho d\rho$ and $r(z) = z = \rho$ for Bessel functions.

It is convenient to use a shorthand notation for the operator in the Liouville equation

$$\mathcal{L}_\lambda(z) \equiv \frac{d}{dz}p(z)\frac{d}{dz} + q(z) + \lambda r(z) \tag{13.154}$$

$$0 = \mathcal{L}_{\lambda_i}(z)\psi_{\lambda_i}(z) \tag{13.155}$$

which is used in the subsequent discussion. The last expression is the eigen-value equation.

13.4. Green's Functions

So far, the discussion has been on the homogeneous Liouville equation, where there is a zero across the equal sign. Now consider the inhomogeneous equation

$$\mathscr{L}_\lambda(z)\Psi_\lambda(z) = f(z) \tag{13.156}$$

The most general case is that λ is not one of the eigenvalues that has an homogeneous solution. The function $f(z)$ can be anything. Again assume that the solution is wanted over the interval $z_1 \leqslant z \leqslant z_2$ and that Ψ obeys boundary conditions of the general form discussed earlier.

The above problem appears to be quite formidable. Actually it can be solved in a rather simple way using Green's functions. Define Green's function $G_\lambda(z, z')$ as the solution to the differential equation

$$\mathscr{L}_\lambda(z)G_\lambda(z, z') = -\delta(z - z') \tag{13.157}$$

The constant z' is contained in the same interval $z_1 \leqslant z' \leqslant z_2$ and G_λ obeys the same boundary conditions as Ψ.

Multiply Eq. (13.156) by G and Eq. (13.157) by Ψ, then subtract the two equations. The result is

$$\frac{d}{dz}[p(z)W(\Psi, G)] = G_\lambda(z, z')f(z) + \Psi_\lambda(z)\delta(z - z') \tag{13.158}$$

Integrate this equation dz between the end points. The first term is a perfect differential and the Wronskian is evaluated at the two end points. This term again vanishes at the two end points for the same reason as in the prior proof. The Wronskian of two functions vanishes at the end points if the two functions obey the same boundary conditions. The resulting integral is simply

$$\Psi_\lambda(z') = -\int_{z_1}^{z_2} dz G_\lambda(z, z')f(z) \tag{13.159}$$

This integral is the reason Green's functions are useful. The solution to Eq. (13.156) is obtained by the integral of Green's function and $f(z)$. Green's function needs only to be found once and this will solve the general inhomogeneous equation [Eq. (13.156)].

THEOREM. *Green's function is symmetric, $G_\lambda(z, z') = G_\lambda(z', z)$.*

The proof of this theorem starts with the equations for two Green's functions

$$\mathscr{L}_\lambda(z)G_\lambda(z, z') = -\delta(z - z') \tag{13.160}$$

$$\mathscr{L}_\lambda(z)G_\lambda(z, z'') = -\delta(z - z'') \tag{13.161}$$

Multiply Eq. (13.160) by $G(z, z'')$ and Eq. (13.161) by $G(z, z')$, then subtract

$$\frac{d}{dz}[pW(G, G)] = -G_\lambda(z, z'')\delta(z - z') + G_\lambda(z, z')\delta(z - z'')$$

Integrate this equation between the end points. The first term again vanishes. The remaining integral gives $G(z', z'') = G(z'', z')$. This completes the proof.

Constructing Green's Function

The general method for constructing Green's function is now described. The process appears to be rather complicated but in practice it is rather easy.

The variable z exists over the range $z_1 \leqslant z \leqslant z_2$. The constant z' is over the same range. It is useful to define two regions:

$$\text{Region I:} \quad z_1 \leqslant z < z' \tag{13.162}$$

$$\text{Region II:} \quad z' < z \leqslant z_2 \tag{13.163}$$

The Liouville equation $\mathscr{L}_{\lambda_i}\phi_{\lambda_i} = 0$ is valid for specific values λ_i of the eigenvalue. Green's function equation is solved for a general value of $\lambda \neq \lambda_i$. This means there is not an eigenfunction $\psi_\lambda(z)$ that satisfies the Liouville equation and both boundary conditions. One can construct a function that satisfies the boundary condition at one end or the other, but not both. Let $\psi_\lambda^{(I)}$ satisfy the boundary condition at the lower value z_1. It is a solution of the homogeneous equation $\mathscr{L}_\lambda\psi_\lambda^{(I)} = 0$ that satisfies the boundary condition at the end in region I.

Similarly, let $\psi_\lambda^{(II)}(z)$ satisfy the Liouville equation and the boundary condition at z_2. Since λ is not an eigenvalue, then $\psi^{(I)} \neq \psi^{(II)}$. Try a solution of the form

$$\text{Region I:} \quad G_\lambda(z, z') = A\psi_\lambda^{(I)}(z)\psi_\lambda^{(II)}(z') \tag{13.164}$$

$$\text{Region II:} \quad G_\lambda(z, z') = B\psi_\lambda^{(I)}(z')\psi_\lambda^{(II)}(z) \tag{13.165}$$

The last step is to match this solution at $z = z'$. There are two conditions:

1. Green's function is continuous at $z = z'$. This condition requires that the constants A and B are equal.

2. The derivatives of Green's functions change value at the match point according to

$$\frac{1}{p(z')} = \left(\frac{d}{dz}G(z, z')\right)_{z=z'-\varepsilon} - \left(\frac{d}{dz}G(z, z')\right)_{z=z'+\varepsilon} \tag{13.166}$$

Where $\varepsilon \to 0$. This constraint on the derivative is derived by starting with the original equation [Eq. (13.157)] and integrating dz over the interval $(z' - \varepsilon, z' + \varepsilon)$. The integral over the delta function gives one. The integral over the derivative terms gives the difference in derivatives. The functions $(q + \lambda r)G$ give negligible contribution as $\varepsilon = 0$.

Green's function $G(z, z')$ has a cusp at $z = z'$. It is continuous, but the slope changes. Note that the derivative term generates the Wronskian

$$\left(\frac{d}{dz} G(z, z')\right)_{z=z'-\varepsilon} - \left(\frac{d}{dz} G(z, z')\right)_{z=z'+\varepsilon}$$

$$= A\left[\left(\frac{d}{dz}\psi_\lambda^{(I)}(z)\right)_{z'}\psi_\lambda^{(II)}(z') - \psi_\lambda^{(I)}(z')\left(\frac{d}{dz}\psi_\lambda^{(II)}(z)\right)_{z'}\right] \tag{13.167}$$

$$= AW(\psi^{(I)}, \psi^{(II)}) = \frac{AC}{p(z')} = \frac{1}{p(z')}$$

This gives $AC = 1$, where C is the constant that is associated with the Wronskian. Green's function has been determined and is actually quite simple.

Example 1

Consider the solution to the differential equation

$$\left[\frac{d^2}{dx^2} + k^2\right]\phi(x) = x \tag{13.168}$$

over the interval $0 \leqslant x \leqslant a$ with the boundary conditions $\phi(0) = 0 = \phi(a)$. The equation will be solved two ways, with or without Green's functions.

Without using Green's functions, the solution is by inspection

$$\phi(x) = \frac{x}{k^2} + A\cos(kx) + B\sin(kx) \tag{13.169}$$

The last two terms are the solution of the homogeneous equation, while x/k^2 is a solution to the inhomogeneous equation. Now apply the boundary conditions. To set $\phi(0) = 0$ requires that $A = 0$ and to set $\phi(a) = 0$ requires that $B = -a/[k^2 \sin(ka)]$. The solution to the equation is

$$\phi(x) = \frac{1}{k^2}\left[x - a\frac{\sin(kx)}{\sin(ka)}\right] \tag{13.170}$$

The next solution uses Green's functions and its equation is

$$\left[\frac{d^2}{dx^2} + k^2\right]G(x, x_0) = -\delta(x - x_0) \tag{13.171}$$

The solution to the homogeneous equations are $\sin(kx + \theta)$, where θ is a constant and it is $\theta = \pi/2$ to get $\cos(kx)$. For region I ($0 < x < x_0$) choose $\psi^{(I)} = \sin(kx)$ that vanishes at $x = 0$. For region II, choose $\psi^{(II)} = \sin[k(a - x)]$ that vanishes at $x = a$. Therefore, Green's function has the form

$$G(x, x_0) = \begin{cases} A\sin(kx)\sin(ka - kx_0), & \text{when } 0 < x < x_0 \\ A\sin(kx_0)\sin(ka - kx), & \text{when } x_0 < x < a \end{cases} \tag{13.172}$$

$$\frac{1}{A} = W = k[\cos(kx_0)\sin(ka - kx_0) + \sin(kx_0)\cos(ka - kx_0)]$$
(13.173)
$$= k\sin(ka)$$

This completes the derivation of Green's function. The final step is to evaluate the integral for $\phi(x)$

$$\phi(x) = -\int_0^a dx' x' G(x, x')$$

$$= -\frac{1}{k\sin(ka)}\left[\sin(ka - kx)\int_0^x dx' x'\sin(kx') + \sin(kx)\int_x^a dx' x'\sin(ka - kx')\right]$$

$$= \frac{1}{k^2}\left[x - a\frac{\sin(kx)}{\sin(ka)}\right]$$
(13.174)

This is the same result as before. Green's function derivation appears to be more complicated. Most of the effort is in finding Green's function.

What if the problem is changed to having x^2 on the right-hand side of Eq. (13.168)? Then a simple ansatz such as Eq. (13.169) is no longer obvious. However, Green's function method still works. The factor of x' in Eq. (13.174) is merely changed to x'^2.

Example 2. *Bessel functions.*

Consider Green's function

$$\left[\frac{d}{dz}z\frac{d}{dz} + k^2 z - \frac{n^2}{z}\right]G(z, z') = -\delta(z - z')$$
(13.175)

The homogeneous equation is the same as for Bessel functions but with solutions $J_n(kz)$, $N_n(kz)$. Here, the boundary conditions are, $G(0, z') = 0$ and outgoing waves as $z > z'$. The former is satisfied by $J_n(kz)$, while the latter by the Hankel function $H_n^{(1)}(kz)$. Green's function must have the form

$$G(z, z') = AJ_n(kz_<)H_n^{(1)}(kz_>)$$
(13.176)

where

	$z_<$	$z_>$
$z < z'$	z	z'
$z' < z$	z'	z

(13.177)

The notation $z_{<,>}$ is often used for Green's functions. The coefficient A is found from the Wronskian

$$\frac{1}{A} = z'W(J, H) = kz'[H^{(1)\prime}J - JH^{(1)\prime}] = ikz'W(J, N) = ikz'\frac{2}{\pi kz'}$$
(13.178)

which gives $A = -i\pi/2$. Green's function is determined.

The meaning of *completeness* is that any function $f(z)$ can be expanded exactly in a series over the eigenfunctions $\psi_{\lambda_i}(z)$. Multiply Eq. (13.188) by $f(z)$ and integrate dz_0 over the interval $z_1 \leqslant z_0 \leqslant z_2$ and find

$$f(z) = \sum_i b_i \psi_{\lambda_i}(z) \tag{13.179}$$

$$b_i = \frac{1}{I_i} \int_{z_1}^{z_2} dz_0 r(z_0) f(z_0) \psi_{\lambda_i}(z_0) \tag{13.180}$$

For any function $f(z)$, there exists a set of coefficients b_i that permit an exact expansion. The value of the coefficients b_i is given by the above integral. Green's function can also be expanded in these functions

$$G_\lambda(z, z_0) = \sum_i a_i(z_0) \psi_{\lambda_i}(z) \tag{13.181}$$

The constant λ does not equal any of the eigenvalues λ_i. Operate on the above equation

$$\mathscr{L}_\lambda G_\lambda(z, z_0) = -\delta(z - z_0) = \sum_i a_i \mathscr{L}_\lambda \psi_{\lambda_i} = r(z) \sum_i a_i(\lambda - \lambda_i) \psi_{\lambda_i}(z) \tag{13.182}$$

Multiply this equation by $\psi_{\lambda_n}(z)$ and integrate dz over the line. The result is

$$-\psi_{\lambda_n}(z_0) = a_n(\lambda - \lambda_n) I_n \tag{13.183}$$

$$I_n = \int_{z_1}^{z_2} dz r(z) \psi_{\lambda_n}(z)^2 \tag{13.184}$$

$$a_n = -\frac{\psi_{\lambda_n}(z_0)}{I_n(\lambda - \lambda_n)} \tag{13.185}$$

$$G_\lambda(z, z_0) = -\sum_i \frac{1}{I_i} \frac{\psi_{\lambda_i}(z) \psi_{\lambda_i}(z_0)}{\lambda - \lambda_i} \tag{13.186}$$

Green's function is formally expressed as a summation over the eigenfunctions. Note that this expression makes no sense if $\lambda = \lambda_n$. A related expression can be obtained by operating on both sides of this expression by $\mathscr{L}$

$$\mathscr{L}G = -\delta(z - z_0) = -\sum_i \frac{r(z)(\lambda - \lambda_i)}{I_i} \frac{\psi_{\lambda_i}(z) \psi_{\lambda_i}(z_0)}{\lambda - \lambda_i} \tag{13.187}$$

$$\delta(z - z_0) = r(z) \sum_i \frac{1}{I_i} \psi_{\lambda_i}(z) \psi_{\lambda_i}(z_0) \tag{13.188}$$

The final expression is called a *completeness relation*. As an example, for Legendre functions $P_n(x)$, $x = \cos\theta$, $r = 1$, and $I_n = 2/(2n + 1)$ so that

$$\sum_n (2n + 1) P_n(x) P_n(x_0) = 2\delta(x - x_0) \tag{13.189}$$

which was given in Chapter 12. If λ_i is a continuous variable the summation over eigenvalues is replaced by a continuous integral. An example is the integral over Bessel functions in Eq. (13.153).

13.5. Singular Integral Equations

The theory of singular integral equations was developed by Muskhel-ishvili.[4] The theory has been used widely to solve a number of problems in science and engineering. An example is the exact solution of the Boltmann equation for the diffusion of photons through disordered systems such as clouds, stellar atmospheres, or human tissue. A brief summary of this theory is presented.

The general problem is to solve an integral equation for the unknown variable $n(x)$ where $A(x)$ and $g(x)$ are known functions

$$g(x) = A(x)n(x) + \int_0^1 du \, \frac{n(u)}{u - x} \tag{13.190}$$

The singular point in the integral for $0 \leqslant x \leqslant 1$ is evaluated using principal parts. The upper and lower limits of the integral can be any value, but the variable x or u can always be rescaled to make them zero and one.

The standard method is reviewed for solving singular integral equations. The notation is identical to Muskhelishvili in order to make it easier for the reader to consult this reference. First, define a function of the complex variable z and then evaluate it for $z = x \pm i\eta$ where x is real and η is infinitesimal. It is assumed that $0 < x < 1$. Frequent usage is made of the identity

$$\frac{1}{u - z} = \frac{1}{u - (x \pm i\eta)} = P\frac{1}{u - x} \pm i\pi\delta(u - x) \tag{13.191}$$

The symbol P denotes principal part and will usually not be written.

$$\Phi(z) = \frac{1}{2\pi i} \int_0^1 du \, \frac{n(u)}{u - z} \tag{13.192}$$

$$\Phi^{\pm}(x) = \Phi(x \pm i\eta) = \pm\frac{1}{2}n(x) + \frac{1}{2\pi i} \int_0^1 du \, \frac{n(u)}{u - x} \tag{13.193}$$

Adding and subtracting the above equations produces two important relations

$$n(x) = \Phi^+(x) - \Phi^-(x) \tag{13.194}$$

$$\int_0^1 du \, \frac{n(u)}{u - x} = i\pi[\Phi^+(x) + \Phi^-(x)] \tag{13.195}$$

The original equation [Eq. (13.190)] can now be cast in the form of

$$g(x) = \Phi^+(x)[A(x) + i\pi] - \Phi^-(x)[A(x) - i\pi] \tag{13.196}$$

Define an angle $\theta(x)$ according to

$$A(x) \pm i\pi = \sqrt{A^2 + \pi^2}\, e^{\pm i\theta} \tag{13.197}$$

$$\tan[\theta(x)] = \frac{\pi}{A(x)} \tag{13.198}$$

$$\frac{g(x)}{\sqrt{A^2 + \pi^2}} = e^{i\theta}\Phi^+ - e^{-i\theta}\Phi^- \tag{13.199}$$

In order to solve this equation we make the exponent a function of z. Define

$$\Gamma(z) = \frac{1}{\pi} \int_0^1 du\, \frac{\theta(u)}{u - z} \tag{13.200}$$

$$\Gamma^\pm(x) = \Gamma(x \pm i\eta) = \pm i\theta(x) + \gamma(x) \tag{13.201}$$

$$\gamma(x) = \frac{1}{\pi} \int_0^1 du\, \frac{\theta(u)}{u - x} \tag{13.202}$$

Multiply Eq. (13.199) by the factor of $\exp[\gamma(x)]$ which brings it to the form

$$\beta(x) \equiv \frac{g(x)e^{\gamma(x)}}{\sqrt{A(x)^2 + \pi^2}} = e^{\Gamma^+}\Phi^+ - e^{\Gamma^-}\Phi^- \tag{13.203}$$

This last equation has the same form as Eq. (13.194). The right-hand side is the difference between a complex function evaluated above and below the branch cut. The difference is the function which is the numerator of the singular integral.

$$e^{\Gamma(z)}\Phi(z) = \frac{1}{2\pi i} \int_0^1 du\, \frac{\beta(u)}{u - z} \tag{13.204}$$

$$\Phi(z) = e^{-\Gamma(z)}\frac{1}{2\pi i} \int_0^1 du\, \frac{\beta(u)}{u - z} \tag{13.205}$$

The above solution is correct when the various functions behave well when $z \to \infty$. In this limit $\Gamma(z)$ and $\Phi(z)$ go to zero as $O(1/z)$. In this case, as proved by Muskhelishvili, Eq. (13.203) has the unique solution of Eq. (13.204). Finally, from Eq. (13.194) derive an expression for $n(x)$

$$
\begin{aligned}
n(x) &= \Phi^+(x) - \Phi^-(x) \\
&= \frac{e^{-\Gamma^+(x)}}{2\pi i}\left[\int_0^1 \frac{du\,\beta(u)}{u - x} + i\pi\beta(x)\right] - \frac{e^{-\Gamma^-(x)}}{2\pi i}\left[\int_0^1 \frac{du\,\beta(u)}{u - x} - i\pi\beta(x)\right] \\
&= \beta(x)e^{-\gamma(x)}\cos(\theta) - e^{-\gamma(x)}\frac{\sin(\theta)}{\pi}\int_0^1 \frac{du\,\beta(u)}{u - x} \\
&= \frac{A(x)g(x)}{A(x)^2 + \pi^2} - \frac{e^{-\gamma(x)}}{\sqrt{A^2 + \pi^2}}\int_0^1 du\, \frac{\beta(u)}{u - x}
\end{aligned}
\tag{13.206}
$$

where we have used identities such as

$$\sin(\theta) = \frac{\pi}{\sqrt{A^2 + \pi^2}} \tag{13.207}$$

$$\cos(\theta) = \frac{A}{\sqrt{A^2 + \pi^2}} \tag{13.208}$$

The solution [Eq. (13.206)] completes the general derivation. An analytical solution has been obtained for $n(x)$. An example is provided below of how to apply these formulas to specific cases.

The example has $g(x) = 1$ and $A(x)$ equal to a constant, which also makes the angle $\theta(x)$ a constant. All functions and integrals are easily evaluated using contour integrals. Define $\alpha = \theta/\pi$

$$\gamma(x) = \alpha \ln\left(\frac{1-x}{x}\right) \tag{13.209}$$

$$\beta(x) = \frac{1}{\sqrt{A^2 + \pi^2}}\left(\frac{1-x}{x}\right)^\alpha \tag{13.210}$$

$$n(x) = \frac{1}{\sqrt{A^2 + \pi^2}}\left(\frac{x}{1-x}\right)^\alpha, \quad \text{for } 0 < x < 1 \tag{13.211}$$

The expression for $n(x)$ is found by evaluating the integral

$$\int_0^1 du \frac{\beta(u)}{u - x} \tag{13.212}$$

Integals of this form are evaluated by considering the contour integral in complex space

$$\oint \frac{dz}{2\pi i} \frac{1}{z - x}\left(\frac{z-1}{z}\right)^\alpha \tag{13.213}$$

The integrand has a branch cut placed between zero and one on the real axis and the contour integral circles this branch cut. The integral can also be evaluated by taking the circular contour at infinity where it becomes $\oint dz/(2\pi i z) = 1$. This step shows that the integral is one if the circle is anticlockwise. Then evaluating it above and below the cut making allowances for the contribution at the pole $z = x$, where $0 < x < 1$, produces the equation

$$1 = \frac{e^{i\pi\alpha}}{2\pi i}\left[i\pi\left(\frac{1-x}{x}\right)^\alpha - \int_0^1 \frac{du}{u-x}\left(\frac{1-u}{u}\right)^\alpha\right]$$

$$+ \frac{e^{-i\pi\alpha}}{2\pi i}\left[i\pi\left(\frac{1-x}{x}\right)^\alpha + \int_0^1 \frac{du}{u-x}\left(\frac{1-u}{u}\right)^\alpha\right] \tag{13.214}$$

$$= \cos(\theta)\left(\frac{1-x}{x}\right)^\alpha - \int_0^1 du \frac{\beta(u)}{u-x}$$

The first square bracket in Eq. (13.214) is from the integral along the top of the branch cut, while the second square bracket is for the integral below the branch cut. Within the brackets, the first term is for the semicircular integral around the pole at x, while the other term is the principle part integral along the cut. The final result is

$$\int_0^1 du \frac{\beta(u)}{u - x} = -1 + \frac{A}{\sqrt{A^2 + \pi^2}} \left(\frac{1 - x}{x} \right)^\alpha \qquad (13.215)$$

Inserting this result in Eq. (13.206) gives the final expression for $n(x)$.

One can show that the original equation is obeyed. The easiest way is to evaluate the integral

$$\int_0^1 du \frac{n(u)}{u - x} = \frac{1}{\sqrt{A^2 + \pi^2}} \int_0^1 \frac{du}{u - x} \left(\frac{u}{1 - u} \right)^\alpha \qquad (13.216)$$

by considering the contour integral

$$\oint \frac{dz}{2\pi i} \frac{1}{z - x} \left(\frac{z}{z - 1} \right)^\alpha \qquad (13.217)$$

Evaluating the circular contour at infinity shows that the integral equals one. It can be evaluated by integrating around the branch cut between zero and one. The steps are the same as for the integral over $\beta(u)$. The phase angle changes sign since the factor of $(z - 1)$ is now in the denominator.

$$1 = \cos(\theta) \left(\frac{x}{1 - x} \right)^\alpha + \int_0^1 du \frac{n(u)}{u - x} \qquad (13.218)$$

$$\int_0^1 du \frac{n(u)}{u - x} = 1 - \frac{A}{\sqrt{A^2 + \pi^2}} \left(\frac{x}{1 - x} \right)^\alpha \qquad (13.219)$$

Now test the original equation [Eq. (13.190)]

$$1 = \frac{A}{\sqrt{A^2 + \pi^2}} \left(\frac{x}{1 - x} \right)^\alpha (1 - 1) + 1 = 1 \qquad (13.220)$$

which proves that the solution satisfies the original equation.

The above solution for $n(x)$ is valid for the range $0 < x < 1$. The final step is to find the solution for x from this range. For example, consider the case that $x > 1$. The starting point is the original equation [Eq. 13.190)] rewritten into the form

$$n(x) = \frac{g(x)}{A(x)} - \frac{1}{A(x)} \int_0^1 du \frac{n(u)}{u - x} \qquad (13.221)$$

The last integral is easily evaluated since $n(u)$ is known over the range

$0 < u < 1$. Again consider the contour integral Eq. (13.217) except that $x > 1$. The identity [Eq. (13.218)] is changed by replacing $\cos(\theta)$ by 1. The factor of $\cos(\theta)$ comes from the phases along the branch cut, but now the pole at x is beyond this cut. For $x > 1$

$$\int_0^1 du \frac{n(u)}{u - x} = 1 - \left(\frac{x}{x - 1}\right)^\alpha \tag{13.222}$$

$$n(x) = \frac{1}{A}\left(\frac{x}{x - 1}\right)^\alpha, \quad \text{for } x > 1 \tag{13.223}$$

Another example is assigned in the problems

References

1. M. Abramowitz and I. A. Stegun, eds., *Handbook of Mathematical Functions*. NBS, 1964.
2. I. S. Gradshteyn and I. M. Ryzhik, *Tables of Integrals, Series, and Products*. Academic Press, (1965).
3. W. B. Jones and W. J. Thron, *Continued Fractions*, in *Encyclopedia of Mathematics*, ed. G. C. Rota. Addison-Wesley, 1980.
4. N. I. Muskhelishvili, *Singular Integral Equations.*, Ch. 10. Dover, 1992.
5. S. G. Mikhlin, *Integral Equations*, Chap III. Pergamon, 1957.

Problems

1. Solve the following differential equations using series methods,

(a) $\left[z \frac{d^2}{dz^2} - a \right] \phi(z) = 0$ \qquad (13.224)

(b) $\left[z^2 \frac{d^2}{dz^2} - a \right] \phi(z) = 0$ \qquad (13.225)

(c) $\left[z^3 \frac{d^2}{dz^2} - a \right] \phi(z) = 0$ \qquad (13.226)

(d) $\left[z \frac{d^3}{dz^3} - a \right] \phi(z) = 0$ \qquad (13.227)

(e) $\left[\frac{d^2}{dz^2} - az \right] \phi(z) = 0$ \qquad (13.228)

2. The function $F(a, b, z)$ is a solution of the confluent hypergeometric equation. Show that the following functions are solutions of the same differential equation

$$y_1 = e^z F(a_1, b_1, -z) \tag{13.229}$$

$$y_2 = z^{1-b} F(a_2, b_2, z) \tag{13.230}$$

$$y_3 = z^{1-b} e^z F(a_3, b_3, -z) \tag{13.231}$$

and express (a_j, b_j) in terms of (a, b).

3. Show that the following integral is a valid representation of the confluent hypergeometric function and that it obeys the differential equation

$$F(a, b, z) = \frac{\Gamma(b)}{\Gamma(a)\Gamma(b-a)} \int_0^1 dt e^{zt} t^{a-1} (1-t)^{b-a-1} \qquad (13.232)$$

4. Use the Kummer transformation to determine the limit as $z \to -\infty$ of $F(a, b, z)$. Can you derive the same result starting from the integral in the previous problem?

5. Start from the integral in Problem 3 and derive the asymptotic expansion of $F(a, b, z)$ as $z \to \infty$.

6. Express the first derivative of the confluent hypergeometric function as another confluent hypergeometric function.

7. Use series to find the solutions of the following differential equation over the interval $0 \leqslant z < \infty$

$$0 = \left[\frac{d^2}{dz^2} - \frac{l(l+1)}{z^2} - z^2 + a \right] \phi(z) \qquad (13.233)$$

Consider how it behaves at small and large z. Find the solution that is well-behaved at $z \to 0$ and also at $z \to \infty$.

8. Find the numerical value of the continued fraction that has a period of two

$$f = \frac{1}{4-} \frac{1}{5-} \frac{1}{4-} \frac{1}{5-} \frac{1}{4-} \cdots \qquad (13.234)$$

9. Derive a continued fraction from the recursion relation for spherical Bessel functions

$$\frac{2n+1}{z} j_n(z) - j_{n+1}(z) = j_{n-1}(z) \qquad (13.235)$$

What is its form for $n = 1$?

10. Derive the polynomials $L_n^{(1)}(x)$ for $n = 0, 1, 2$. Do it two ways: (1) use the definition [Eq. (13.115)] in terms of confluent hypergeometric functions, and (2) use Eq. (13.122).

11. Use generating functions to evaluate the following two integrals,

(a) $\int_0^\infty dx e^{-x} L_n(x) L_m(x) \qquad (13.236)$

(b) $\int_{-\infty}^\infty dx e^{-2x^2} H_n(x) H_m(x) \qquad (13.237)$

12. Verify that the form of $p(z)$, $r(z)$ for $L_n(z)$ in Table 13.1 gives the correct differential equation for Laguerre polynomials. What are the correct forms for $p(z)$, $r(z)$ to give the differential equation for $L_n^{(\alpha)}(z)$?

13. Find the following Green's function over the interval $0 \leqslant r \leqslant \infty$ with the boundary condition that it is zero if $r = 0$ and it is an outgoing wave at infinity.

$$\left[\frac{d}{dr} r^2 \frac{d}{dr} + k^2 r^2 - l(l+1) \right] G(r, r_0) = -\delta(r - r_0)$$

14. Solve the following Green's function

$$\left[z \frac{d^2}{dz^2} + (1 - z) \frac{d}{dz} + a \right] G(z, z_0) = -\delta(z - z_0) \tag{13.238}$$

where $0 \leqslant z, z_0 \leqslant \infty$ and a is not an integer.

15. Solve the following differential equation

$$\left[\frac{\partial}{\partial \rho} \rho \frac{\partial}{\partial \rho} + \rho k^2 - \frac{n^2}{\rho} \right] \phi(\rho) = 0 \tag{13.239}$$

for $a \leqslant \rho \leqslant b$ subject to the boundary conditions that $\phi(a) = 0 = \phi(b)$. Find the general form of $\phi(\rho)$ and derive the equation that determines the eigenvalues k_α. Next, prove that the solutions for different α are orthogonal.

16. Solve Eq. (13.190) and obtain $n(x)$ for $x > 0$ for the case $A = $ constant, $g(x) = x/(x^2 + a^2)$.

Index